AF606581

SCIENCE NOW!

ACTIVITIES AND ASSESSMENT PACK

Ann Fullick

Ian Richardson

David Sang

Martin Stirrup

Heinemann Educational Publishers
Halley Court, Jordan Hill, Oxford OX2 8EJ
A Division of Reed Educational & Professional Publishing Ltd

OXFORD MELBOURNE AUCKLAND
JOHANNESBURG BLANTYRE GABORONE
IBADAN PORTSMOUTH (NH) USA CHICAGO

First published 1995

02 01 00 99
10 9 8 7 6 5 4 3

ISBN 0 435 50684 6

Designed and typeset by **AMR,** Basingstoke

Illustrated by Art Construction, Phillip Burrows, Jon Davis and Jane Jones

Cover design by Miller, Craig and Cocking

Cover photo by Images

Printed and bound in Great Britain by Athenæum Press Limited, Gateshead, Tyne & Wear

Contents

Activity worksheets

Contents

Contents

How this pack supports the pupil book and your scheme of work

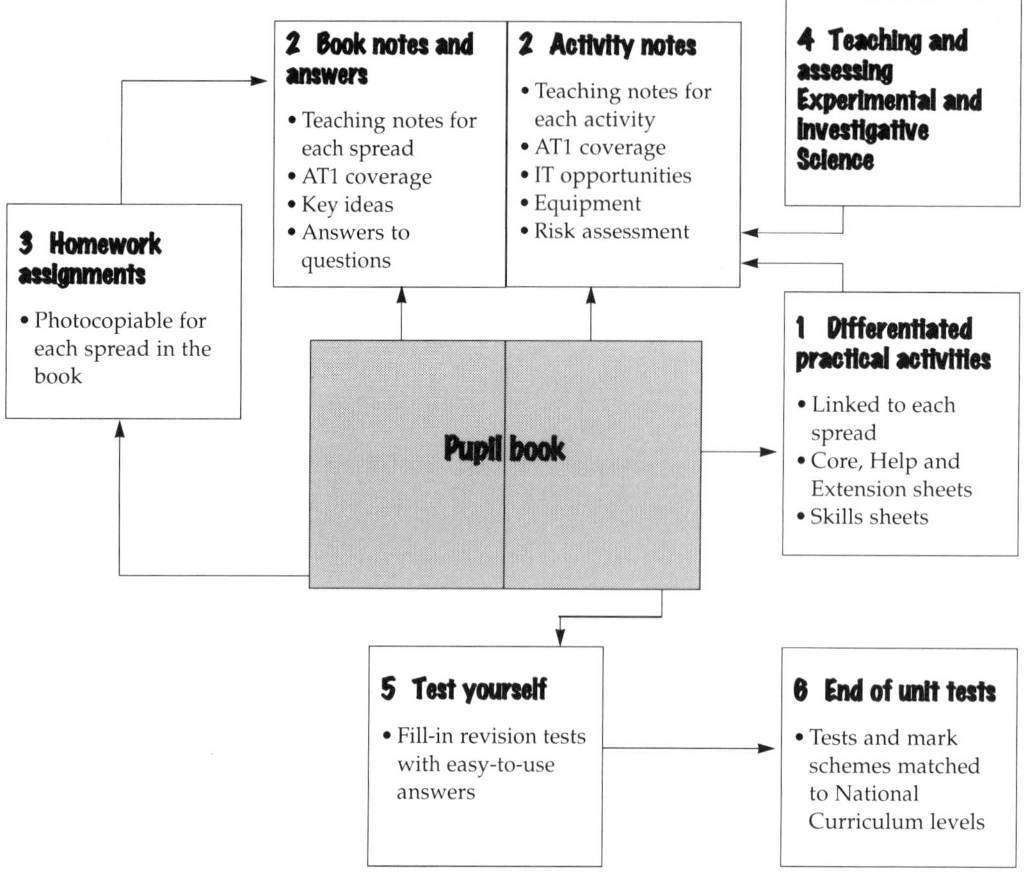

For ease of use, this pack is organised into six main sections.

1 Practical activities

These are photocopiable worksheets for use by pupils.

Relation to the book

The activities are linked to the main double-page spreads in the book in the sense that they are based on the key concepts developed there. Very occasionally, pupils will need to refer to the book for information (e.g. on classification) to help them carry out the activity.

The vast majority of spreads in the book have one linked activity and some have two. The relationship of the activities to the book spreads is denoted by their letter and number code. So, activity 1a is linked to book spread 1a, activity 4c to book spread 4c and so on.

Differentiation – Core, Help and Extension sheets

Every activity has a Core sheet which describes the activity, provides context and presents the main goals.

In addition, many have an accompanying Help sheet. This can be used in one of two ways:
1 It can be used alongside the Core sheet by those needing extra help as they go along. This type of sheet might be put out in class for pupils to select for themselves.
2 The Help sheet may be an alternative which can be handed out to weaker pupils instead of the Core sheet.

Many activities also have an Extension sheet. These are also of two types:
1 One type extends the activity to challenge pupils who finish the Core early.

2 The other type may be used as an alternative to the Core sheet for more able pupils who can tackle the activity in a more open-ended way with less guidance, or who can handle more sophisticated forms of recording and measurement.

Skills sheets

These are provided to support pupils where necessary in specific skills such as drawing graphs or using microscopes.

2 Teacher and technician notes

There are two types of notes pages, Book notes and answers, and Activity notes.

Book notes and answers

These notes give the coverage of ATs 2, 3 and 4 and the key ideas in the related book spread.

They also give answers to:

- all the in-text activities and the 'What do you know?' questions in the related book spread
- questions in any related 'Extras' sections.
- all homework assignments for that book spread.

Activity notes

There is a page of these notes for each activity, offering ideas on running the activity and on differentiation. The worksheets available are denoted by the following codes which show how they might be used:

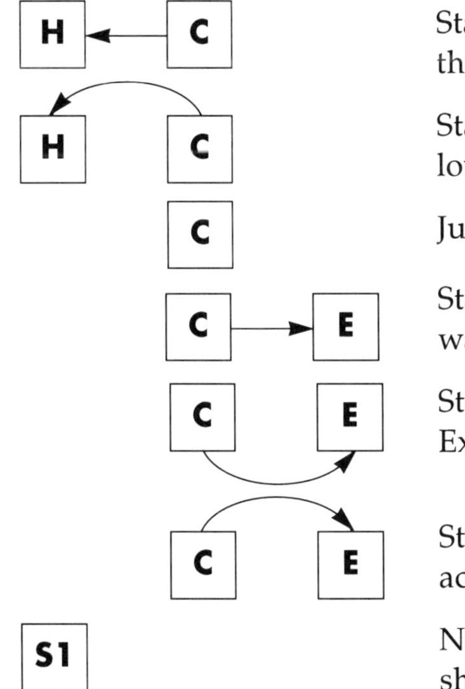

Start with Core and then go to Help part of the way through for support for low achievers.

Start with Help for a complete alternative to Core for low achievers.

Just Core for all pupils.

Start with Core and then go on to Extension part of the way through for high achievers.

Start with Core and those who finish fast move on to Extension.

Start with Extension for a complete alternative for high achievers.

Numbered Skills sheet to support a particular activity sheet.

Resource sheet for use in the activity itself.

The notes also show the activity's coverage of the AT1 programmes of study. Risk assessments and opportunities for IT are also given where appropriate.

At the bottom of each page an 'Equipment needed' box shows clearly what equipment is required to run the activity.

3 Homework assignments

These are free-standing assignments which draw on the concepts covered on double-page spreads in the book, but not on particular information. They are intended to be used for homework to save sending the book home, but could also, of course, be used in class if required.

The assignments are linked to double-page spreads in the book and arranged under the book spread code. They have been arranged by unit so that it is possible, if desired, to make a homework booklet for each book unit. The answers to all assignments are provided on the 'Book notes and answers' pages.

4 Teaching and assessing Experimental and Investigative Science

This section provides guidance on the requirements of this AT in the 1995 Order. It shows how activities in the pack can be used in a teaching programme for AT1.

It also exemplifies level judgements for the outcomes of a range of activities in the pack to help in assessing experimental and investigative science.

5 Test yourself

These are revision tests for each unit in the book. They are intended for pupils to fill in and keep as a record of the key ideas in the unit. Page numbers showing where answers can be found are given beside each question. This helps pupils answer the question and encourages them to consolidate their understanding of the unit.

The answers are given in straightforward filled-in form for handing out to pupils to mark their own tests.

6 End of unit tests

These are based on individual units, although they could also be put together to provide end of term or end of year tests. They are at two levels: Core and Extension. A mark scheme is provided for each test which matches ranges of aggregate scores to National Curriculum levels.

The tests are not for filling in. The text, however, is also provided on a disk which enables you to add spacing easily to produce disposable versions, or to add and alter questions as you wish. (For technical reasons the pictures are not included on the disk, but these can easily be scanned, or cut and pasted in.)

How Science Now! matches the curriculum for England and Wales

Life Processes and Living Things

Programme of study		Book 1	Book 2	Book 3
1 Life processes and cell activity	a	Unit 2	Unit 1	Unit 2
	b	Unit 2		
	c	Unit 2		Unit 2
	d	Unit 2		
	e	Unit 2	Unit 7	Unit 2
2 Humans as organisms	a	Unit 5	Unit 7	
	b	Unit 5	Unit 7	
	c	Unit 5	Unit 7	Unit 2
	d		Unit 7	
	e		Unit 7	Unit 2
	f		Unit 1	
	g		Unit 1	
	h		Unit 1	
	i	Unit 9		Unit 5
	j	Unit 9		Unit 5
	k	Unit 9		
	l		Unit 1	
	m		Unit 1	Unit 5
	n			Unit 2
	o			Unit 2
	p			Unit 2
	q			Unit 5
	r			Unit 5
	s			Unit 5
3 Green plants as organisms	a	Unit 5	Unit 4	Unit 2
	b	Unit 5		Unit 2
	c			Unit 2
	d	Unit 5		
	e	Unit 5		
	f	Unit 9		
	g			Unit 2
4 Variation, classification and inheritance	a	Unit 2	Unit 4	Unit 9
	b			Unit 9
	c	Unit 2		
	d	Unit 2		
	e			Unit 9
5 Living things in their environment	a		Unit 4	Unit 9
	b		Unit 4	Unit 9
	c	Unit 5	Unit 4	
	d	Unit 5		
	e		Unit 4	
	f		Unit 4	Unit 9
	g		Unit 4	Unit 9

Materials and their Properties

Programme of study		Book 1	Book 2	Book 3
1 Classifying materials	a	Unit 3		Unit 1
	b	Unit 3		Unit 1
	c	Unit 3		Unit 1
	d		Unit 2	Unit 3
	e		Unit 2	Unit 3
	f		Unit 2	Unit 8
	g		Unit 2	Unit 8
	h	Unit 6	Unit 2	
	i	Unit 6		
	j	Unit 3	Unit 2	
	k	Unit 3		
	l	Unit 3		
2 Changing materials	a			Unit 8
	b			Unit 8
	c	Unit 8		Unit 8
	d		Unit 1	Unit 1
	e	Unit 8	Unit 1	Unit 1
	f	Unit 11		
	g	Unit 11		
	h	Unit 11		
	i			Unit 8
	j		Unit 2	
	k		Unit 6	Unit 8
	l	Unit 8	Unit 9	Units 7/3
	m		Unit 9	
	n	Unit 8	Unit 9	
	o	Unit 8	Unit 9	
	p		Unit 9	
3 Patterns of behaviour	a		Units 6/9	Unit 7
	b		Unit 9	Unit 7
	c			Unit 7
	d			Unit 7
	e		Unit 6	
	f		Unit 6	
	g		Unit 6	Unit 7
	h		Unit 6	
	i		Unit 6	

Notes: Units from more than one book are frequently listed against one teaching point from the programme of study. This is because the concepts described are developed progressively over the three years of the course. Full coverage of the programme of study at all levels is achieved by the end of book 3.
The numbering of the units in book 3 may change. Final numbering will be included in Activities and Assessment Pack 2.

Physical Processes

Programme of study		Book 1	Book 2	Book 3
1 Electricity and magnetism	a			Unit 4
	b			Unit 4
	c	Unit 10		
	d	Unit 10		
	e	Unit 10		Unit 4
	f			Unit 4
	g		Unit 3	
	h		Unit 3	
	i		Unit 3	
	j		Unit 3	
2 Forces and motion	a		Unit 8	
	b		Unit 8	
	c	Unit 1	Unit 8	Unit 11
	d	Unit 1	Unit 8	Unit 11
	e	Unit 1	Unit 8	
	f	Unit 1		Unit 11
	g			Unit 11
	h			Unit 11
	i			Unit 11
3 Light and sound	a	Unit 7		Unit 6
	b	Unit 7		
	c	Unit 7		
	d	Unit 7	Unit 5	Unit 6
	e			Unit 6
	f			Unit 6
	g			Unit 6
	h			Unit 6
	i			Unit 6
	j	Unit 7		
	k	Unit 7		
	l			Unit 6
	m			Unit 6
	n			Unit 6
4 The Earth and beyond	a		Unit 5	
	b	Unit 7	Unit 5	
	c		Unit 5	
	d	Unit 7	Unit 5	
	e		Unit 5	
5 Energy resources and energy transfer	a	Unit 4	Unit 10	Unit 10
	b		Unit 10	
	c	Unit 4	Unit 10	
	d		Unit 10	Unit 10
	e			Unit 10
	f	Unit 4		Unit 10
	g			Unit 10
	h		Unit 10	Unit 10

Contexts and content in P7 to S2	Book 1	Book 2	Book 3
Variety and characteristic features	Unit 9	Unit 4	Unit 9
The processes of life	Unit 2 Unit 5 Unit 9	Unit 1	Unit 2 Unit 5
Interaction of living things with their environment	Unit 5	Unit 4	Unit 9
Forms and sources of energy	Unit 4	Unit 10	Unit 10
Properties and uses of energy	Unit 10	Unit 3	Unit 4
Conversion and transfer of energy	Unit 4	Unit 10	Unit 10
Forces and their effects	Unit 1		Unit 11
Earth in space		Unit 5	Unit 9
On planet Earth	Unit 3 Unit 11	Unit 2	Unit 1
Materials from Earth	Unit 3 Unit 6 Unit 8	Unit 2 Unit 9	Unit 1 Unit 7 Unit 8

Notes: Units from more than one book are frequently listed against one area of study from the science guidelines. This is because the concepts and related contexts are developed progressively over the three years of the course. Coverage of the areas of study at all levels is achieved by the end of book 3.
The numbering of the units in book 3 may change. Final numbering will be included in Activities and Assessment Pack 2.

Attainment target	Book 1	Book 2	Book 3
2 Living things	Unit 2 Unit 4 Unit 5 Unit 9	Unit 1 Unit 4 Unit 8	Unit 2 Unit 5 Unit 9
3 Materials	Unit 3 Unit 6 Unit 8	Unit 2 Unit 6 Unit 9	Unit 1 Unit 3 Unit 7 Unit 8
4 Forces and Energy	Unit 1 Unit 5 Unit 7 Unit 10	Unit 8 Unit 10	Unit 2 Unit 4 Unit 6 Unit 10 Unit 11
5 Environment		Unit 10 Unit 5 Unit 9	

Notes: Units from more than one book are frequently listed against an attainment target. This is because the concepts described are developed progressively over the three years of the course. Coverage of the attainment targets at all levels is achieved by the end of book 3.
The numbering of the units in book 3 may change. Final numbering will be included in Activities and Assessment Pack 2.

Safety informati

The following symbols are used in the pupil activity sheets to help pupils work safely:

Take care

Wash your hands after handling lead.

This warning is accompanied by suitable explanatory text.

We have attempted to identify any activities that might present some risk, and suggest appropriate safe approaches. Most educational employers have adopted various national publications such as *Topics in Safety* (ASE, 2nd edition 1988), *Hazcards* (CLEAPSS School Science Service, 1995), *Safeguards in the School Laboratory* (ASE, 9th edition, 1988) as the basis for their general risk assessments, and the proposed activities should be compatible with advice contained therein. In all cases, however, teachers must follow the guidance or local rules produced by their employer.

It is assumed that good laboratory practice is observed throughout, for example:

- Eye protection is worn both by pupils and teacher whenever the risk assessment requires it.
- Other protective or control equipment (e.g. safety screens, efficient fume cupboards) is similarly used when the risk assessment requires it.
- Long hair is tied back, pupils do not wear 'wet-look' hair preparations, and ties, scarves and cardigans are not allowed to hang freely.
- Pupils are trained in how to heat chemicals safely, and reminded frequently of the technique.
- Pupils are taught how to smell gases safely.
- Containers of chemicals are clearly labelled, with an appropriate name, and any hazards.
- Eating, drinking and chewing are not permitted.
- Electrical and other equipment is well maintained and subject to regular checks.

Measuring forces

You can use a push or a pull to start something moving. A push or a pull is a force. We use newtonmeters to measure forces. Forces are measured in newtons (N).

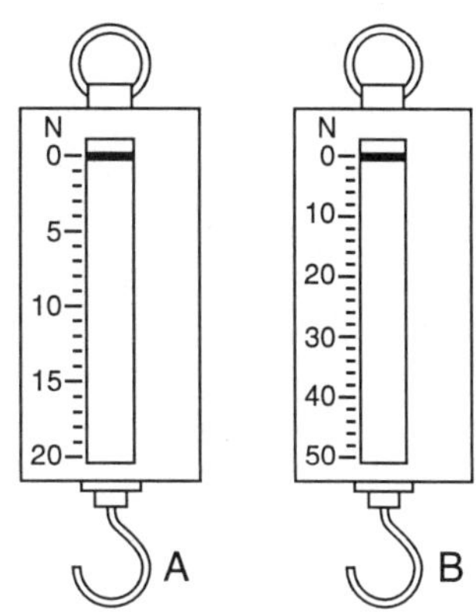

Newtonmeter B looks the same as newtonmeter A, but it can measure bigger forces than A. B has a bigger **range** than A. Its range is 0–50 N.

1 What is the range of newtonmeter A?

2 You are going to measure these forces using a newtonmeter:

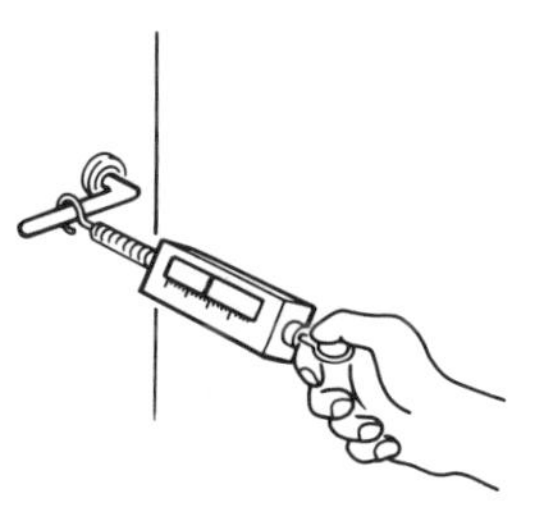
pulling the lab door open

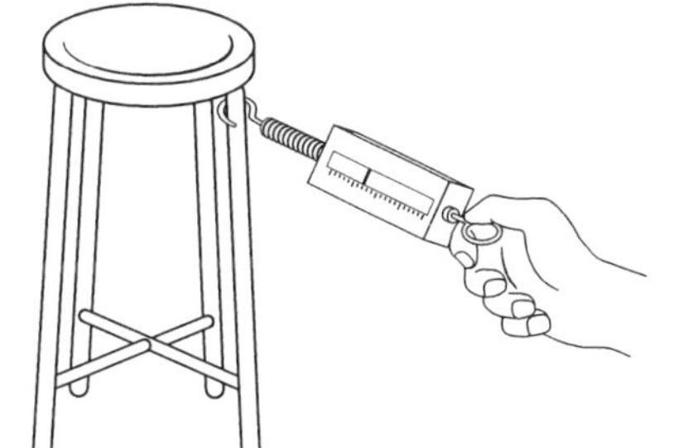
pulling a stool along the floor

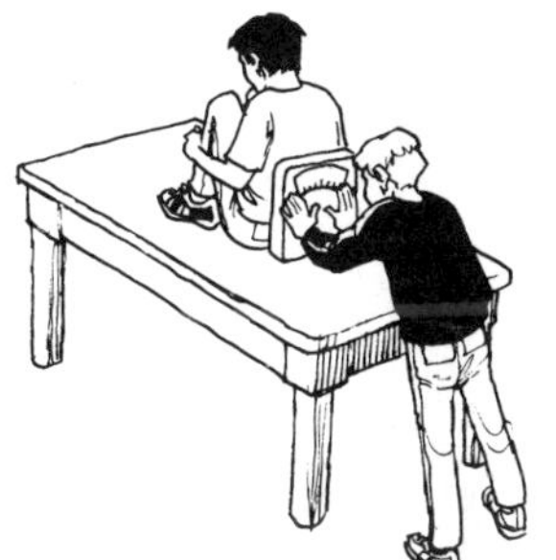
pushing a friend along the bench

pushing your hardest against the wall

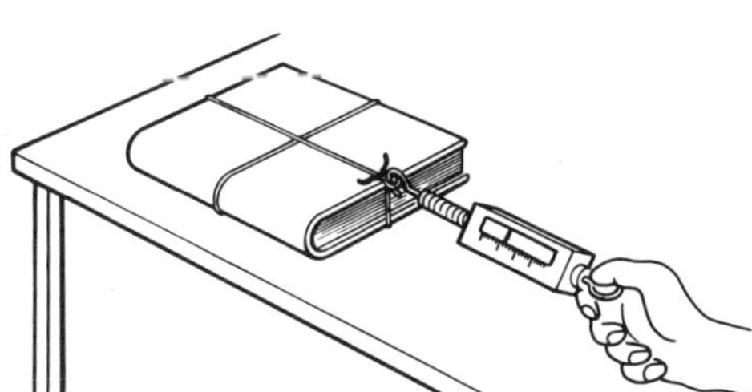
pulling a book along a bench

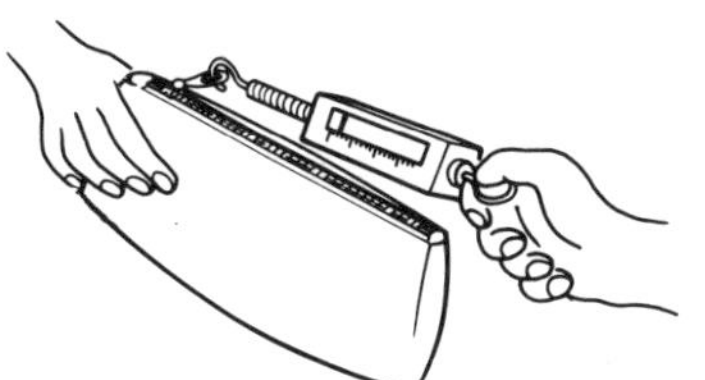
opening a pencil case

a Before you start, make a **prediction** (say what you think will happen):

- Which will be the biggest force?
- Which will be the smallest?

b Prepare a table to record your results.

c Now measure each of the forces. Make sure you use the best newtonmeter for the job. Choose the one with the best range.

d Record your results in your table. Were your predictions correct?

e Think of some more pushes and pulls to measure. Add the results to your table.

Activity 1a Help

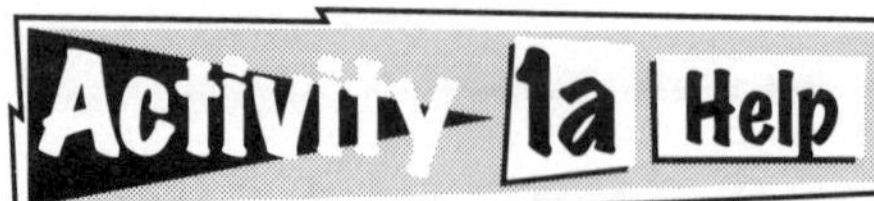

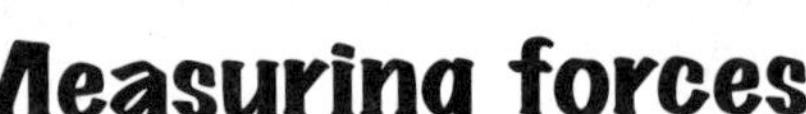

Measuring forces

1 These newtonmeters are measuring forces.
How many newtons is each force? Write your answers under each picture.

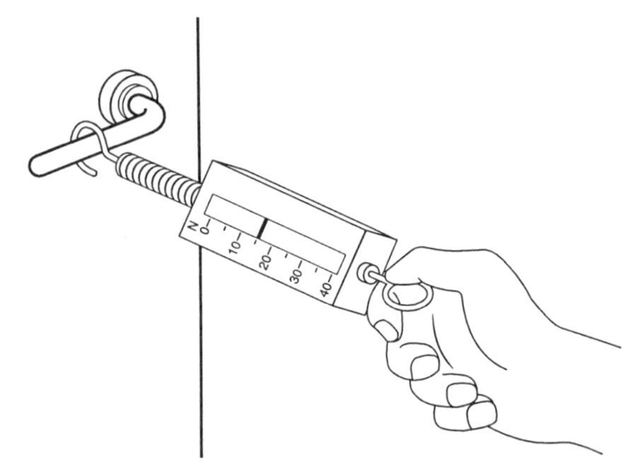

This newtonmeter is measuring a force of 15N.

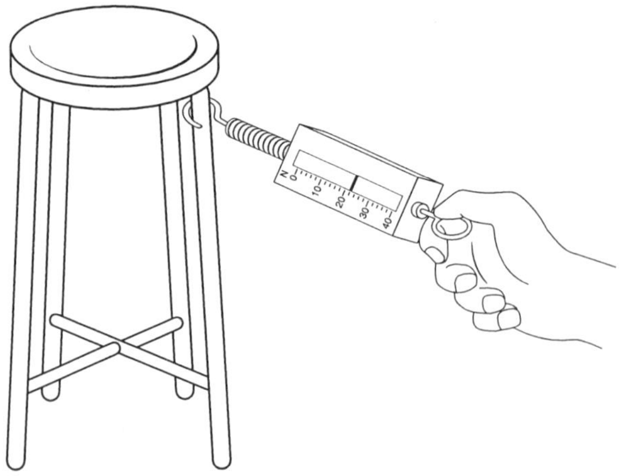

Force = N

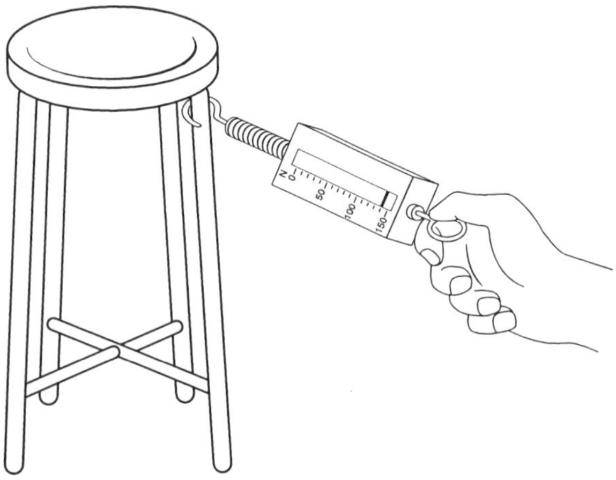

Force = N

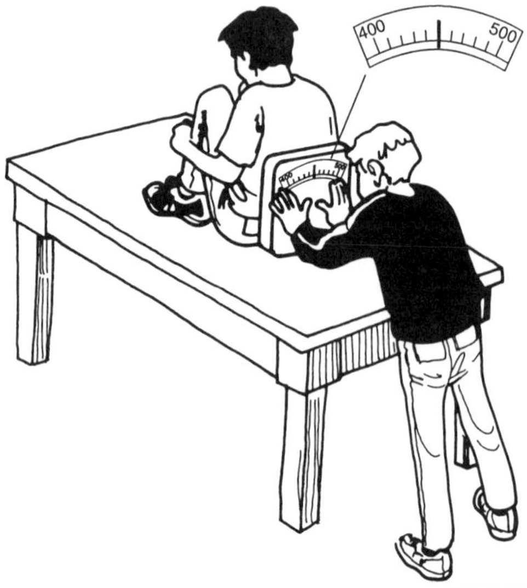

Force = N

2 Now measure some forces of your own.
Write them in the table.

Push or pull	Force
pulling open a door	N
pulling a stool along the floor	N
pushing a person along the bench	N
pushing your hardest against a wall	N
opening a pencil case	N
pulling a book along the bench	N

Light and heavy

Can you tell (predict) which of these objects is the heaviest? Which is the lightest? Do this activity to find out.

1 Look at the objects you are given. Don't touch! Which do you think is the lightest? Which is the heaviest? Copy the table, and put your predictions in the first empty column.

Object	Score (by looking)	Score (by hand)	Weight (N)	Score (by weighing)
A				
B				
C				
D				
E				

Results table: score 1 for lightest, 2 for next lightest, and so on.

2 Now hold one object in your left hand and one in your right. Which is heavier? Judge the objects in pairs like this, and fill in the next column.

3 Weigh each object with a newtonmeter. Fill in the last two columns.

4 How good were your predictions of the weights of the objects by looking at them?

5 How well did you judge their weights by holding them?

6 Which is the best method for finding the lightest and the heaviest?

Easy going

Friction is the force that can make it difficult for one surface to slide over another. It is easier to start things moving without friction. Things move faster without friction.

Smooth surfaces have less friction than rough surfaces. You can use a liquid such as oil on a rough surface, to reduce friction.

You are going to find out the best ways to reduce friction.

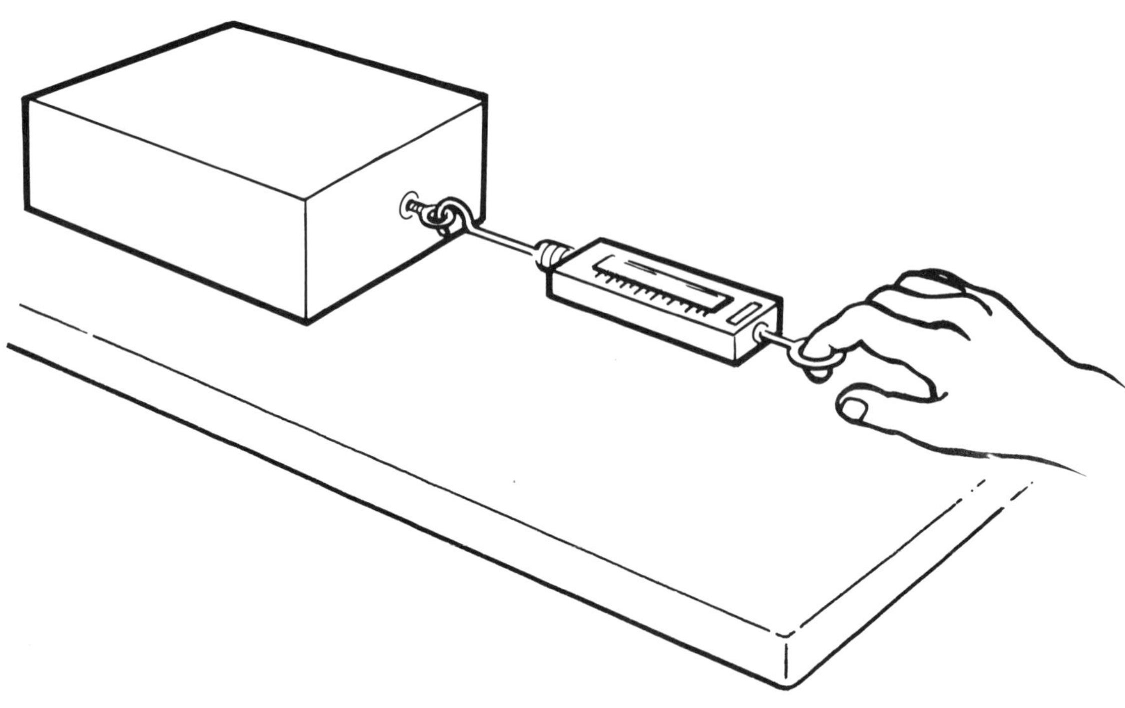

1 Before you start, read the instructions below and decide how you will record your findings.

2 Start by finding the force of friction on a plain wooden block. Pull the block along with a newtonmeter. The force when the block starts to move tells you the force of friction.

3 Plan an investigation of how the force of friction depends on the surface of the block and the surface it moves along. You could:

- change the surface of the block by wrapping something around it
- change the surface it moves along. You can pull it along a wooden tray. You can lubricate the tray with oil or water. You can cover the tray with small polystyrene beads.

4 **Once your plan has been approved**, try it out.

5 What reduced friction the most out of all the things you tried? What reduced it the least?

Take care

Be careful not to spill any beads or oil as they could make the floor dangerously slippery.

Easy going

Activity 1c Help

⚠ Take care

Be careful not to spill any beads or oil as they could make the floor dangerously slippery.

1 Measure the force of friction in each test:

wooden block	on oily tray	force = N
wooden block	on wet tray	force = N
wooden block	on tiny beads	force = N
wooden block	on dry tray	force = N

The easiest to pull has the least friction, and the hardest to pull has the most friction.

2 Which was easiest to pull?

____

3 Which was hardest to pull?

____

4 Which was better at helping to reduce friction, oil or water?

____

5 Now try these:

wooden block	on wooden tray	force = N
polythene-covered block	on wooden tray	force = N
sandpaper-covered block	on wooden tray	force = N

6 Which surface has most friction?

____

How do you know?

____

7 Which surface has least friction?

____

How do you know?

____

Easy going

The force of friction depends on whether the surfaces are rough or smooth. What else might it depend on?
You are going to investigate two factors.

Factor 1: By turning the wooden block on its side, you can change the area of the block that is rubbing on the bench.

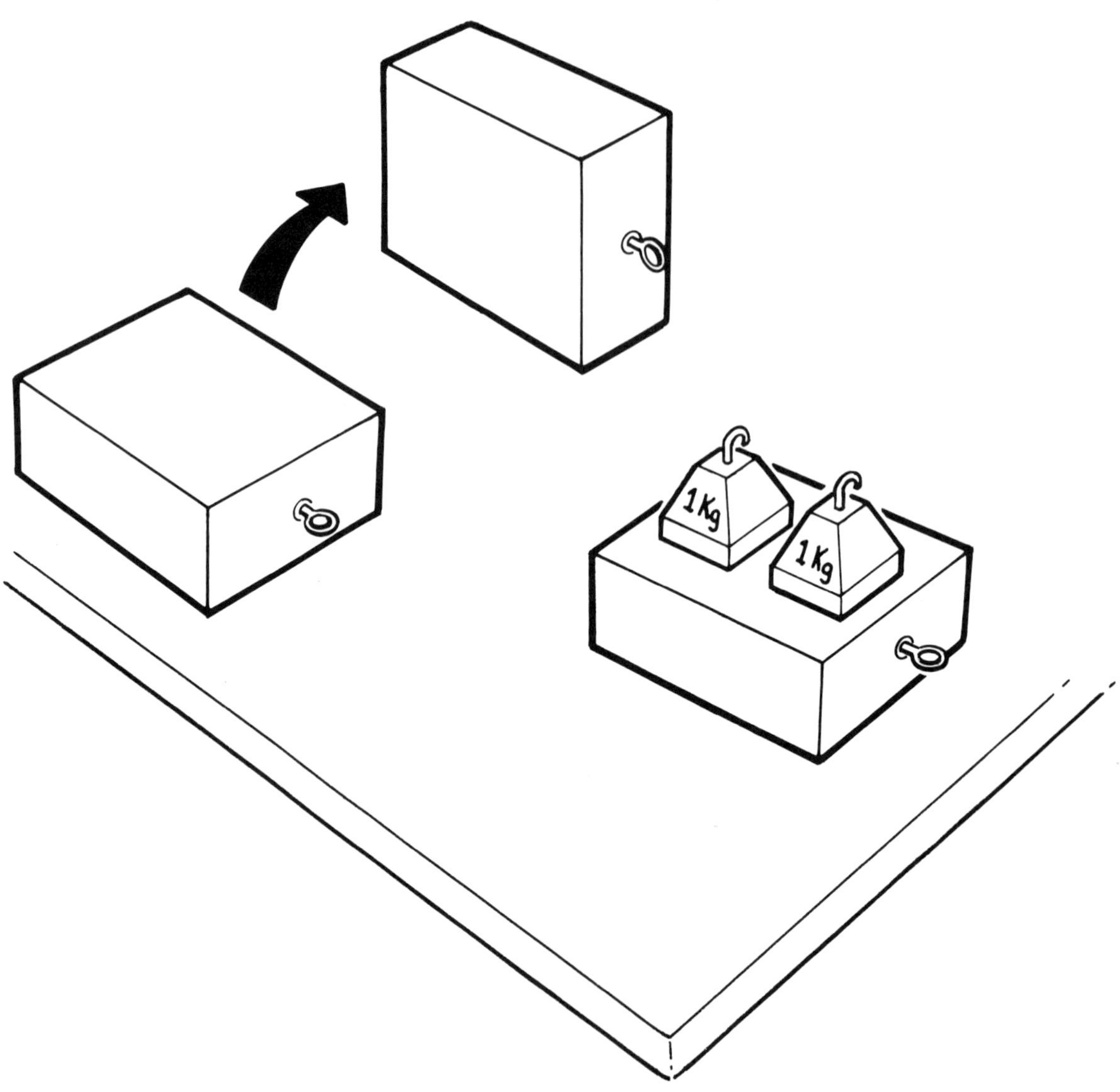

Factor 2: By putting 1 kg masses on the block, you can increase its mass.

1 Plan an investigation of how the force of friction depends on the area of contact and the mass of the block.
Start by making a prediction about the following:

- If the wooden block has a bigger area rubbing on the bench, will the force of friction be bigger, or smaller, or will it stay the same?
- If you increase the mass of the block, will the force of friction be bigger, or smaller, or will it stay the same?

2 **Once your plan has been approved**, try it out.

Down the tube

A streamlined shape is good for travelling fast through water or air. A diver shapes her body to be very streamlined when she enters the water. Many fishes have a streamlined shape so that they can swim fast.

Can you design a good streamlined shape? In this experiment, you have to make a piece of Plasticine into the best shape for falling fast through water.

1. Before you start, think about these questions:
   - How will you record your results?
   - Why must all the pieces of Plasticine you use have the same weight?
2. Mark a ring around the cylinder, 10 cm from the top. This is the starting line. Mark a finishing line, 10 cm from the bottom. Fill the cylinder right to the top with water.
3. Shape your first piece of Plasticine. Hold it just above the water and let it drop in.
4. Use a stopclock to time the Plasticine as it falls. Start the clock as it passes the starting line. Stop the clock as it reaches the finishing line.
5. Experiment with different shapes and measure the time each takes to get to the finishing line.
6. Which shape reaches the finishing line most quickly?
7. Can you explain why this is the fastest shape?
8. Which shape reaches the finishing line most slowly?
9. Can you explain why this is the slowest shape?

Down the tube

1. Fill two cylinders with water.
2. Put them side by side.
3. Take two pieces of Plasticine that weigh the same.
4. Make them into different shapes.

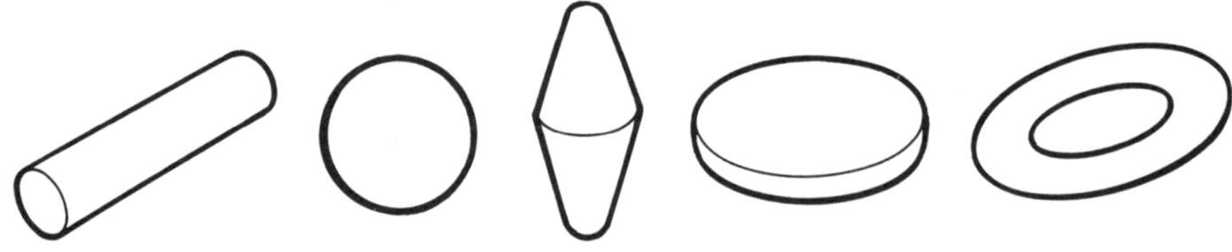

5. Drop one into each cylinder at the same time.
6. Which one reaches the finishing line first?
7. Which is the most streamlined shape?
8. Do the same again with the winning shape and another shape that weighs the same.
9. Carry on until you have the most streamlined shape possible. What shape is it?

Down the tube

Speed

You can work out the **speed** of the falling Plasticine like this:

- Measure the distance from the starting line to the finishing line, in centimetres.
- Calculate the speed by dividing the distance by the time taken for the Plasticine to fall:

$$\text{speed (in cm/s)} = \frac{\text{distance (in cm)}}{\text{time taken (in s)}}$$

1 Copy and fill in a table like this to work out the speed. Here is an example of how to do this:

$$\text{speed} = \frac{22\,\text{cm}}{2.9\,\text{s}} = 7.6\,\text{cm/s}$$

Shape of Plasticine	Distance it fell (cm)	Time taken (s)	Speed (cm/s)
sphere	22	2.9	7.6

Size

2 Break up a piece of Plasticine into smaller pieces. Make several different-sized balls of Plasticine. Which will fall fastest? Which will fall slowest? Your answers to these questions will help you to make a prediction such as:

- A smaller ball of Plasticine will fall **faster/slower** than a larger ball.

3 Decide on your prediction, and carry out an investigation to test it. You should test at least five different Plasticine balls.

4 Try to explain your results. Think about the forces involved: friction and weight.

Teeter totter

A see-saw is a kind of lever. By pressing down on one end of the see-saw, you can lift a weight on the other end.

You can investigate a see-saw like this:

1. Make a model see-saw by balancing a ruler on a triangular wooden block. The block is the pivot.
2. Put an object on one end. This is the load.
3. Put weights on the other end until they are heavy enough to tip the see-saw. How much weight does it take?
4. Remove the weights. Move the load closer to the pivot. Now make a prediction:
    - How much weight will you need to tip the see-saw?
5. Test your prediction.
6. Try changing the load. Make it heavier. How much weight do you need now?

Teeter totter

1 You can investigate the force needed to tip the see-saw by copying and completing the table.
What pattern can you see in the results?

Distance from load to pivot	Force needed to lift load
10 cm	N
20 cm	N
30 cm	N
40 cm	N
50 cm	N

2 A line graph can help to show the pattern in your results.

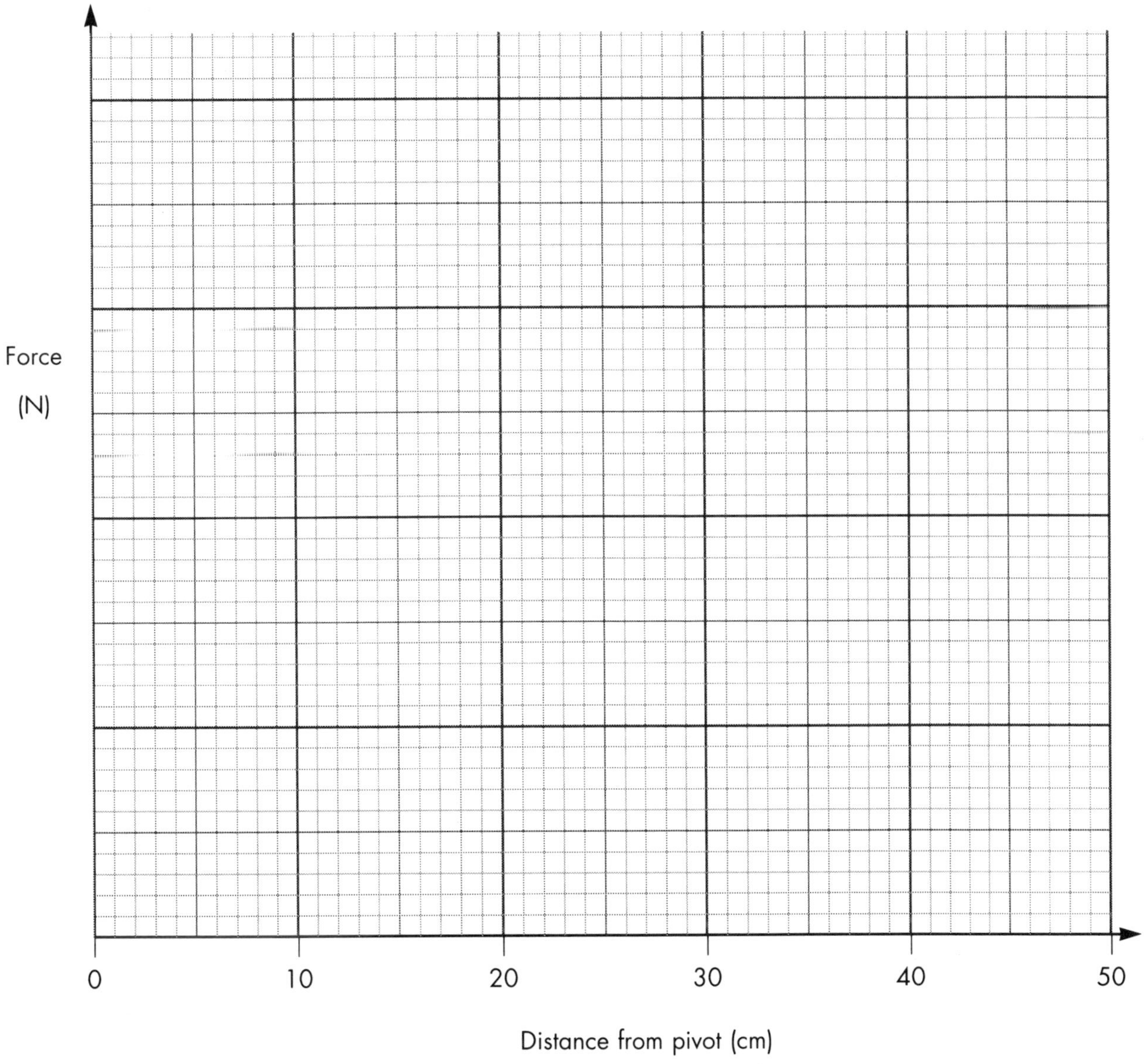

Force (N)

0 10 20 30 40 50

Distance from pivot (cm)

3 Now repeat your experiment with a different load.

Is it living? Does it respire?

All living things respire. They use oxygen to get energy from their food. As living things respire, they produce a waste gas called carbon dioxide. There is a simple test for carbon dioxide using limewater.

Limewater is a clear liquid which turns cloudy when carbon dioxide bubbles through it. Think about this question:

- How could you use limewater to show that living things really do respire?

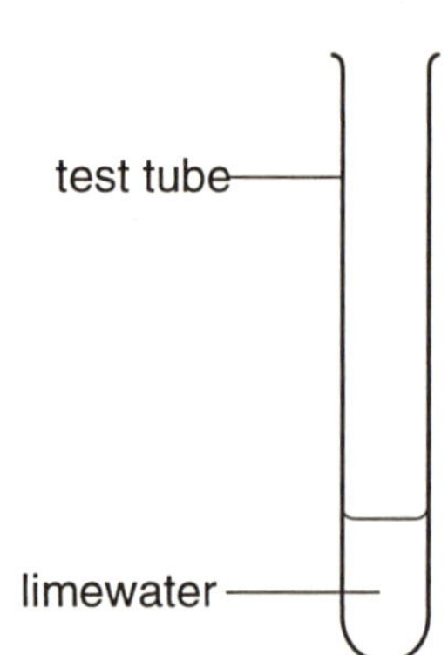

This activity uses limewater to see if people produce carbon dioxide when they respire.

1. Collect two test tubes and pour about 3 cm^3 of limewater into each one.
2. Is the limewater clear or cloudy?
3. Collect a straw. Breathe out **gently** through the straw into one of the test tubes, blowing bubbles through the limewater. Keep bubbling for about a minute.

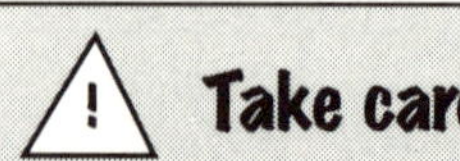
Take care

Be careful not to suck on the straw.

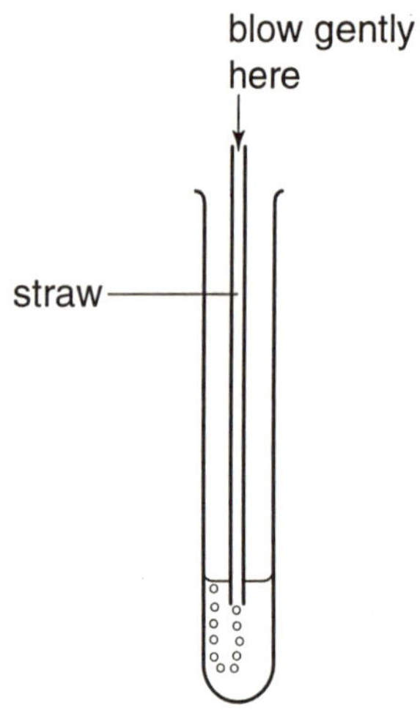

4. Is the limewater clear or cloudy now?
You can compare it with the limewater that has not been used. This is your **control.**
5. Make a record of your limewater test. Say what you did, what happened and what the experiment tells you.

Is it living? Does it respire?

It is easy to use the limewater test to show that there is carbon dioxide in the air people breathe out, because people can blow through a straw. It is not quite to easy to use it for other animals or plants.

1 Plan an experiment using limewater to show that simple animals, such as maggots or woodlice, respire and produce carbon dioxide. Here is the apparatus you can use:

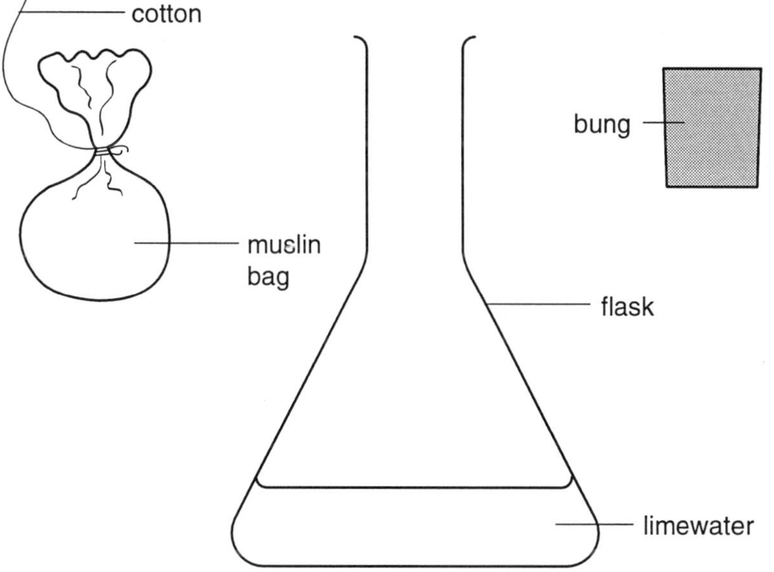

Think about these things:

- Your experiment will need to be as kind as possible to the animals. Think carefully about how you can set it up without drowning them.
- How will you make sure it is a fair test?

2 How could you show that plants respire too?

3 Discuss your ideas with your teacher.

Do plants move?

It is easy to see animals moving. Because plants don't move their whole bodies, and because they move very slowly, we often don't notice their movements.

Plants need light. You are going to plan and carry out an experiment to show that plants move towards the light.

You can use this apparatus:

- beakers or Petri dishes
- cress seeds
- scissors
- cotton wool
- Sellotape
- black paper

1 Design your experiment.

- Think carefully about how you can find out whether plants move towards the light.
- Remember that you must make it a fair test. You will need a control plant which gets light from all around it.
- Cress seeds need plenty of water if they are going to grow.

2 **Once your plan has been approved**, set up your experiment and leave it until your next lesson.

3 What has happened?

4 Make a record of your experiment. Say what you did, what happened and what the experiment tells you.

5 Have you any suggestions for improvements to the experiment if you did it again?

Do plants move?

Plants need light. You are going to carry out an experiment to show that plants move towards the light.

1 Put some cotton wool in the bottoms of two beakers and wet it well with water.

2 Sprinkle some cress seeds on the wet cotton wool in both beakers.

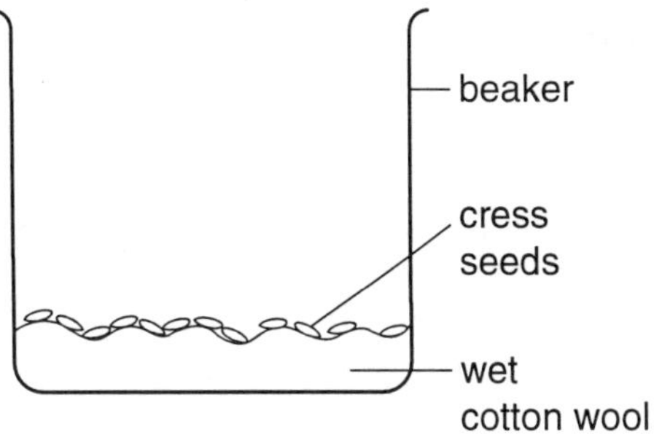

3 Use black paper to make a cover for one of the beakers. Cut a hole in one side of the cover and put the cover over your beaker. In this beaker, light will only reach your seeds from one side.

In the other beaker, light will reach them from all sides. This is your control.

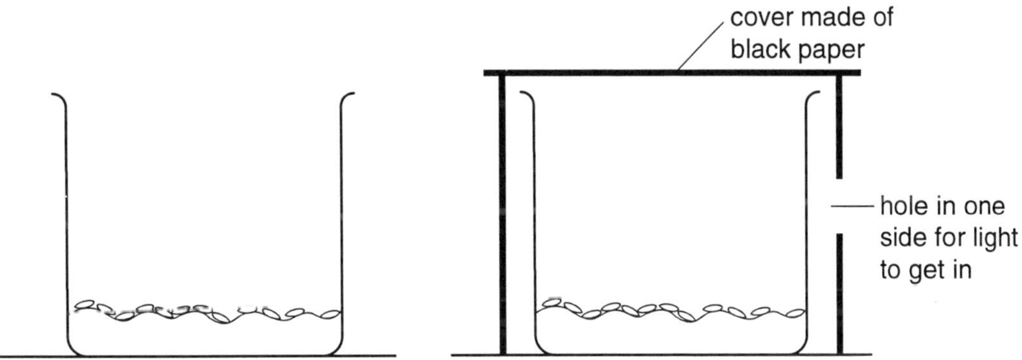

4 Leave both beakers in a light place for a week. Then take the cover off the beaker and compare the two lots of plants.

a What has happened to the plants under the cover?

b What does this tell you about movement in plants?

5 Make a record of your experiment.

- Say what you did.
- Draw a picture to show what your two lots of cress plants looked like.
- Explain what happened.

Looking at cells

You will need the sheets **What is a microscope?** and **How to use the microscope**. Read these sheets and remind yourself of what to do before you carry on.

Plant cells

Plant cells are easier to see than animal cells.

To see plant cells under the microscope, we need a very thin piece of plant material which is only one cell thick. Your teacher will give you some thin bits of plant from onion or rhubarb.

Making a slide

1 Place a single drop of water on a microscope slide.

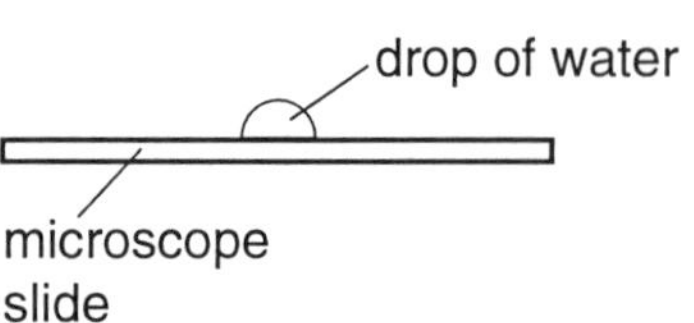

2 Collect your piece of plant material and place it on the drop of water.

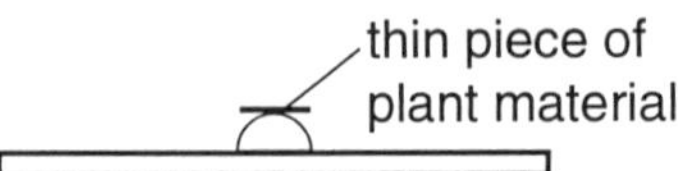

3 Lower a coverslip down carefully. Try hard not to get any air bubbles.

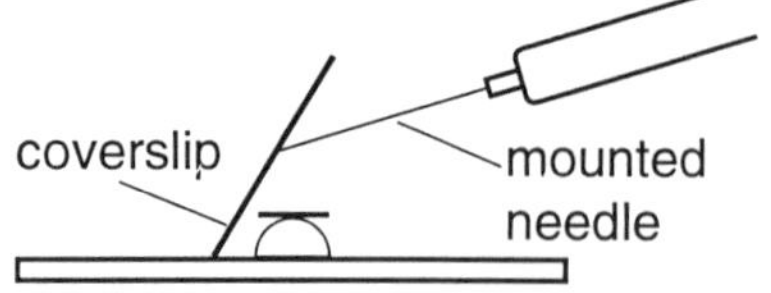

4 Use a piece of tissue paper to dry any extra water off your slide and then place it on the microscope stage.

Looking at the cells

Remember that plant cells are rather like shoeboxes containing plastic bags full of jelly!

5 Use the low power of the microscope to look for lots of plant cell boxes. If you use a higher powered lens of the microscope, you may be able to see the cell walls and the nuclei. Circles with thick black walls are not cells, they are air bubbles!

6 Use a pencil to make two drawings:

a a plan of what you can see

b one or two cells, drawn much bigger.

Label any of the details you can see.

Looking at cells

Animal cells

Animal cells are not as easy to see as plant cells. They are almost invisible under the microscope. To see them, they are treated with special dyes called **stains.** These colour parts of the cell, often the nucleus and the cell membrane. When you look at your slide of animal cells, note down what colour they are stained.

You will be given a slide of human cheek cells. These are scraped gently from the inside of the mouth and added to a drop of stain on a microscope slide.

1. Look at your slide under the microscope and make a careful drawing in pencil of one cell. You should be able to see the cell membrane, the nucleus and the cytoplasm.
2. How are animal cells different from plant cells?
3. Why does this make them so much harder to see?

Looking at cells

Green plant cells

The cells from a plant leaf look more like the diagrams in your textbook. They are much more complicated than simple onion or rhubarb cells. It is difficult to get very thin slices of leaf to look at, so you will look at ready-made slides. You will be using a different sort of microscope.

1 Look at some leaf cells under a higher power. This will be set up ready for you, so focus carefully. Look for the nucleus, the cell wall, the chloroplasts, the cytoplasm and the vacuole. (You might not be able to see them all.)

2 Make a large drawing of a single cell and label as much as you can on it.

3 The onion or rhubarb cells you looked at were not green. Why not?

Activity 2d,e,f Core

All creatures great and small

On spreads 2d, 2e and 2f in your textbook you can see how scientists divide living organisms into groups to identify them. In this activity you will have a go at doing this.

You will have a selection of animal specimens. Some of your specimens may be alive, but many of them will be dead and preserved in some way. This makes it easier to have lots of different types of creature in the same place at the same time.

1 Look carefully at the different animals. They all belong to the same group. Decide which group they belong to.

2 Prepare a two-minute talk on your group of animals and then present it to the rest of the class.

- Show the special features of the group which can easily be seen from looking at the animals.
- Use your textbook to find out more about your group of animals. Your teacher may give you other places to look for information as well.
- Remember that you need to tell your classmates as much as possible about your animal group in an interesting way, so that they remember it.

All creatures great and small

Here are some questions to help you plan your talk:

1 What is the name of your group of animals?

2 What are the similar things that all members of your group have? (Look for things like fur, or hard outside skeletons, or feathers and wings.)

3 Look at the animals. Can you see the special features easily? Find one which is very clear to hold up during your talk.

4 What does it say in your textbook about animals like these?

5 Are all the animals you have got the same sort of size? Find out whether they are usually small, or big, or if they range from small to big.

6 Where do your animals usually live?

7 Can you find out any more about these animals using other information? Ask your teacher to help you decide where to look for information.

The key to the class

1. Plan a key which would identify every member of your class. You need to begin with everyone, and then think of questions which will break you down into smaller and smaller groups.

   Think carefully about the differences you will use to separate people. The similarities and differences you choose must be there next week as well as today, so avoid things like hair styles and clothing because they can change from day to day.

2. Compare your ideas for a key with others in your group.

3. Decide on the best ways of grouping everyone, and produce one big class key. Everyone must produce a drawing of themselves for the final lines of the key.

4. Persuade someone who doesn't know your class well, such as a different teacher or the lab technician, to come and try out your finished key.

The key to the class

Making and using keys can be rather confusing. Here is a simple example.

Make a key to identify these things from Sofie's pencil case.

Every time you make a group, all the things in it must be similar, and quite different from the other groups.

Here is one way of doing it:

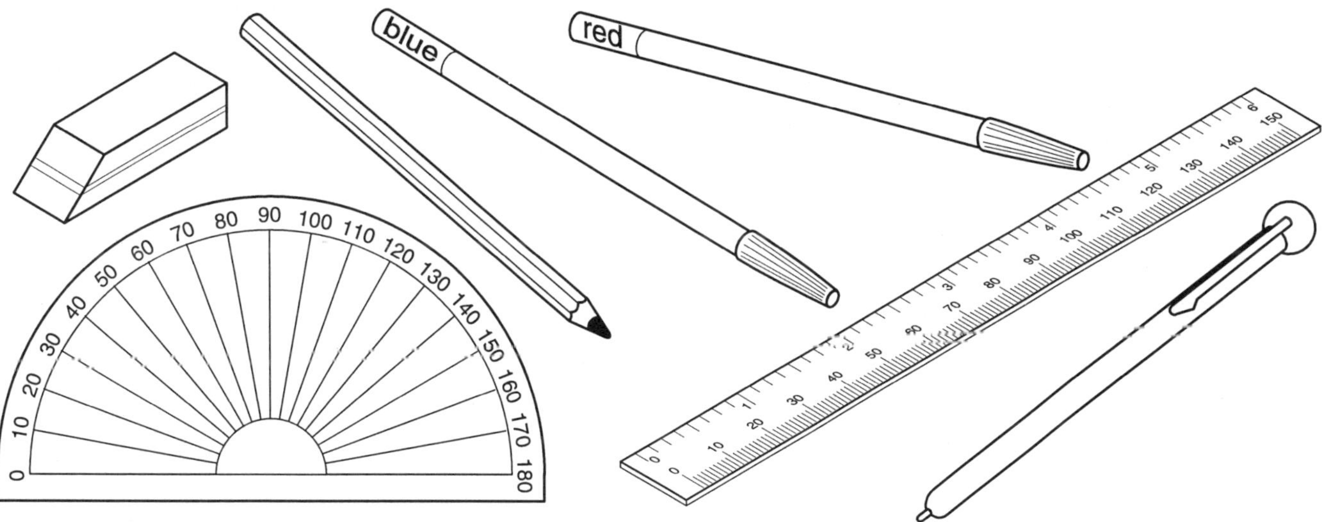

1 Split the pencil case contents into two groups. You could start with things you can write with and things you can't write with.

1 Can you write with it?
yes — go to question **2**
no — go to question **5**

2 Now look just at the writing things and split the group again.

2 Does it write using ink?
yes — go to question **3**
no — pencil

3 Now concentrate on the pens.

3 Does it have a felt tip?
yes — go to question **4**
no — biro

4 Finally in this group, separate the felt-tip pens.

4 Does it use red ink?
yes — red felt-tip pen
no — blue felt-tip pen

5 Now go back to the things you don't write with, and separate that group too.

5 Can it be used to rub out mistakes?
yes — rubber
no — go to question **6**

6 Now there are only two things left.

6 Is it used for measuring angles?
yes — protractor
no — ruler

7 If you have time, make a key for the things in your own pencil case, or your lunch box or school bag.

Pouring liquids

Estimating volumes

How good are you at estimating the volumes of different containers? Do this activity to find out.

1 Look at the collection of liquid containers provided. Think about the volumes of the containers. Arrange them in order, from the one you think has the smallest volume to the one you think has the largest volume.

2 Number the containers from 1 (smallest) upwards. Draw up a simple table like this:

Container number	Volume (cm^3)

3 Fill container 1 with water. Tip this water into a measuring cylinder. Record the volume in your table. This is the volume of container number 1.

4 Repeat step **3** for the other containers in turn.

5 If the measured volumes get bigger each time, you are very good at estimating volumes!

6 If you got a container in the wrong order, try to think why you thought it was bigger (or smaller) than it was.

Pouring liquids

How runny is it?

Liquids can flow and change their shape. But different liquids flow in different ways. Some liquids, like water, are very runny. They flow easily and quickly. Water will pour out of a jar in a fraction of a second. Other liquids, like honey or syrup, are 'thick' and move much more slowly. Syrup might take several seconds to pour from its jar.

The scientific term for 'runniness' is **viscosity**. Runny liquids like water have a **low viscosity**. 'Thick' liquids like syrup have a **high viscosity**.

Investigating 'runniness'

You can investigate how runny a liquid is by timing how long it takes to flow through a filter funnel. A runny liquid will flow through faster than the same volume of a 'thick' liquid.

1 Plan your experiment.
You will need:

- a stopclock
- a filter funnel
- some different liquids (such as water, glycerine, meths, oil, ...)
- a measuring cylinder
- some beakers
- a stand and clamp.

Think about these questions as you plan your experiment:

a How much liquid will you use each time?

b How will you use the stopclock?

c How will you make sure that it is a fair test?

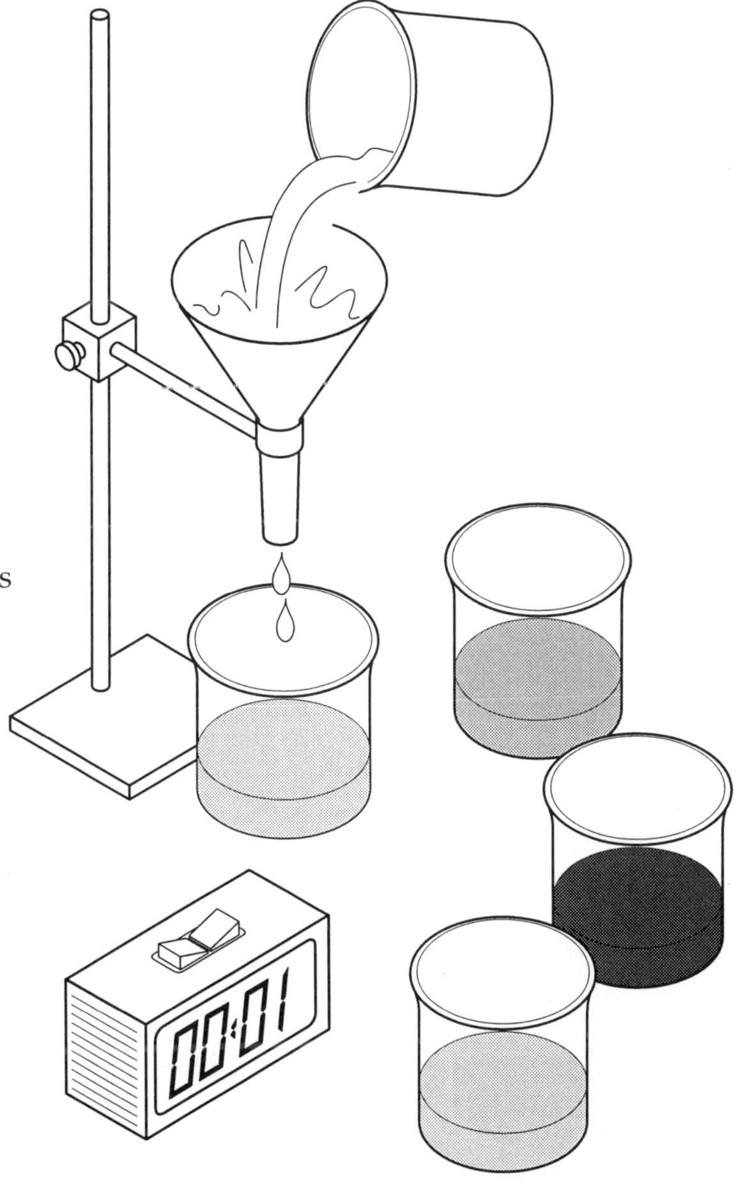

Activity 3b Core

Which bag is best?

You are the manager of a supermarket. You want to order some plastic carrier bags to give to your customers. You want good strong bags that won't break.

Some suppliers have sent you samples of the bags they make. Plan an experiment to test the strength of the sample bags. Here are some ideas to help you:

a John filled his bags with tins of baked beans until the handles broke.

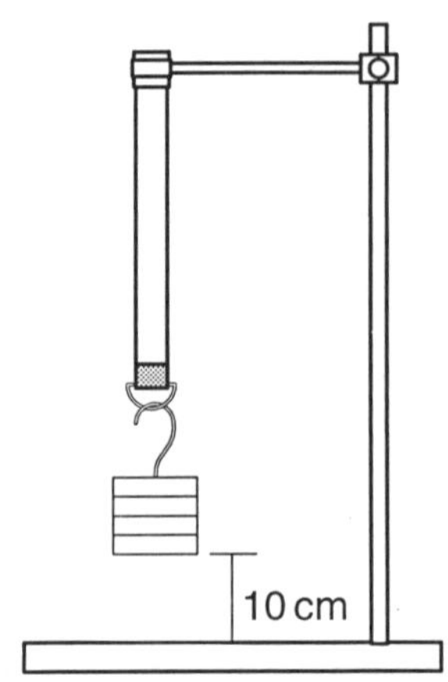

b Sally cut her bags into thin strips and tested these by hanging masses from them.

c Zung hooked the handles of his bags around a door knob and wrote down how hard he had to pull to break them.

1 Which is the best idea, **a**, **b** or **c**?

2 Explain why you think that this is the best idea.

3 Plan your own experiment using this idea.
Think about these questions:

- **a** How could you measure the load needed to break the plastic?
- **b** How could you make sure the test was fair and safe?

4 **Once your plan has been approved**, try it out.
Record your results in a table as you go.

5 Display your results in a bar chart. Choose the best bag for the supermarket, explaining your choice.

6 Compare your results with those from other groups.

- **a** Did you all get the same results?
- **b** If not, whose experiment do you trust the most? Explain your answer.

7 The best bag from your experiment costs twice as much as the second best bag. Would this affect your choice of bag to buy for the supermarket? Explain your answer.

Which bag is best?

Activity 3b Help

1 Cut a strip of plastic 20 cm long and 1 cm wide from each bag. Number each one as you cut it, so that you know which is which.

2 Fold over the bottom 2 cm of strip and tape it as shown in the diagram here.

3 Set up a stand and clamp. Clamp the top of strip 1 between a split cork as shown in the diagram below.

4 Hang a mass holder onto the bottom of the strip as shown. Make sure that the base of the holder is about 10 cm from the bench.

5 Add 100 g masses one at a time until the plastic strip snaps.

6 Use the table to record the strip number and the number of 100 g masses needed to snap it.

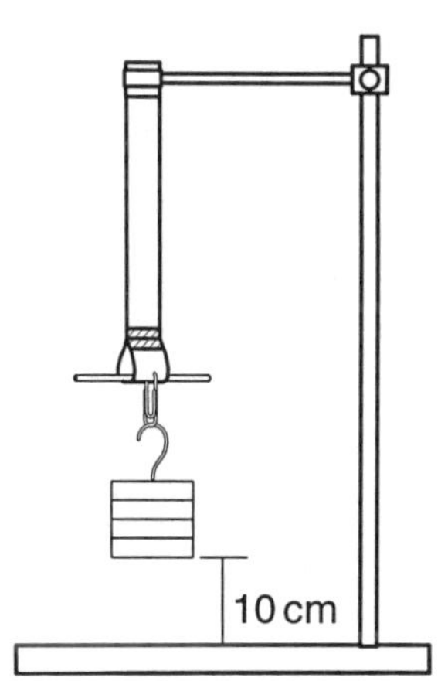

Plastic bag number (and name)	Number of 100 g masses needed to snap it
1	
2	
3	
4	
5	
6	

7 Repeat steps **2** to **6** for the second strip, and the other strips one by one.

8 When your table is completed, answer these questions:

a Which strip was the strongest?

____

b Which bag would you buy?

____

c Why would you buy that bag?

____

Metals and non-metals

Can you tell a metal from a non-metal? Your teacher has set out some displays to help you sort out your ideas. Each display shows just one common difference between metals and non-metals.

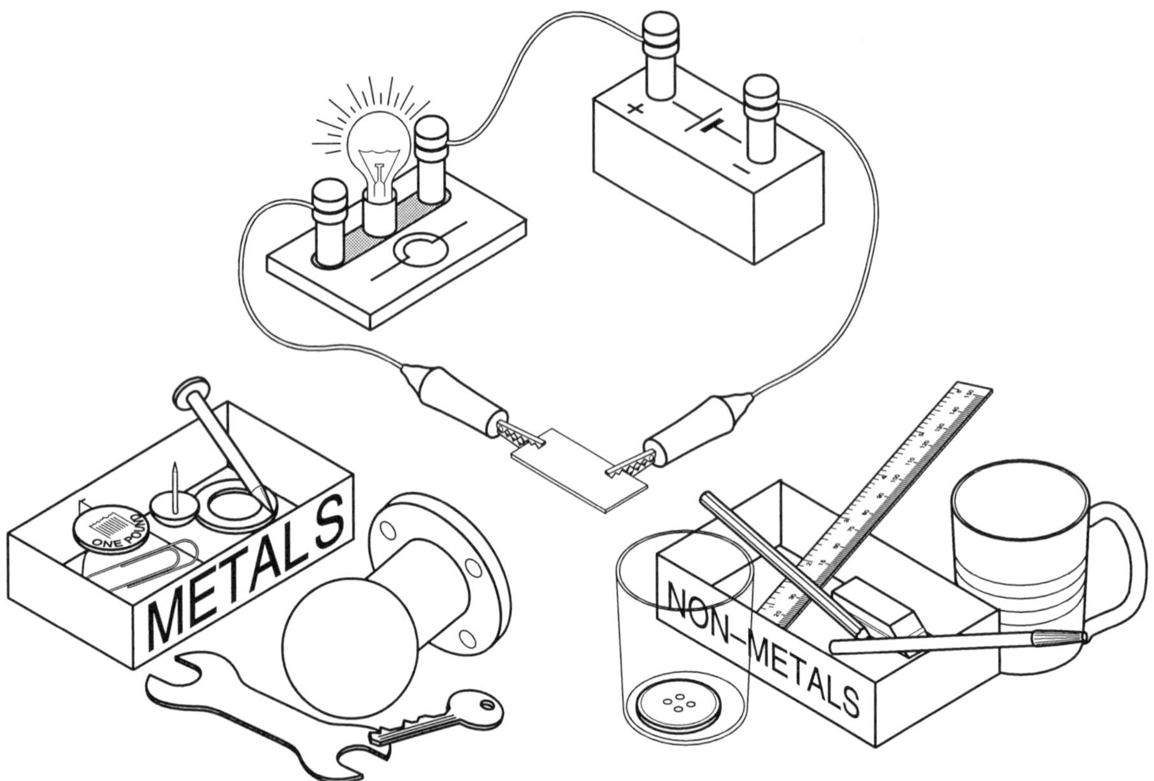

Does it conduct electricity?

Your teacher will demonstrate one example.

1. Draw up a table with four large columns headed **Display number, Property, Metals** and **Non-metals**.
2. Look at one display and see if you can decide what property is being shown. (If you are not sure, ask.) Write this in the **Property** column.
3. In the **Metals** column, describe what is the same about this property for all the metals.

   In the **Non-metals** column, compare the non-metals (these may be much more variable).
4. Go to the other displays in turn and complete your table.
5. Look at your completed table. What does it tell you about the properties of a typical metal?
6. Now look at the **Non-metals** column. Is it so easy to describe the properties of a non-metal?
7. Write a brief report for the class, explaining how you could tell metals and non-metals apart.

Take care

Wash your hands after handling lead.

Metals and non-metals

1 Complete this table about metals and non-metals.

Display number	Property	Metals	Non-metals
demonstration	What happens when you heat it?		
1	What does it look like?		
2	Does it conduct electricity?		
3	Can you bend it?		
4	Is it solid, liquid or gas?		
5	How strong is it?		
6	How 'heavy' is it?		

2 List all the properties that metals have in common.

____

____

____

____

3 List all the properties that non-metals have in common.

____

____

____

Metals and non-metals

1 Look at the data in this table.

2 For substances **a**–**e**, decide whether it is a metal or a non-metal.

Substance	What does it look like?	Does it conduct electricity?	Can you bend it?	Is it solid, liquid or gas?	How strong is it?	How 'heavy' is it? (water = 1)
a	silvery grey	yes	yes	solid	strong	7
b	purple-black	no	no	solid	weak	5
c	shiny white	yes	yes	solid	fairly strong	19
d	dark grey	no	no	solid	weak	3
e	brown	no	–	liquid	–	3
f	silvery	yes	–	liquid	–	14
g	silvery white	yes	yes	solid	weak	0.5

3 Substances **f** and **g** are both metals, but they show some unusual properties. For each, write down the properties that are unusual in a metal.

'Heaviness' and volume

A first try

> **Take care**
> Wash your hands after handling lead.

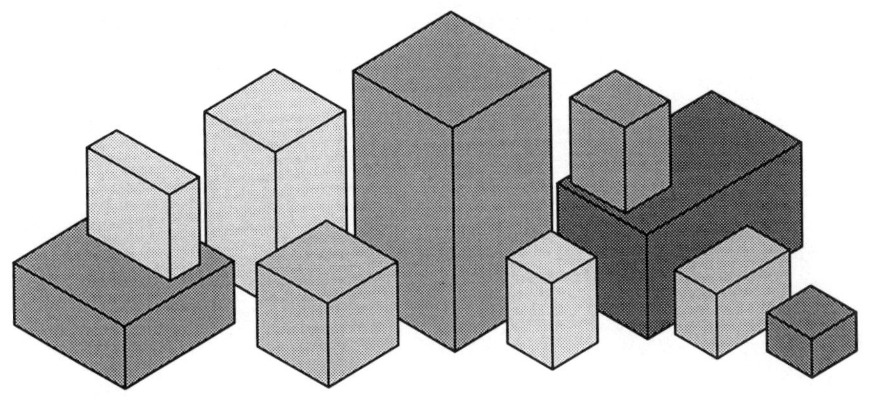

1 Without picking them up, look at the collection of numbered blocks you have been given. Which would you expect to have the larger mass, a small block or a large block?

2 Plan an experiment to test your idea. You will need to think about these points:

- How will you find the volume of each block? (You may be given the volumes.)
- You will need to draw up a table to record your results.

3 **Once your plan has been approved**, try it out.

4 Can you see a clear pattern in your results? If not, was your first idea wrong, or do you just need to improve your experiment?

A fair test

The problem with your first experiment is that volume is not the only factor that affects how heavy something is. The blocks were made of different materials, which have different densities. Changing two factors like this, volume and density, confuses the results.

To make a fair test, you need to look at blocks made of the same material, so that only the volume changes. This will show you whether there is a pattern between the volume and the mass of a block.

5 Five of the blocks are made of the same material.

a Transfer the results for these blocks to a new table, arranging them in order of increasing volume.

b Can you see any relationship between volume and mass now? Describe the relationship you see.

6 Plot a line graph of your results. Put the volume in cubic centimetres as the horizontal (x) axis and the mass in grams as the vertical (y) axis.

'Heaviness' and volume

A first try

1 Look at the blocks without picking them up.

a Which block do you think is the heaviest?

____

b Which block do you think is the lightest?

____

2 Weigh each block in turn and write its mass in the table. Your teacher will tell you the volumes.

3 Look at your results. Were you right about the lightest and the heaviest?

____

Block number	Mass (g)	Volume (cm^3)
1		
2		
3		
4		
5		
6		
7		
8		
9		
10		

Table 1

This first experiment was not a fair test. The blocks are made of different materials. So big ones can be lighter than small ones.

A fair test

4 Five of the blocks are made of the same material.
Put a star (*) by these blocks on table 1.

5 Copy the information about these five blocks into table 2. Change the order so that the smallest block is first.

6 Can you see any pattern?

7 Complete these sentences.

a If you weigh blocks of the same material, the mass of the block depends on its ______________.

b As the volume goes up, the mass __________ __________ too.

Block number	Mass (g)	Volume (cm^3)

Table 2

⚠ **Take care**

Wash your hands after handling lead.

Making sandcastles

Solids keep their shape because all their particles are stuck together. The hardness of a solid depends on how well the particles are stuck together.

Dry sand flows like a liquid, because all the sand grains can slip over one another. Dry sand is useless for making sandcastles.

Damp sand makes good sandcastles. The water helps to stick the sand grains together, so they act like the particles in a solid. But this water 'glue' is not very strong, so the sandcastles are not very hard. They are easily broken.

If you mix cement with your damp sand, the grains are stuck together very strongly when the cement sets. A sandcastle built this way would be very hard. This is a good model for a hard solid. Concrete is made like this.

What's the best mix?

You could make some hard sandcastles by mixing wet sand with cement. But how much cement should you add? You want to use as little as possible, as cement is expensive.

Your teacher will give you plaster of Paris to use, as this is safer than cement.

1 Plan a series of experiments to find the best mix. You will need:

- sand and plaster of Paris
- yoghurt pots or plastic cups
- a spoon or spatula.

Think about these questions:

a How will you measure your sand and plaster?

b How will you make your sandcastles, and how long will you leave them to set?

c How will you measure how hard or strong they are? (You could load them with weights, or drop balls on them from different heights, ...)

d How will you make sure your tests are fair and safe?

2 **Once your plan has been approved**, make your sandcastles and test them. Display your results in a bar chart or graph.

Making sandcastles

How to make your sandcastles

1. Collect six yoghurt pots and enough sand to fill them, a mixing bowl, some plaster of Paris and a spoon. Number the pots 1 to 6. (Plaster of Paris is safer to use than cement.)
2. Tip one yoghurt pot-full of sand into the mixing bowl. Slowly mix in water until the sand is very wet. (If you add too much water, you can carefully tip a little away.)
3. For the first sandcastle, add one spoonful of plaster of Paris and mix in well.
4. Pour the mixture back into the pot and leave it to set.

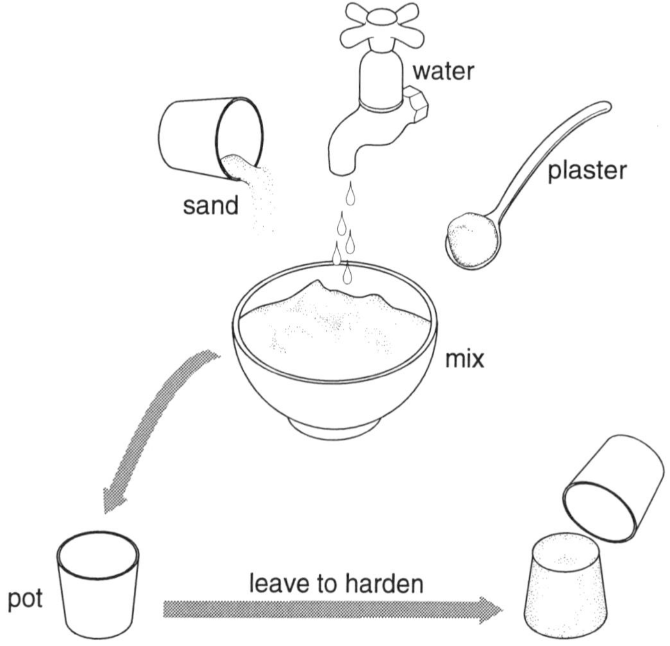

5. Repeat steps **2** to **4** for the next five pots, adding 2 spoonfuls of plaster of Paris to the second, 3 to the third, and so on.
6. Leave your pots to harden till the next lesson.

Activity 3e Help

Making sandcastles

Testing for strength

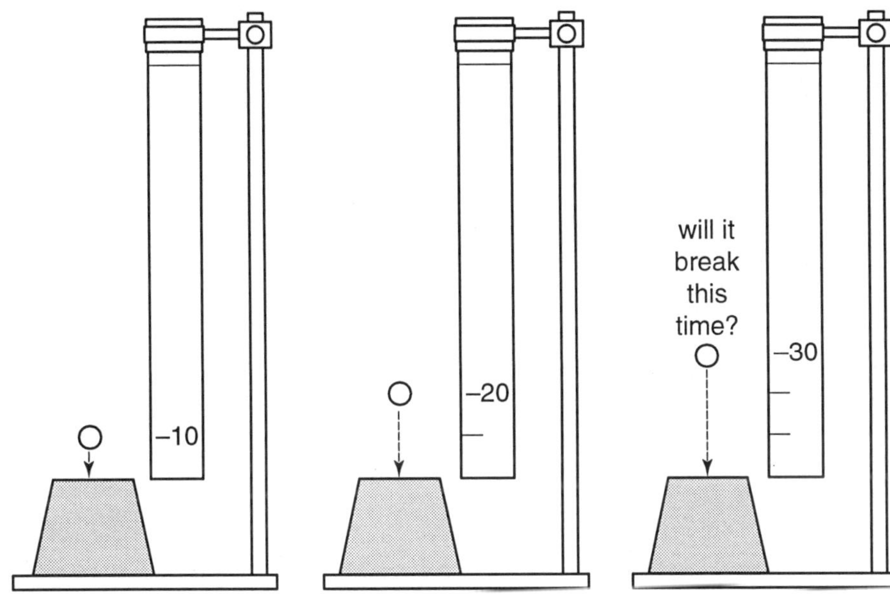

1 Collect a large steel ball bearing.

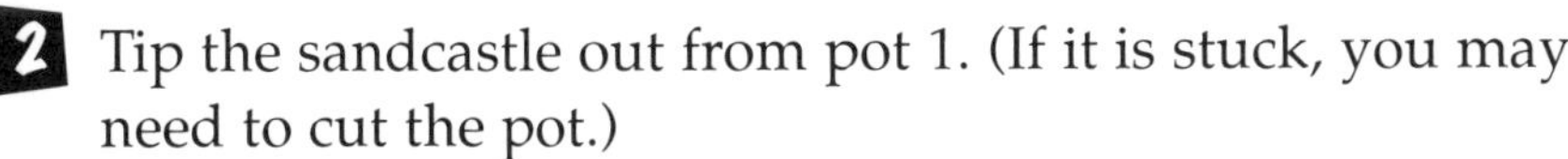

2 Tip the sandcastle out from pot 1. (If it is stuck, you may need to cut the pot.)

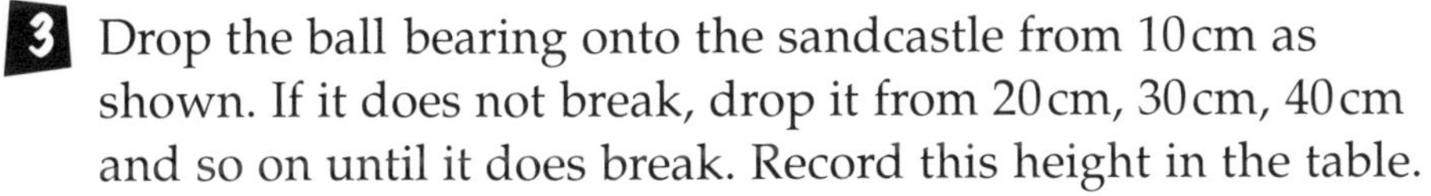

3 Drop the ball bearing onto the sandcastle from 10 cm as shown. If it does not break, drop it from 20 cm, 30 cm, 40 cm and so on until it does break. Record this height in the table.

4 Repeat steps **2** and **3** for pots 2 to 6, recording your results in the table.

5 Display your results in a bar chart.

6 Which mixture gave you the strongest sandcastle?

7 Which mixture would you use to make strong sandcastles if the plaster of Paris was very expensive?
Explain your answer.

Pot number	Spoonfuls of plaster	Height needed for ball to break it
1		
2		
3		
4		
5		
6		

Capturing kinetic energy

Hot air rises. It is moving, so it must have movement energy. The scientific name for movement energy is kinetic energy.

You can capture some of the kinetic energy of hot air by making your own piece of spinning kinetic sculpture.

1 Start with a circle of card. Design your own kinetic sculpture. One idea is shown here.

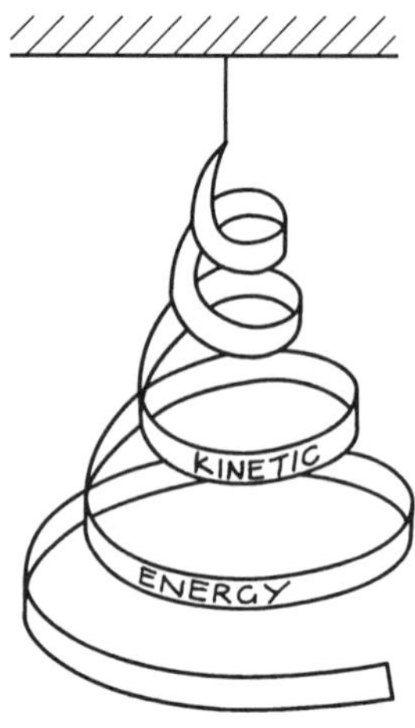

2 Use a pin to support your sculpture, or hang it from a thread. Place it where is can spin around in the hot air rising from a radiator.

3 How could you adjust your sculpture so that it spins faster?

4 Whose sculpture spins the fastest?

Fuels: chemical energy stores

Here are some different fuels. They all store chemical energy. If you burn them, their chemical energy becomes heat energy.

1 Look at these fuels. Which do you think is the best store of energy? Give a reason to support your guess.

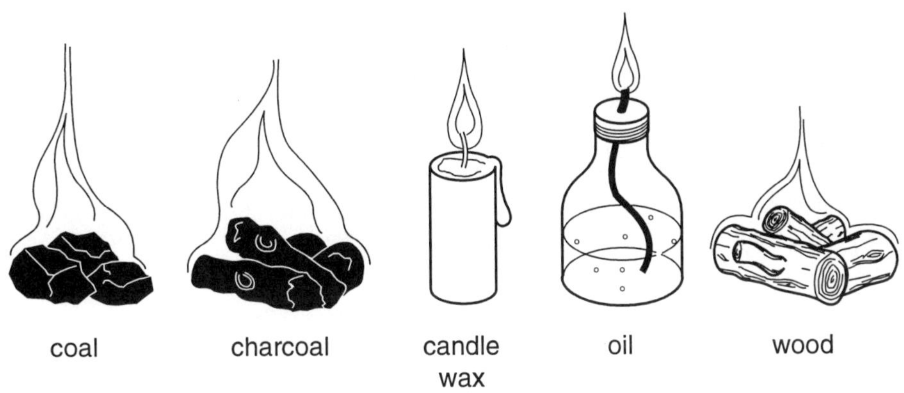

Here are some ideas about how you could find the answer to the question:

- Which fuel stores the most energy?

2 Which do you think would be the best way to compare the different fuels?

3 How would you make it a fair test?

4 What would you measure?

5 How would you decide which fuel stores the most energy?

Camping supplies

You are going camping, and you have to take your own food. You need to take enough to give you the energy you will need for a very active 24 hours. You will need food for:

- supper, breakfast and lunch, and any snacks between meals.

Because you will be walking, climbing and swimming, you will need a lot of energy:

15 000 kJ

1. Make up a shopping list from the following table.
2. Write a menu for your camping trip.
3. Give the energy value of each meal.

Food	Quantity	Energy value
milk	1 pint	1100 kJ
bread and butter	1 slice	200 kJ
peanuts	1 packet	600 kJ
chocolate	1 bar	1200 kJ
cornflakes	1 serving	500 kJ
crisps	1 packet	200 kJ
baked beans	1 small tin	800 kJ
lemonade	1 small bottle	200 kJ
soup	1 tin	1200 kJ
spaghetti	1 small tin	800 kJ
eggs	1 size 2	400 kJ
cheese sandwich	1	800 kJ
flapjack	1 slice	500 kJ
yoghurt	1 pot	400 kJ
rice pudding	1 small tin	1200 kJ
apple	1	100 kJ
banana	1	100 kJ
orange	1	100 kJ

Activity 4c Help

Camping supplies

1 On a sheet of squared paper, draw a rectangle 10 cm × 15 cm.

2 Decide what foods you want to take when you go camping. Cut them out and stick them inside your rectangle. (Each 1 cm square represents 100 kJ.)

3 Have you filled your rectangle? If you don't, you will be very hungry!

milk

bread and butter

peanuts

chocolate

cornflakes

crisps

baked beans

lemonade

soup

spaghetti

eggs

cheese sandwich

flapjack

yoghurt

rice pudding

apple

banana

orange

Electrickery

Electricity is a good way of transferring energy from place to place. Lots of things need electricity to make them work.

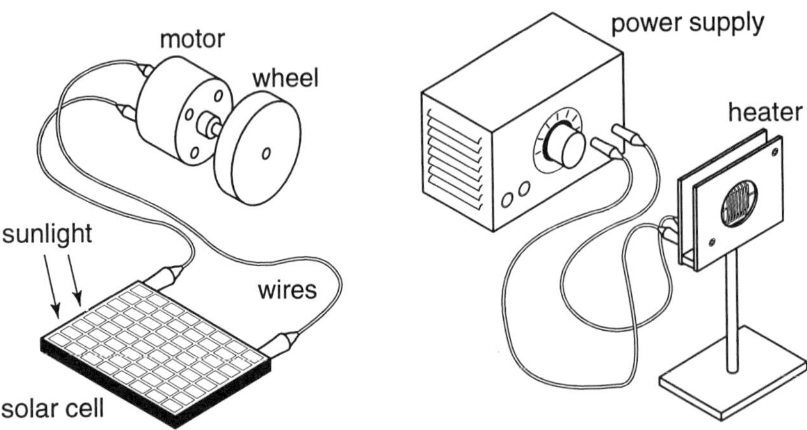

There are different ways to make electricity. You can use a battery, which has a store of chemical energy. You can use a dynamo.

1. Look at the different demonstrations which show electricity at work.
2. For each one, look for the wires that are carrying the electrical energy.
3. Decide on the different forms of energy involved.
4. Draw an energy transfer diagram for each demonstration. Here is a list of forms of energy which you may find useful:
   - electrical energy
   - chemical energy
   - kinetic energy
   - heat energy
   - light energy
   - sound energy.

Electrickery

Here are some energy transfer diagrams.
Which one belongs to which demonstration?

Complete the table.

Demonstration	Energy transfer diagram
	chemical energy → (electrical energy) → light and heat
	light → (electrical energy) → kinetic energy
	chemical energy → (kinetic energy, electrical energy) → light and heat
	electrical energy → heat and light
	electrical energy → kinetic energy
	chemical energy → (electrical energy) → sound energy

Jumpers

Have you ever used a pogo stick? When you squash its spring, the spring stores elastic energy. Then the spring pushes you up in the air.

You can make a 'jumper' using a stick and a cotton reel. It has a rubber band to store elastic energy. You are going to find out how high it will jump.

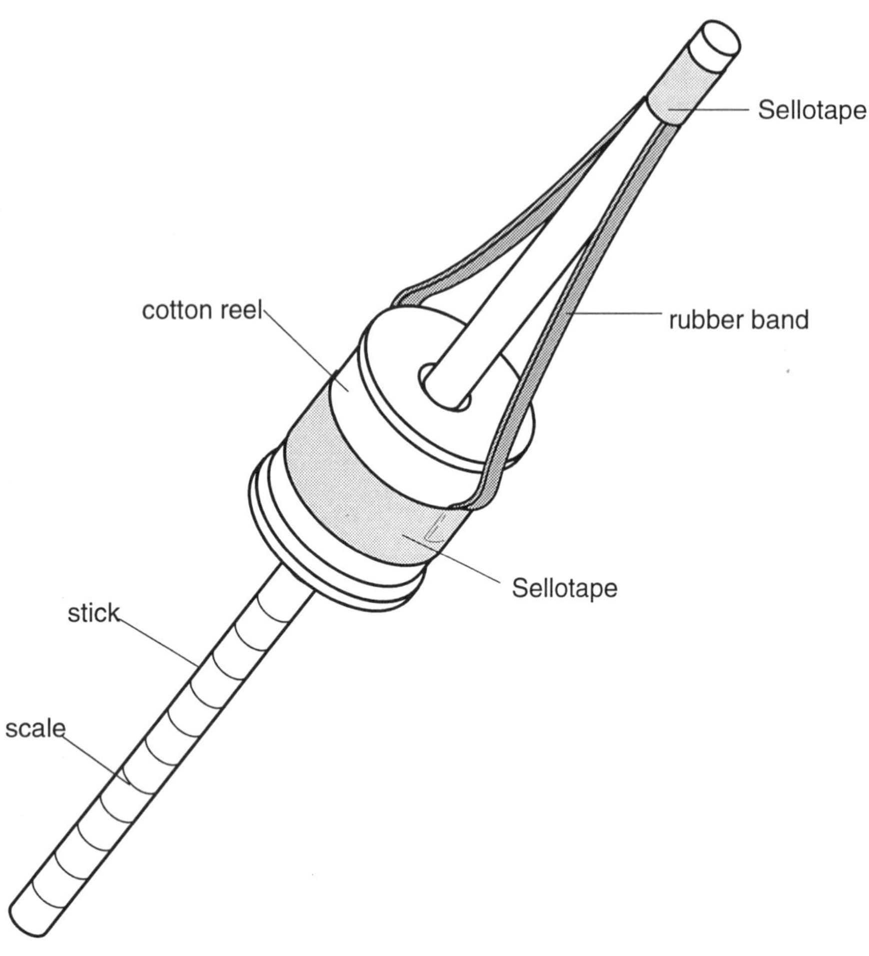

1. Make a jumper.
2. See how high it will jump.
3. How can you make your jumper jump higher?
4. Try pulling the cotton reel down by different amounts. Does your jumper go to different heights?
5. Mark a scale on the stick of your jumper.
6. How will you measure how high your jumper goes?
7. For each mark on the scale, see how high the jumper jumps. Do this three times, and work out the average height.
8. Record your results in a table.
9. Try a different rubber band, and repeat your investigation.

Jumpers

How to mark a scale on the stick of your jumper:

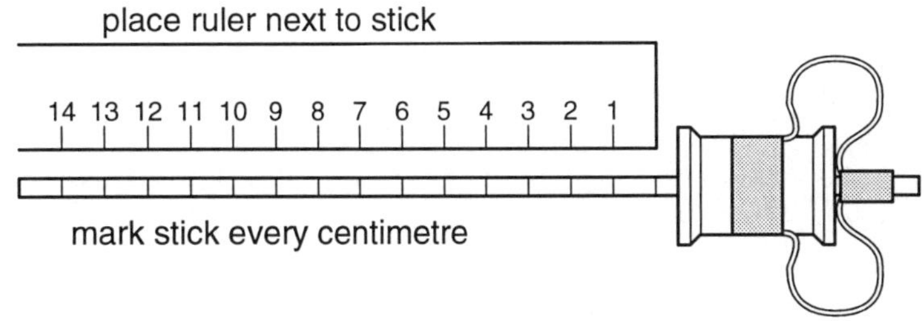

How to measure how high it jumps:

Wear eye protection

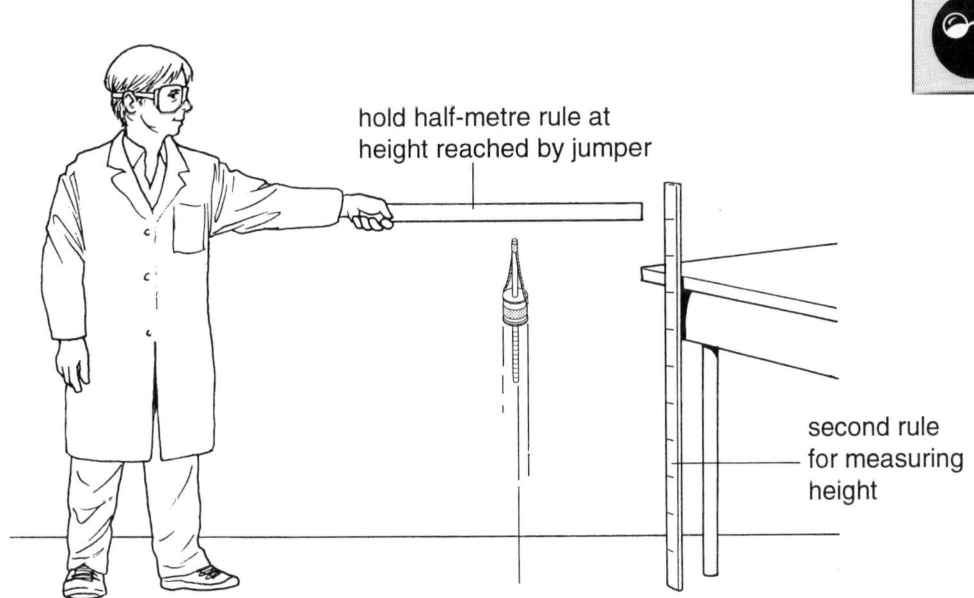

Repeat three times for each scale mark, and record your results:

Scale mark	Height of jump 1	Height of jump 2	Height of jump 3	Average height of jump
1				
2				
3				
4				
5				
6				
7				

Jumpers

You need to use a force to pull down your jumper before you let it go. You can measure the force using a newtonmeter.

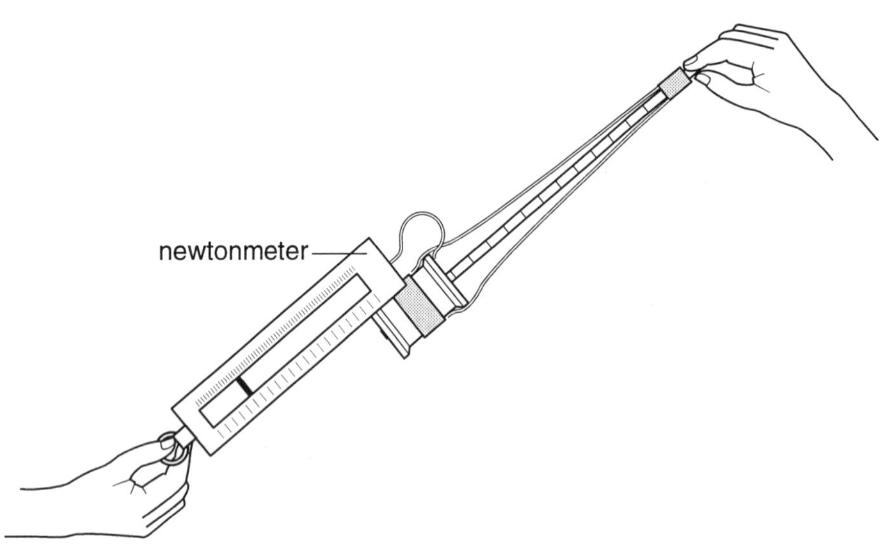

Some rubber bands need a bigger force than others to stretch them.

You are going to investigate how the height of the jump depends on the force needed to stretch the rubber band.

1 Select some different rubber bands.

2 Attach them in turn to your jumper.

3 For each band, measure the force needed to pull the reel down to the bottom of the stick.

4 For each band, measure the height to which your jumper jumps.

5 What factors are you keeping the same each time?

6 Put your results in a table.

7 Draw a line graph of your results.

8 What can you say about how the height depends on the force?

Testing for starch

To make food, a plant needs moist soil, light, air and green leaves. You can see whether a plant has these things, but how can you find out whether it has been making food or not?

Plants trap light energy and use it to turn carbon dioxide and water into carbohydrates. One of the main carbohydrates they make is starch.

Here is a useful test for starch. Try it out to see how it works.

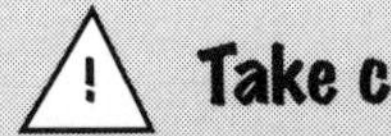

Take care

Do not eat the bread.
It may be contaminated.

The iodine test for starch

1 Collect these things:

- a bottle of iodine solution
- a test tube containing $3\,cm^3$ of starch solution
- a test tube containing $3\,cm^3$ of water
- a white tile
- a piece of bread.

a What colour is the iodine solution?

2 Add 5 drops of iodine solution to the test tube of starch solution and shake it gently. Do the same to the test tube of water.

a What colour is the iodine solution in the starch solution?

b What colour is the iodine solution in the water?

c What happens when iodine reacts with starch?

3 Place the piece of bread on a white tile and add 5 drops of iodine solution to the bread.

a What happens to the iodine solution?

b What does this tell you about bread?

4 The plants in your lab have all had plenty of light, air and water. Plan how you could show that a plant in your lab has made food.

5 **Once your plan has been approved**, take one leaf from a plant and try it out.

6 Does your experiment work as you expected?
Try and explain what has happened.

Has your plant made some food?

Just putting iodine on the leaf of a plant will not tell us whether it has made any starch. This is because the leaf has a waterproof covering which keeps the iodine out. Also, the starch is stored safely inside the plant cells. So to find the starch, we have to give the leaf some special treatment to get rid of the waterproof layer and break open the cells. This is explained on the sheet **Testing a leaf for starch,** which your teacher will give you.

In your lab there are:

- plant A which has been kept in the light
- plant B which has been kept in a dark cupboard for several days
- plant C which has been kept in the light, but with parts of its leaves covered up.

1. Plan an experiment using iodine to show that plants need light to make starch.
2. **Once your plan has been approved**, carry out the experiment.
3. Make a record of your experiment. Say what you did, what happened and what the experiment tells you.

Why roots?

Activity 5b Core

Plant roots hold the plant in the soil. They are also important for getting water into the plant from the soil. You are going to investigate this question:

- Where does the water go as it travels around the plant?

You can use this apparatus:

- a beaker
- blotting paper
- coloured water
- a bean seed which has grown a root system and shoots.

Take care

Do not eat the beans. Some beans have a coat which is toxic until it has been boiled.

1 Plan how you will set up your experiment.

How will you find out where the water has gone in the plant?

2 **Once your plan has been approved**, set up your investigation. Remember that the roots of the growing bean are quite fragile, so handle them carefully.
Label your apparatus with your name and leave it until your next lesson.

3 Next lesson, make a record of your experiment. Say what you did, what happened and what the experiment tells you.

Why roots?

Take care

Do not eat the beans. Some beans have a coat which is toxic until it has been boiled.

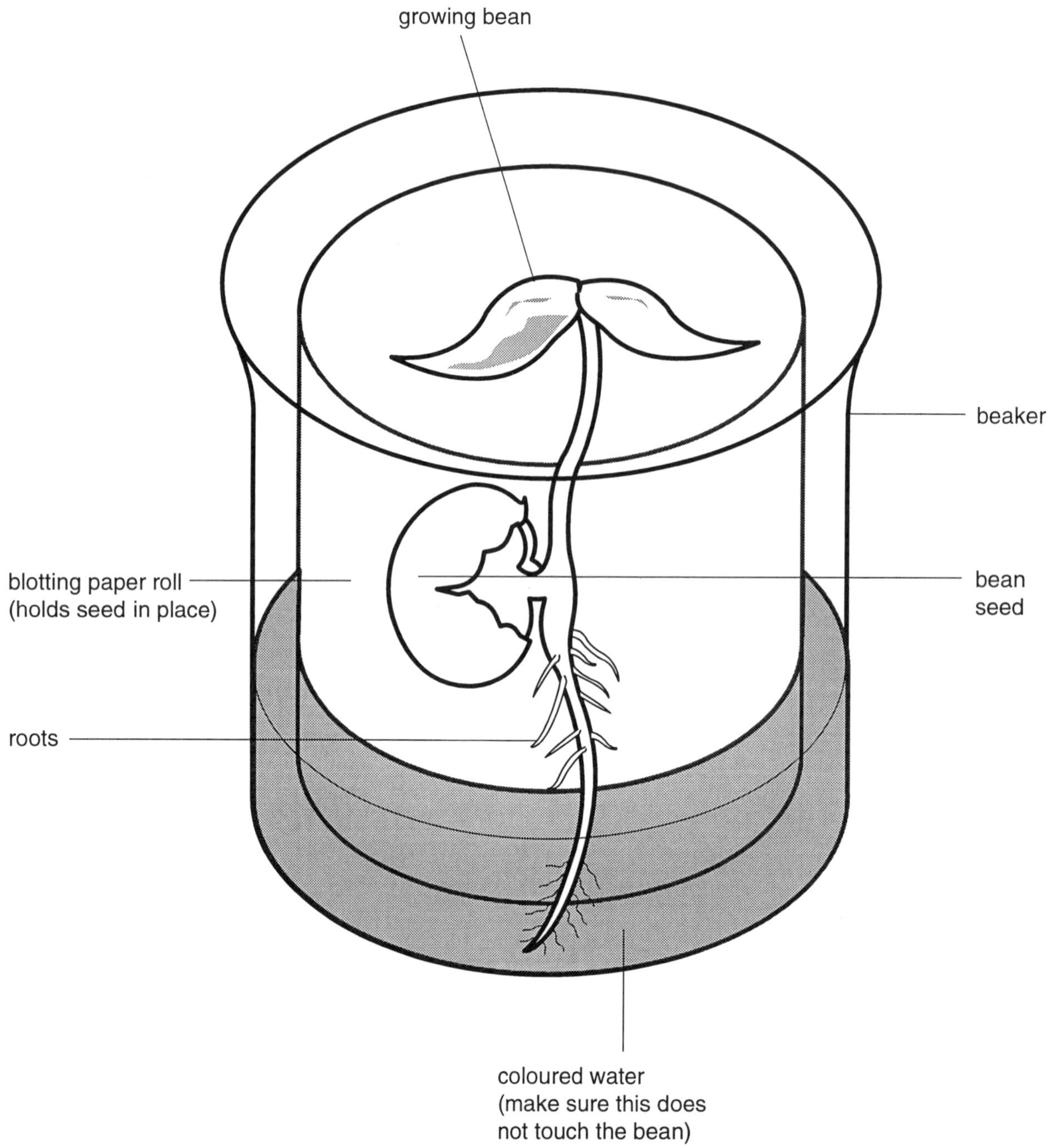

growing bean

beaker

blotting paper roll (holds seed in place)

bean seed

roots

coloured water (make sure this does not touch the bean)

Activity 5c,d Core

Looking at labels

A balanced diet is very important to stay healthy. A balanced diet means eating the right amounts of different types of food.

Food manufacturers put information about their foods on the food packets. A lot of the information is about what the food is made up of, such as the amounts of carbohydrates, fats and proteins. These are measured in grams. There is also information about the amount of energy the food provides. This is measured in kilojoules.

1 Your teacher will give you some food containers. Look at the labels to find out what is in them.

2 Make a table like the one below, and fill in the details of as many foods as you can.

Type of food	Carbohydrate (per 100 g)	Fat (per 100 g)	Protein (per 100 g)	Energy (per 100 g)
baked beans	14.4 g	0.3 g	5.0 g	341 kJ

3 Why do you think you looked at the contents in every 100 grams of the food, rather than the amount in each packet or tin?

4 Look at some of the foods which provide the highest amounts of energy. Are they high in carbohydrate, protein or fat?

5 Which of the foods you have looked at would you choose if:

a you wanted to build up lots of muscles as you train for sports day

b you wanted lots of energy to climb a mountain

c you needed to put on weight?

6 The ingredients in 100 g of dried soup or dehydrated mashed potato usually add up to 100 g. But if you look at your table, you will often find the ingredients do not add up to 100 g. What is the rest made up of? Read the ingredients on some of your labels and see if you can work it out.

Looking at labels

Proteins, carbohydrates and fats are not the only things in a balanced diet. Minerals, vitamins and fibre are very important too. Some packets give you lots of information about all these things, others do not. Look again at your food packets.

1. Which foods show you the vitamins and minerals they contain? Why do you think they do this?

2. Which vitamins and minerals appear most often on the labels?

3. Which foods give information about the amount of fibre in them?

4. An adult should have about 30 g of fibre every day. Which of the foods do you think need to be eaten regularly to get this amount of fibre?

For most of us, the information on food labels is interesting, but it doesn't really matter. But when people have to eat special diets to remain healthy, it is important to know exactly what's in the food.

5. Look at as many of the food packets in the lab as possible, and decide which give:

   - full information (ingredients, energy, carbohydrates, fats, proteins, fibre, vitamins and minerals)
   - just the main ingredients and energy
   - just the ingredients or nothing at all.

   Make a bar chart with the headings **Very full information, Main ingredients and energy** and **Little information**.

6. Why do you think some food manufacturers give so much information on their labels, and others give so little?

7. What information do you think should be on all food labels?

What the owl had for dinner

Activity 5e,f Core

Owls eat a varied diet which contains lots of bits that are difficult to digest. Bones, teeth, fur, feathers and the hard outer skeletons of insects would give the owl indigestion. So the owl gets rid of these bits through its mouth as an **owl pellet**. Owl pellets are dry and do not smell, but they are full of interesting information about the food chains and webs an owl belongs to. Your teacher will give you a sheet showing the remains found in an owl pellet.

1 Find out what the owl has been eating. You might find it helpful to cut the remains out and sort the ones that look the same into piles.This table will help you identify the remains:

Type of animal	What does it eat?
Shrews are the smallest British mammals so their skulls are very small.	insects, worms, slugs and spiders
Mice are a little bigger than shrews.	grass and fruit
Voles are 10–11 cm long and have more rounded skulls than mice.	grass
Rabbits, even baby rabbits, are much bigger than voles. Their skulls have gnawing front teeth and very large eye sockets.	grass and small flowering plants
Bird skulls have beaks instead of jaws.	some eat buds, leaves, fruits or seeds of plants, some eat insects, slugs and worms
Large beetles have indigestible outer skeletons.	some eat plants, others eat smaller beetles
Frogs have very flat skulls with large eyes.	insects, slugs, small fish

2 Make up food chains for the owl and each of the animals in your owl pellet.

3 If you have time, try and link as many of the animals as possible into a food web. Your food web can include organisms which are not found in the owl pellet.

What the owl had for dinner

A food web shows how the animals and plants in the countryside are all linked together.

Use this diagram to help you make a food web for the owl.

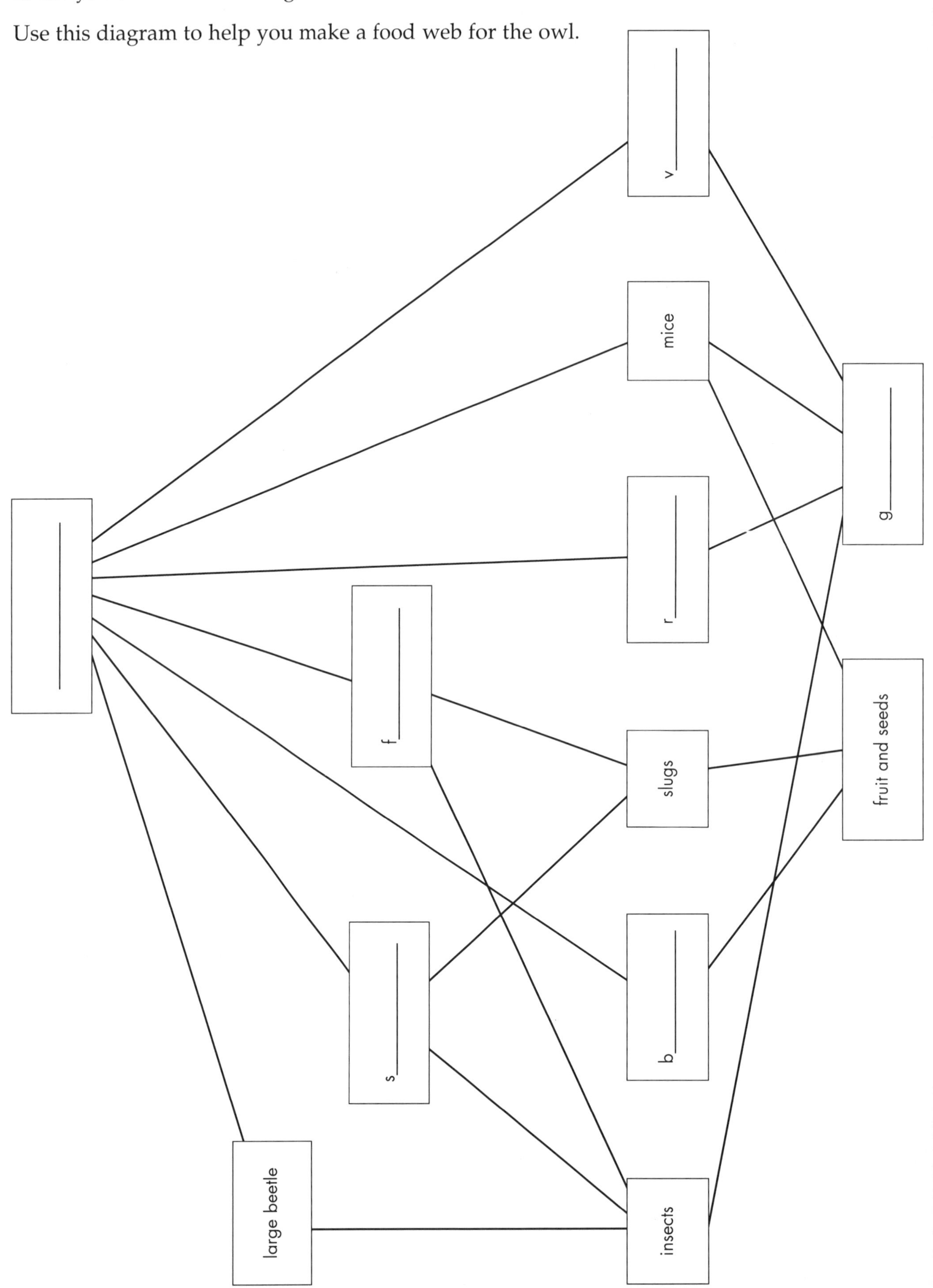

What the owl had for dinner

You have looked at paper samples of the sorts of things that are found in owl pellets. Now think about how useful looking at real owl pellets can be.

1 From the remains in your owl pellet, which animals do owls seem to eat most?

2 Is it very good science to base your ideas on the diet of an owl by looking at one pellet?

3 Your teacher has some more sheets showing what was found in other owl pellets. How can you use these to get a better idea of what owls eat?

4 All the owl pellets were found under the same tree. What are the problems with using these pellets for finding out what all owls eat?

5 James Thomas is a farmer with a number of barn owls nesting on his farm. He is pleased that there are fewer mice and similar animals attacking his crops and stores. He wants to persuade his neighbouring farmers to put up 'owl boxes' to encourage more barn owls to nest.

How would you use evidence from the owl pellets to help James convince the other farmers? Remember that:

- different owls living in different places might eat different things
- the same owl might eat different food on different days.

Pellet sheet 1

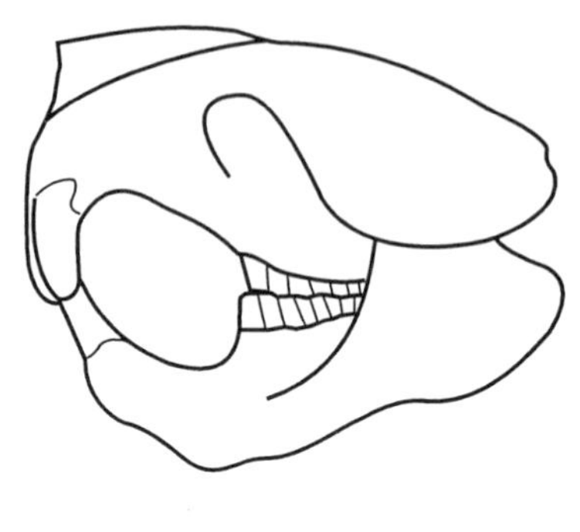

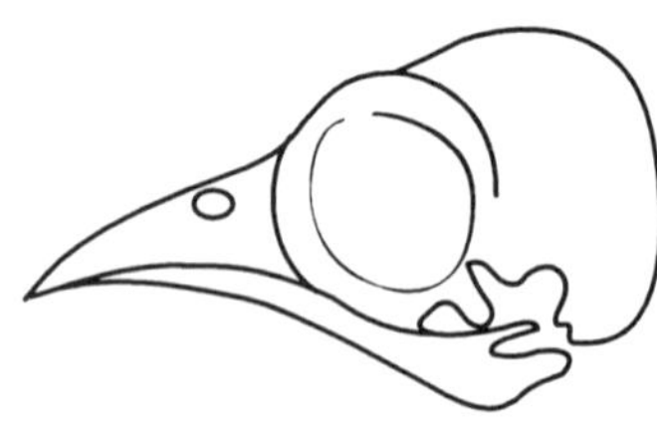
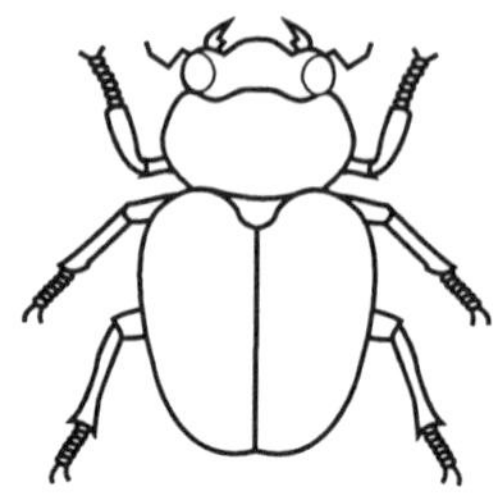
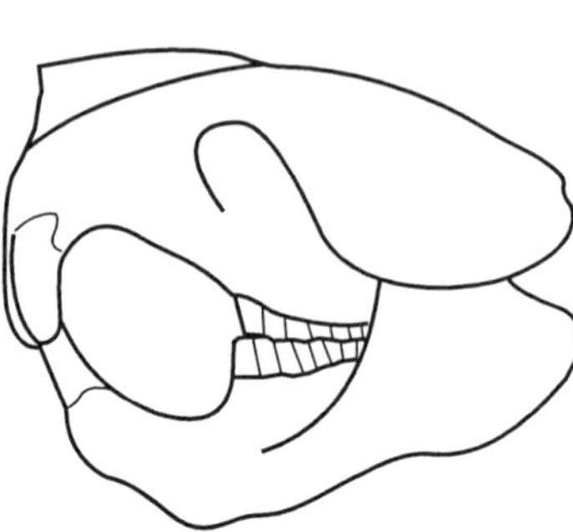
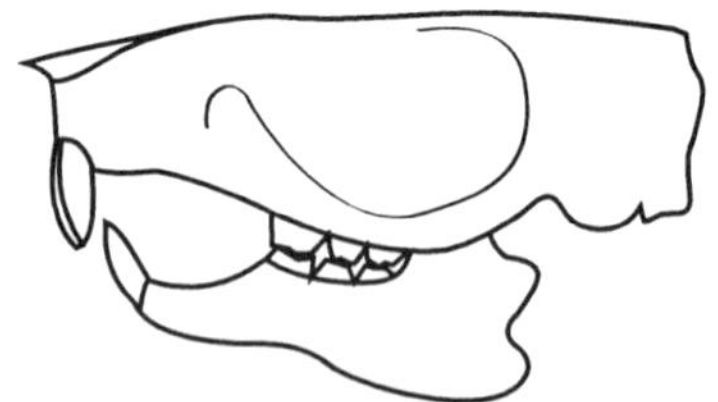

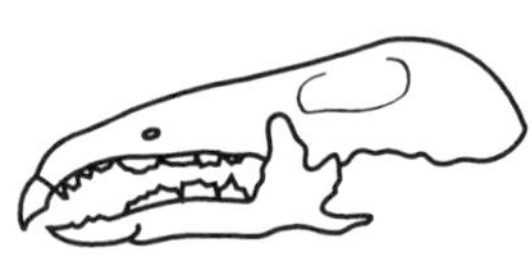
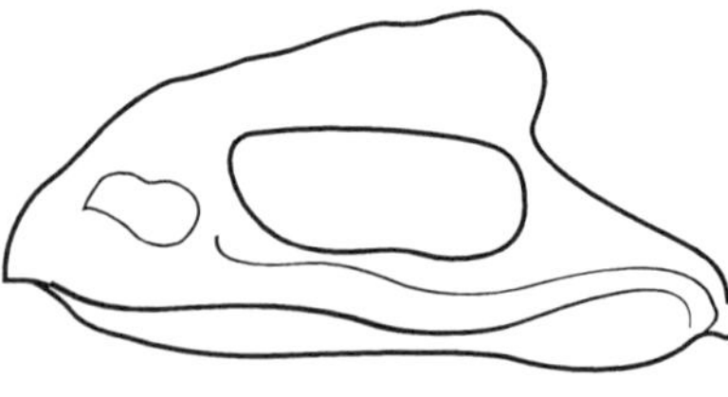
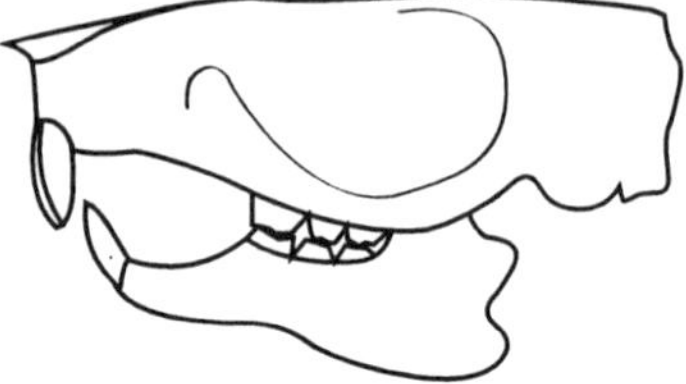
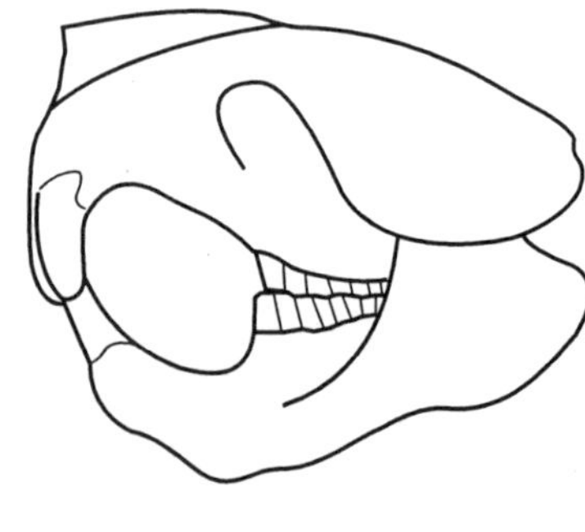
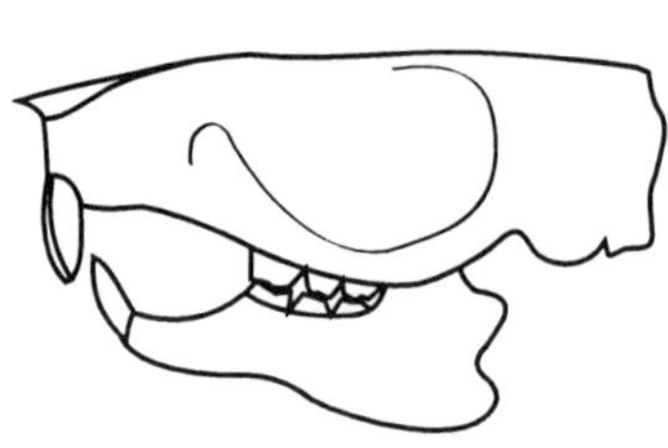
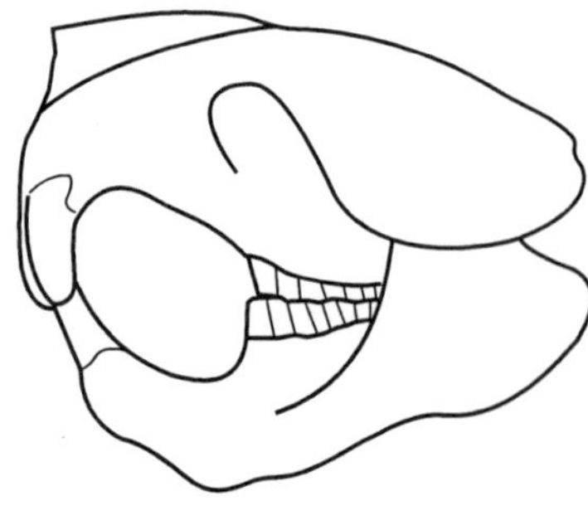

Pellet sheet 2

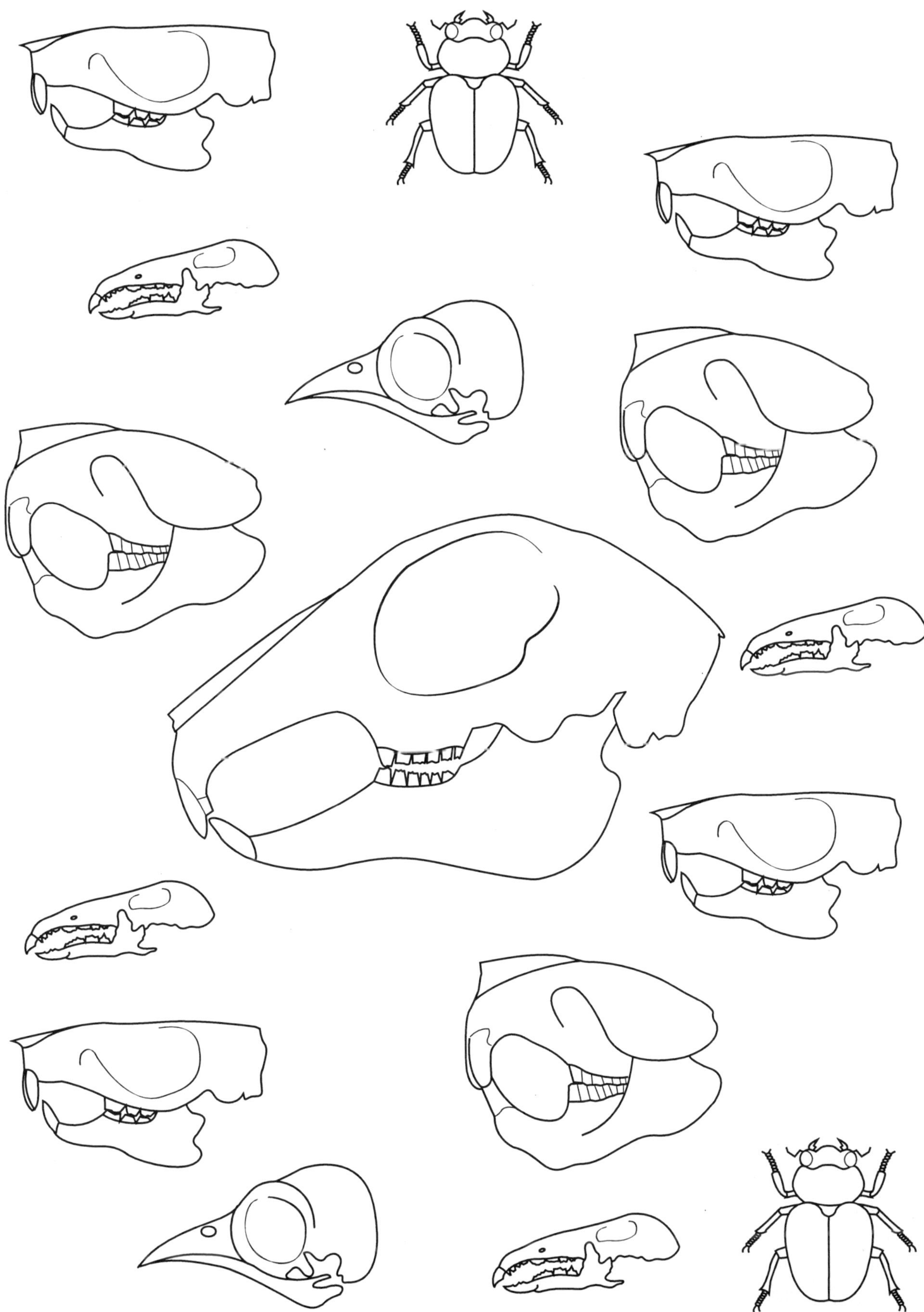

Pellet sheet 3

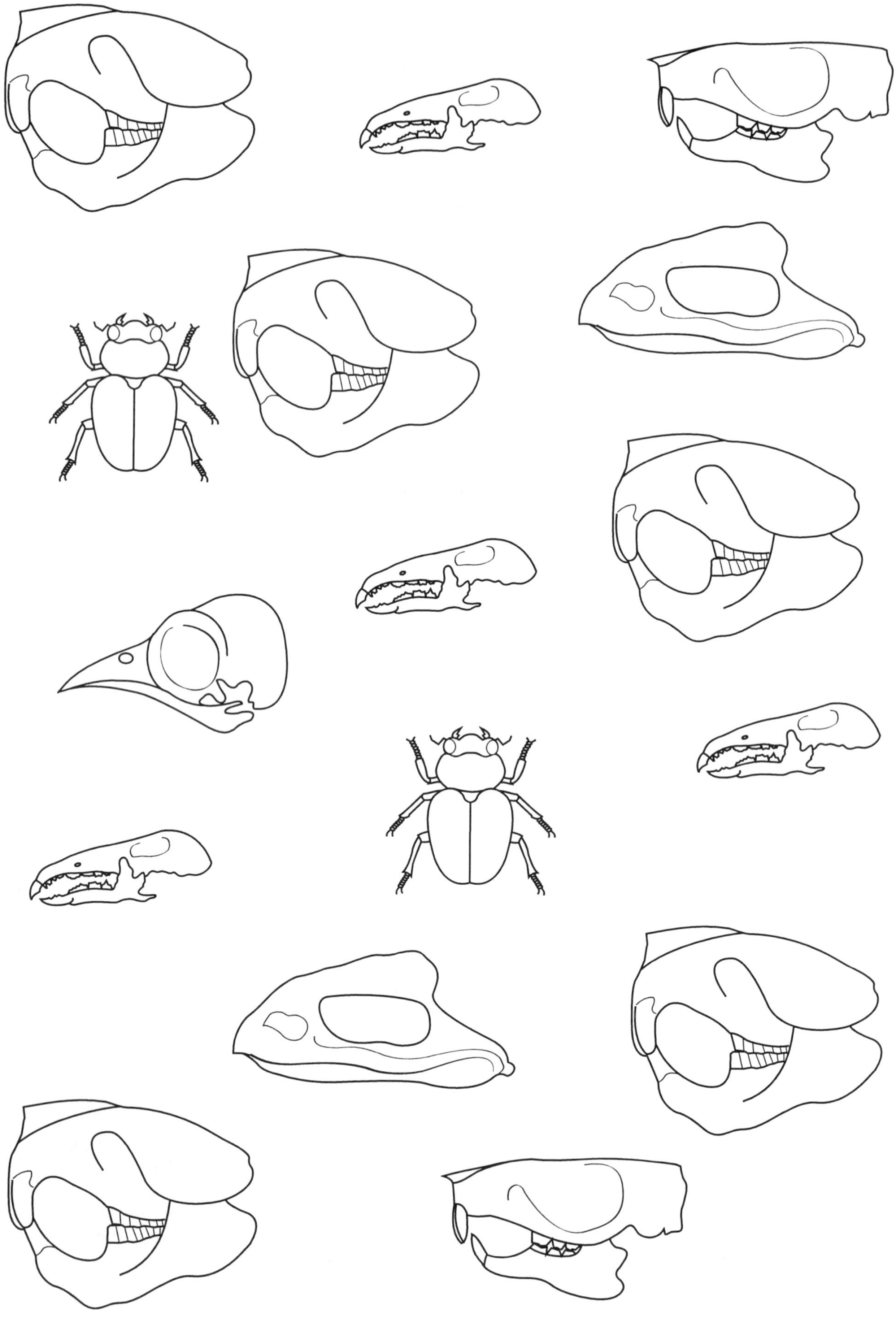

Pellet sheet 4

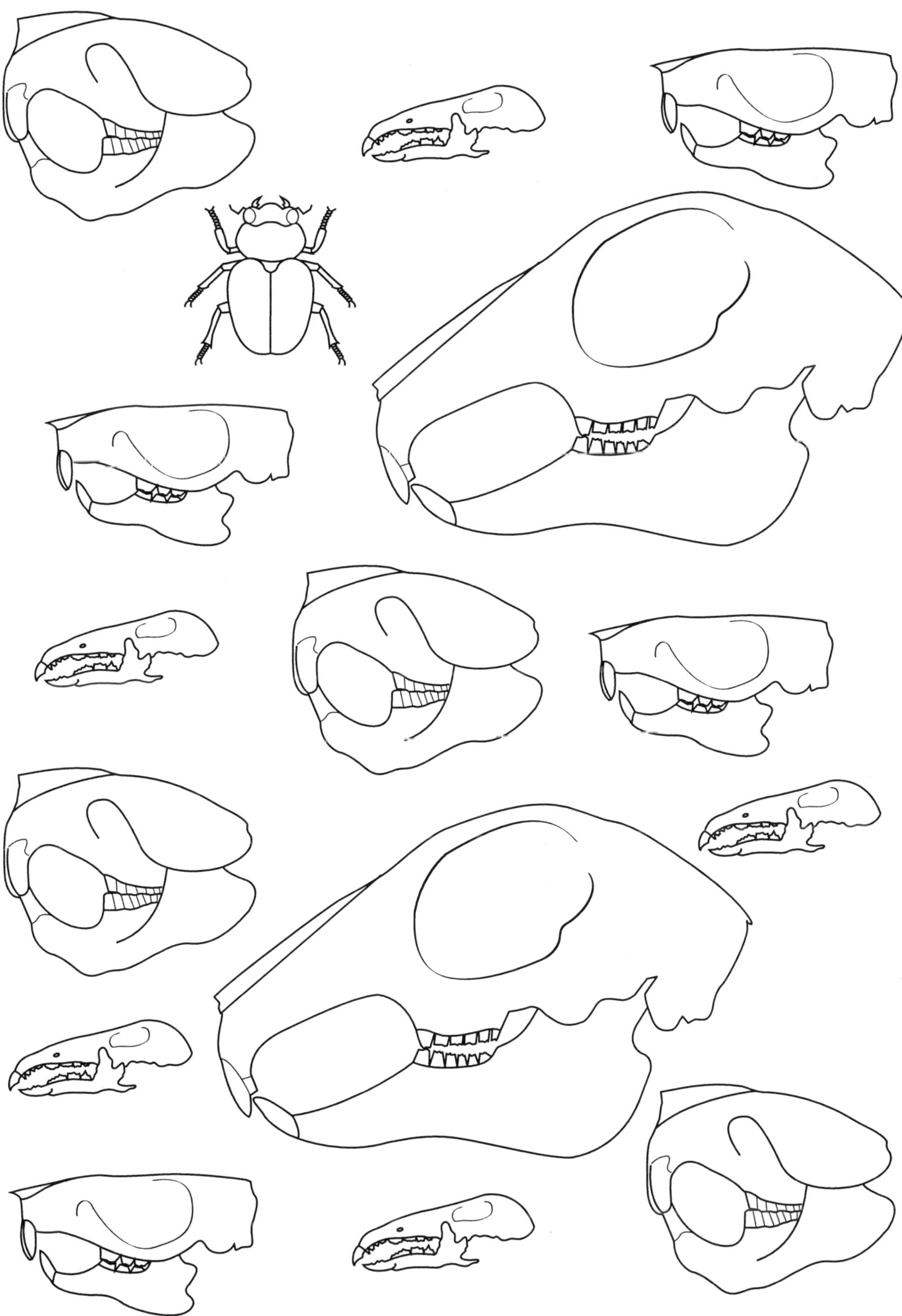

Calamine alert!

You are a technician at a laboratory. You have been asked to investigate a new brand of calamine lotion that seems to be causing problems.

Calamine lotion is a suspension of a light-coloured powder in water. The lotion takes the sting out of sunburn. But Brand X calamine lotion seems to bring some people out in a bad rash.

Brand X must contain something other than calamine and water, which is causing the problem. But is this something in the solid part or the watery part? Your job is to separate the solid from the liquid, so that each part can be tested on its own, to see where the problem lies.

- You will be given a sample of Brand X, and a collection of equipment that you can use to separate it into its solid and liquid parts.
- You must end up with a sample of the clear liquid and a sample of the dry solid.

1. Plan how you will do this safely.
2. Check your plan with your teacher.
3. **Once your plan has been approved**, carry out your experiment.

Calamine alert!

Wear eye protection

1 Fold a piece of filter paper to make a cone:

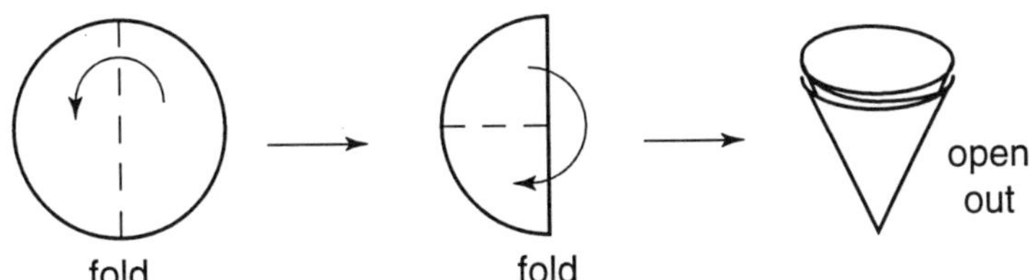

2 Put the paper in a filter funnel. Set the funnel above a beaker.

3 Pour the sample of calamine lotion into the filter paper cone. Do not overfill it. The liquid must not get above the paper. Pour it in a bit at a time if necessary. Be patient!

4 Collect the liquid that comes through the paper (the **filtrate**) in the beaker. When all the filtrate has gone through, remove the beaker.

5 Put a new beaker in place and wash the stuff left in the paper (the **residue**) clean with water. Let this filter through, and then throw it away.

6 Carefully remove the filter paper from the funnel. Open it out and leave it somewhere safe to dry.

7 What did the original calamine lotion look like?

___

8 What does the filtrate look like?

___

9 Write a few sentences to explain how you filtered the calamine lotion.

___

___

___

Calamine alert!

Testing filter papers

Not all filter papers are the same. Some cheap filter papers are like blotting paper. The fibres are packed very loosely. Others have their fibres more tightly packed. These are usually more expensive. You are going to find out what difference this makes to how the filter papers work.

1 Look at some different types of filter paper under a microscope. Note down your observations.

2 Design an experiment to test the properties of a range of different filter papers. Here are some ideas:

- You could use the calamine lotion as a test liquid.
- You could time how long a set volume of liquid takes to pass through each paper.
- You could look at the filtrate to see how clear it is.

3 Rashik found that the expensive filter paper gave a very clear filtrate, but it took a very long time to drip through. The cheap filter paper was much quicker, but the filtrate was still a little cloudy, with some very fine solid still in suspension.

a What are the advantages of the expensive filter paper?

b What are the advantages of the cheap filter paper?

4 Which type of filter paper would you use for:

a filtering calamine lotion

b filtering coffee?

Give reasons for your answers.

Evaporation

When washing dries on the line, the water evaporates into the air. What controls how long it takes to evaporate? Here are some thoughts:

Sunny days are best, because the clothes dry well in the heat.

My granny says a good blowy day is the best time to dry the washing.

I folded the clothes double to get them all on the line. My dad said they had to be opened out so that air got to all the surfaces.

1 Each of these ideas gives one factor that helps water evaporate. Work out what each factor is and write it down on a diagram like this. These are called the **input factors** for evaporating.

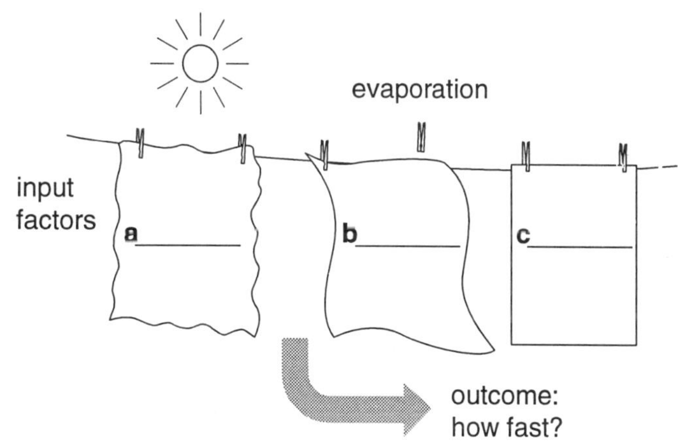

2 Select one input factor to investigate.

3 Plan an experiment to investigate the effect of your chosen input factor on how long it takes for water to evaporate. Here are some hints:

- **a** To make it a fair test, you will need to control the other input factors (keep them the same).
- **b** What other factors will need to be kept constant?
- **c** How will you vary your chosen input factor?
- **d** How will you measure the outcome?
- **e** How will you record and display your results?

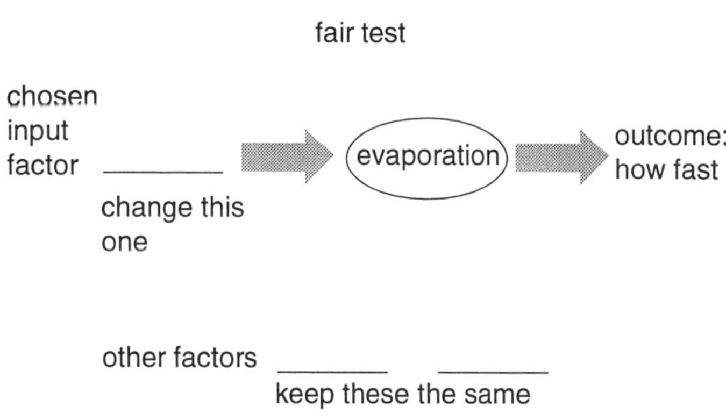

4 Make a prediction about your experiment (say what you think will happen).

5 **Once your plan has been approved**, try it out. Was your prediction correct?

6 Write a report on your experiment. Share your results with others who have looked at different factors.

Evaporation

Water can take quite a while to evaporate. These experiments use propanone (nail varnish remover) to study evaporation. Propanone evaporates faster than water.

Take care

Propanone catches fire easily. It should never be heated directly with a flame or used near a flame.

Temperature

1. Collect three beakers. Half-fill one with very hot water, one with warm water, and one with cold water.
2. Place a watch-glass over each beaker.
3. Put one drop of propanone onto each watch-glass. Start the stopclock.
4. Use the table to record how long it takes for the propanone to evaporate away in each case.

Temperature	Time taken
hot	
warm	
cold	

5. What effect does temperature have on the time it takes for propanone to evaporate?

____

____

Moving air

1. Collect two watch-glasses and put them at least half a metre apart on the bench.
2. Place one drop of propanone on each. Start the stopclock.
3. Leave one of the watch-glasses alone. Fan the other one with a book, or blow it with the cold jet from a hair drier.
4. Use the table to record how long it takes for the propanone to evaporate away in each case.

Air	Time taken
still	
fanned	

5. What effect does air movement have on the time it takes for propanone to evaporate?

____

Pure water from ink

Here's your chance to make pure water from ink.

1 Put some ink into the flask. Do not overfill it or it will boil over and spoil the experiment. A few pieces of broken tile will help the ink to boil evenly.

2 Connect up the rest of the apparatus as shown in the diagram on the Resource sheet. Light the Bunsen burner and set it to a medium flame. Start to heat the flask.

3 As the ink starts to boil, adjust the flame so that it simmers gently. If uncondensed steam starts to come out of the delivery tube, turn the flame down more.

4 When nearly all the ink has boiled away, turn off the gas. Remove the collecting tube and beaker of water from the apparatus. You should have crystal clear distilled water.

5 Label the diagram of your simple distillation apparatus. Add a few words next to each label to show what happened.

6 Did you have any problems with your apparatus? Was it easy to get all of the steam to condense?

7 Look at the diagram of the improved distillation apparatus. Your teacher will demonstrate this to you.

8 How does this apparatus make sure that all of the steam condenses?

9 Label the diagram, and add a few words next to each label to show what happened.

Take care

Use the Bunsen burner safely and correctly.

Make sure that steam can escape easily through the delivery tube.

Steam can burn you very badly, so make sure you keep away from the steam coming out of the delivery tube.

Pure water from ink

Simple distillation apparatus

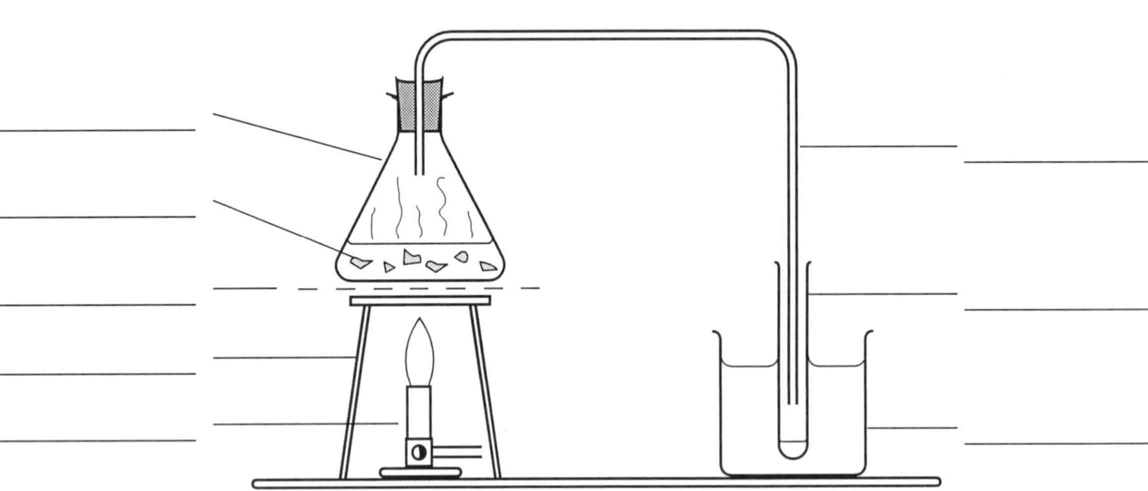

Improved distillation apparatus

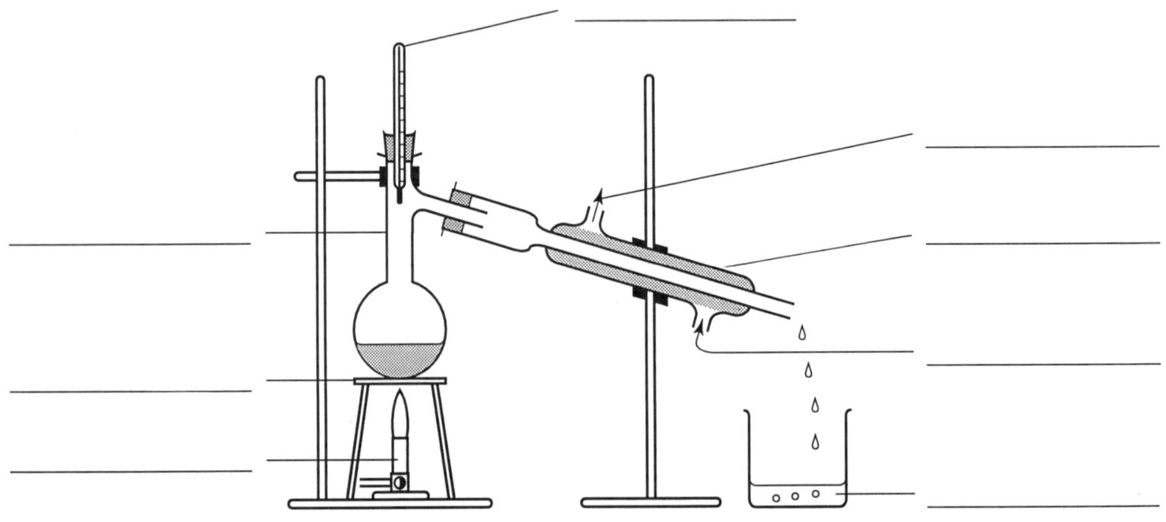

What dyes are used in Smarties?

Take care

Do not eat the sweets. They may be contaminated.

1 Collect six watch-glasses and put a different coloured Smartie onto each one.

2 Add one drop of water to each watch-glass using a glass rod.

3 Cut a piece of chromatography paper to fit snugly in a beaker.

4 Draw a pencil line 1 cm up from the base of the paper.

5 Mark seven pencil dots along this line. Label them with the colours of your Smarties. Label the seventh dot X.

6 Using a thin glass tube, transfer one spot of liquid from the red Smartie onto the dot marked 'red'. Wash the tube and repeat for the other colours.

7 Put a spot of Sunset Yellow food dye on the dot marked 'X'.

8 Carefully put a little water into the beaker so that it is just half a centimetre deep. If any water gets onto the side of the beaker, dab it off with tissue paper.

9 Carefully lower the paper into the beaker. Make sure that the water doesn't splash up above the pencil line.

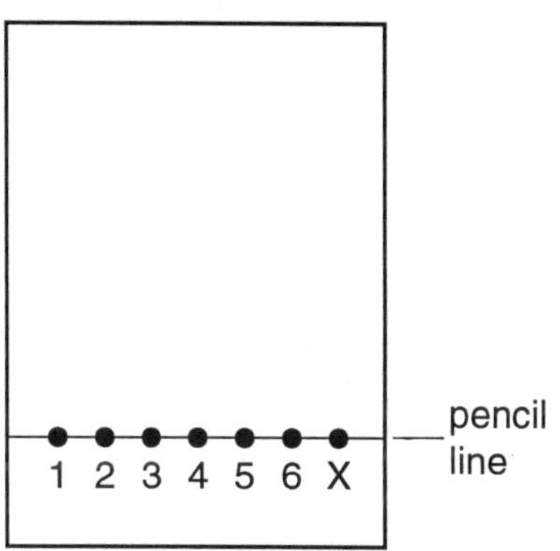

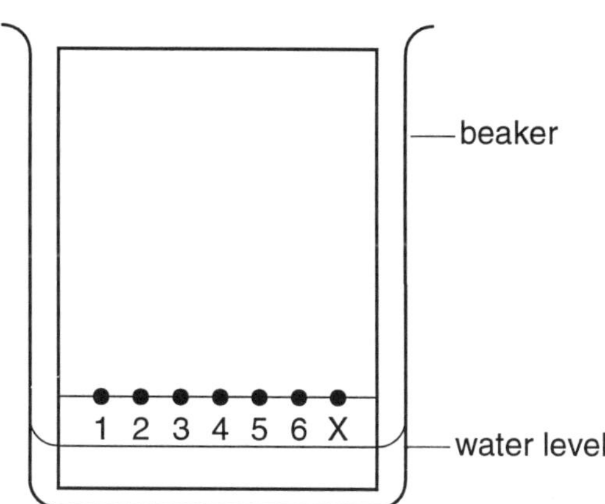

10 Wait while the water soaks up through the paper, carrying the dyes with it.

11 When the water gets near the top, take the paper out and let it dry.

12 Which colours have separated out into different dyes?

13 Did any Smarties contain Sunset Yellow?

Now you hear it, now you don't

Some people have to be able to hear very faint sounds. A doctor uses a stethoscope to hear your heart beating. A water engineer uses a long metal rod to hear the sound of water in pipes under the road.

You are going to investigate the question:

- What materials will sounds travel through?

1 How do you know that sounds will travel through air?

2 Tap the bench very gently with the end of a pencil. Can you hear the sound it makes? Now place your ear on the bench and tap gently. What do you observe?

3 Stroke the bench very gently with one finger. Rub so gently that you can't even hear it. Now place your ear on the bench and stroke again. What do you observe?

4 (You could try this at home.) Put a squeaky toy on the side of the bath. Lie back in the water, so that your ears are submerged. Squeeze the toy so that it makes a sound. What do you observe?

5 Your teacher will show you the bell-in-the-belljar experiment. What do you observe?

6 For each experiment, write down:

a what you observe

b what this tells you about sounds.

Activity 7a Help

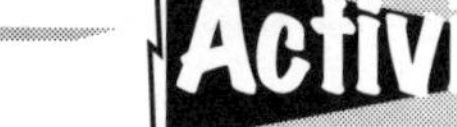

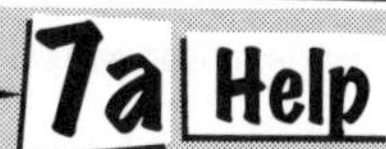

Now you hear it, now you don't

Fill in your answers in the table.

Experiment	Observation	Conclusion
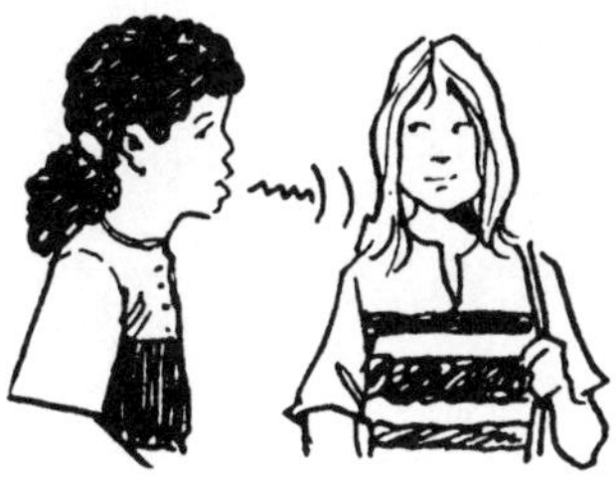	Can you hear the sound of someone's voice travelling through the air?	Do sounds travel through air?
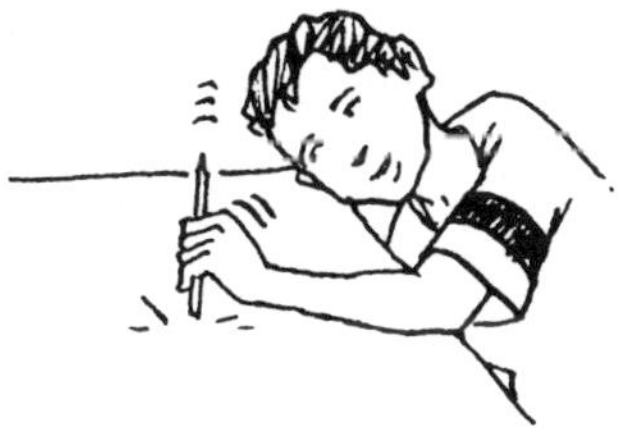	Can you hear the tapping through the bench?	Do sounds travel through wood?
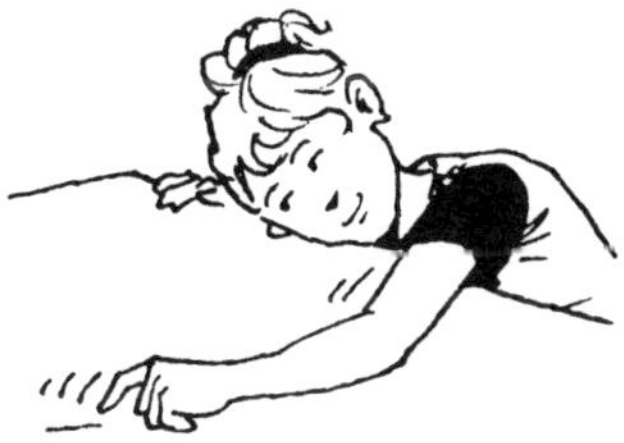	Can you hear the sound of the bench being stroked?	Do sounds travel better through air or through wood?
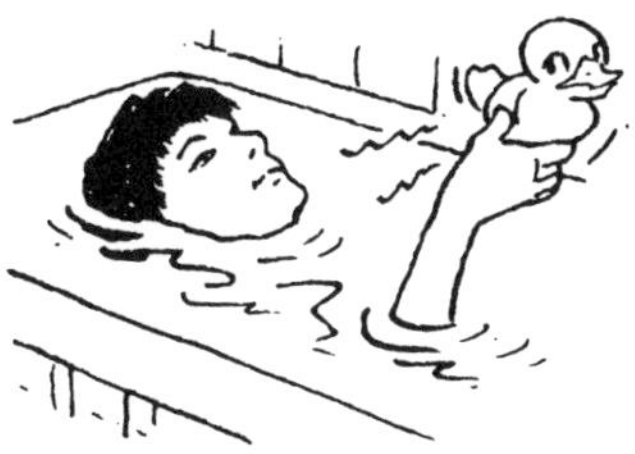	Can you hear the sound of the toy through the water?	Do sounds travel through water?
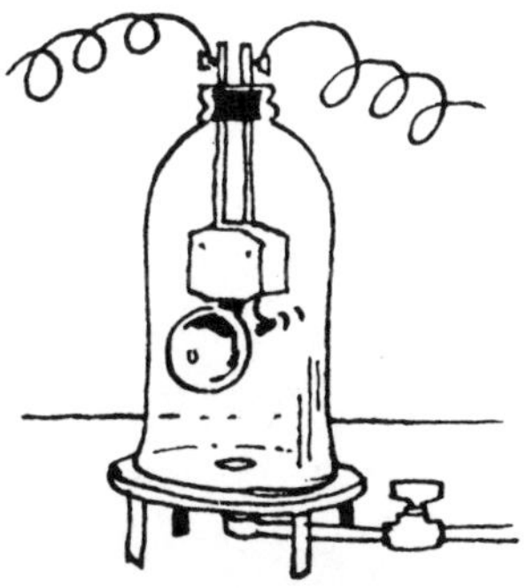	What happened when the air was pumped out of the belljar?	Do sounds travel through a vacuum?

Hearing high and low

Lots of animals can hear sounds that we can't hear. Dogs can hear very high notes. You may have seen a dog whistle which makes a sound too high for humans to hear. Bats use high-pitched sounds to find their way around in the dark.

You are going to find out how high you can hear. A signal generator can make very high and very low sounds. The sounds come out of a loudspeaker.

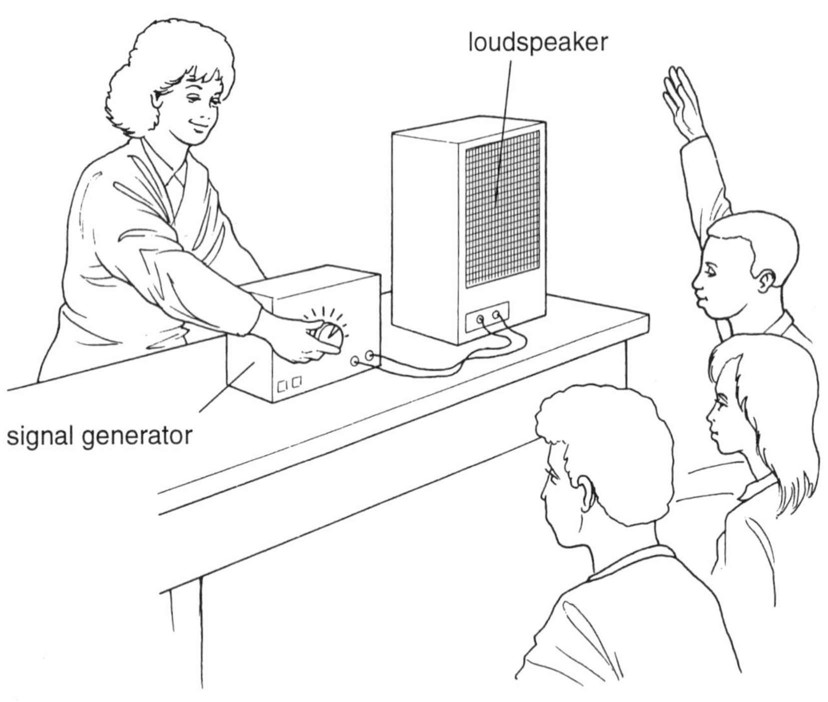

1. Your teacher will make some high-pitched sounds. Can you hear 10 kHz? (10 kHz means 10 kilohertz, 10 000 vibrations every second.)
2. Put your hand up if you can hear the sound. Put your hand down when you can't hear it any more.
3. Make a table to show how many people could hear each note.
4. Draw a bar graph to show your results.
5. Someone might be cheating. They might say that they can hear a high note when they can't. How can you test whether they are telling the truth?
6. Now repeat the investigation to find out who can hear very low notes.

Hearing high and low

Here are the results that one class found, together with their bar graph:

Pitch of sound	How many could hear it?
10 kHz	28
11 kHz	28
12 kHz	27
13 kHz	20
14 kHz	3
15 kHz	1
16 kHz	0

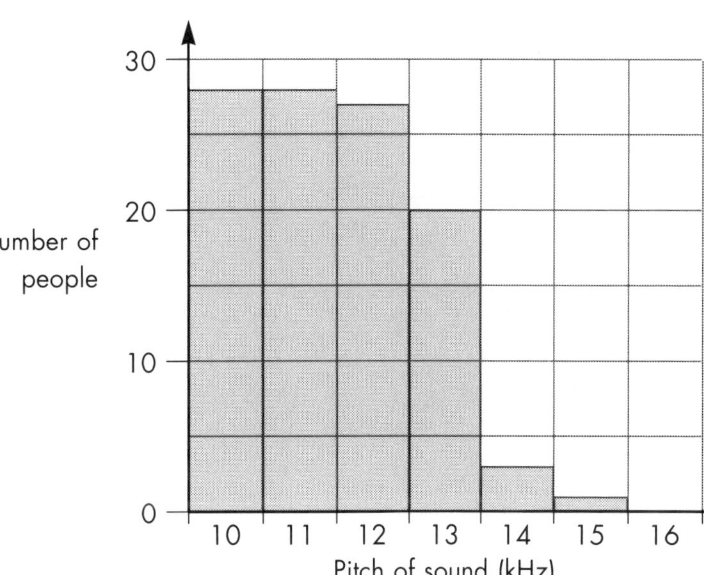

1 Use this table for your results, and draw a bar graph.

Pitch of sound	How many could hear it?
10 kHz	
11 kHz	
12 kHz	
13 kHz	
14 kHz	
15 kHz	
16 kHz	

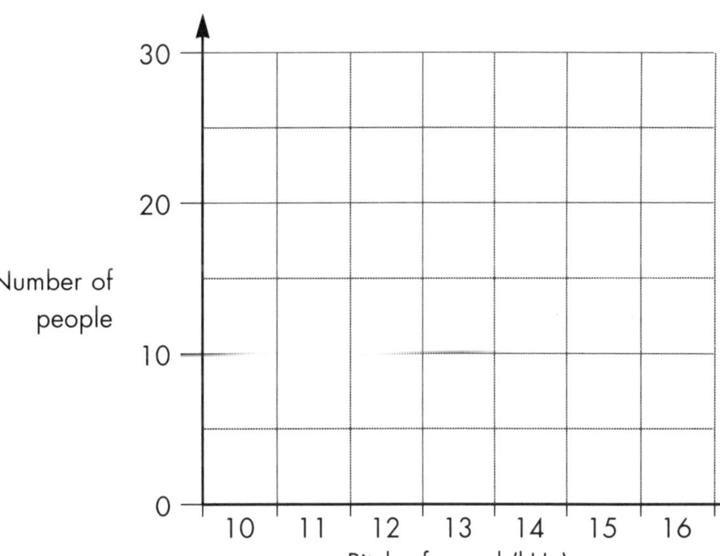

2 Who could hear the lowest notes? Use this table for your results, and draw a bar graph.

Pitch of sound	How many could hear it?
30 Hz	
25 Hz	
20 Hz	
15 Hz	
10 Hz	

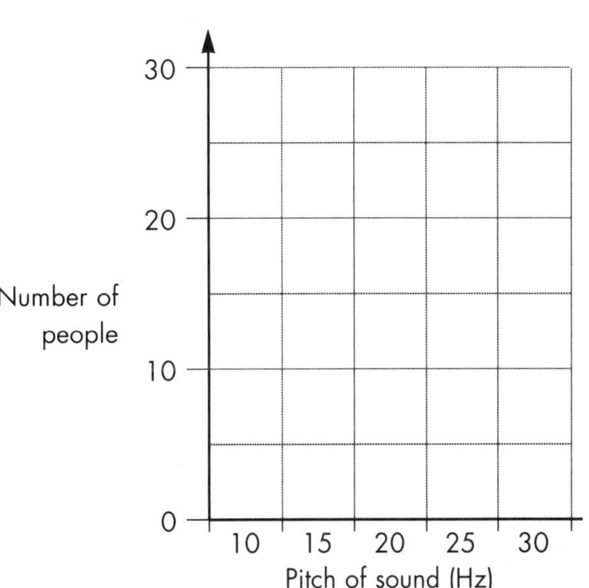

Shadow challenge

When the Sun shines, you have a shadow on the ground. If the Sun is high in the sky, your shadow is short. In the evening, when the Sun is lower, your shadow is longer. You are going to make some cut-out people and look at their shadows.

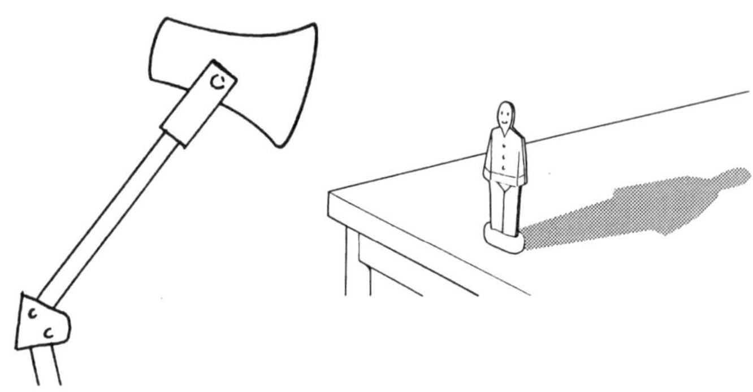

1. Make a cut-out person, about 10 cm high. Use Plasticine so that it will stand up straight.
2. Use a lamp for the Sun. Shine it so that your person's shadow falls on the bench.
3. Think about how the Sun moves across the sky during the day. How can you move the lamp so that it is like the Sun?
4. How does your person's shadow change as the Sun moves? Be prepared to give a demonstration to others in the class.
5. Record how the shadow changes. Think about its length, its width and its direction. (It may help if you make some measurements.)
6. You can make a shadow on a greaseproof paper screen. When you turn the Plasticine shape around, its shadow changes. This shape has a circular shadow, but when you turn it, its shadow becomes a square.

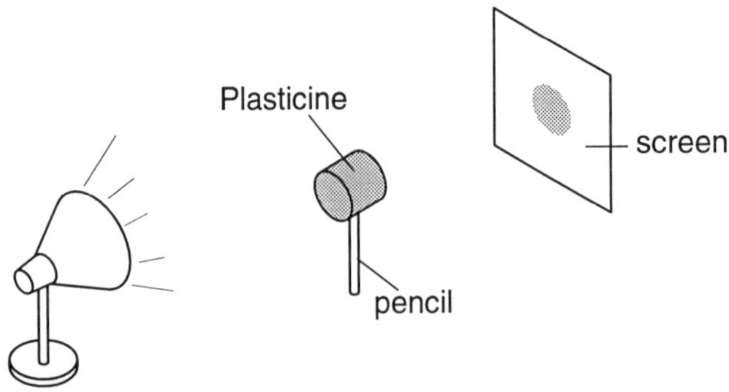

7. Design a Plasticine shape of your own. Don't let anyone else see it. Now, challenge someone. Let them see the shadow of your shape as you turn it round. They are not allowed to look behind the screen. Can they guess what shape your Plasticine is? Can you guess what shape theirs is?

Shadow challenge

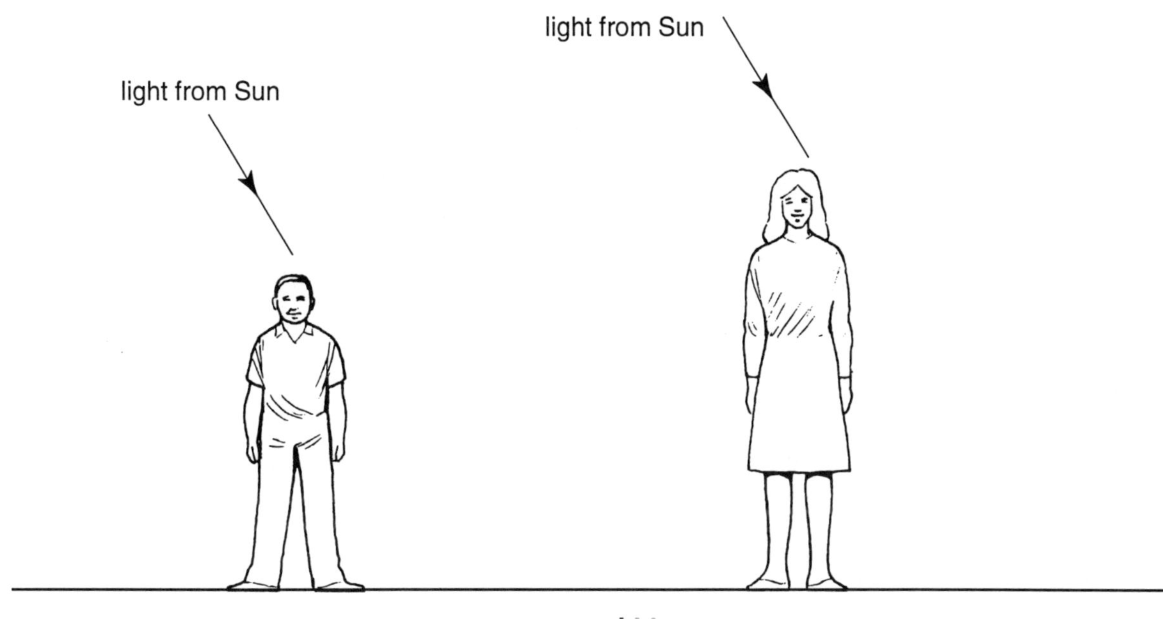

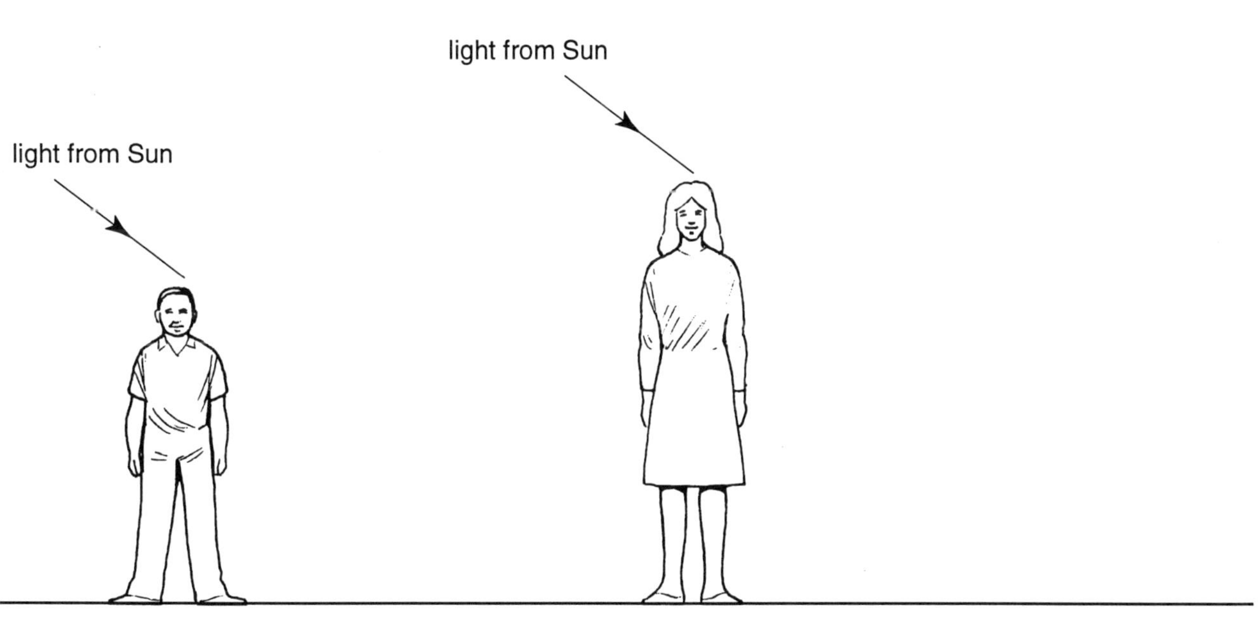

1 These diagrams show Mr Short and Mrs Tall at midday and in the evening. Copy and complete the diagrams to show where their shadows will be on the ground.

2 Mrs Tall is 2.0 m high. How long is her shadow at midday?

3 How tall is Mr Short? How long is his shadow in the evening?

A message to Planet Mirth

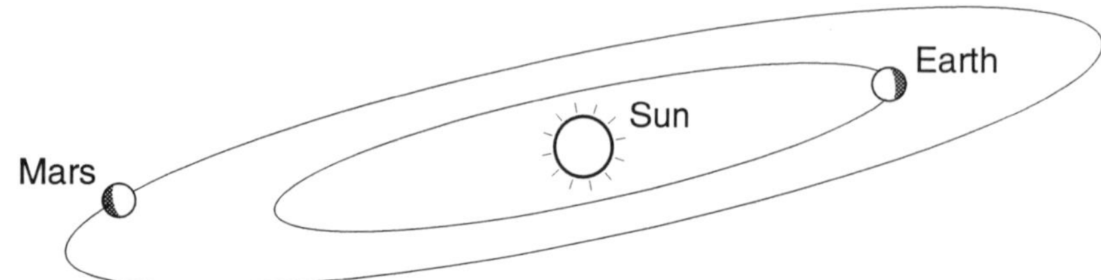

Our home is the Earth. The furthest people have ever travelled from the Earth is to the Moon. One day, astronauts may travel to Mars.

Mars is a long way from the Earth. The closest it ever comes is 80 million kilometres away. It takes about 5 minutes for light to travel this distance. This means that, when the astronauts get to Mars, it will take about 5 minutes for a message to reach them from the Earth. It will take another 5 minutes for their reply to get back to the Earth.

Recently, space explorers have discovered a new planet called Mirth. It takes 1 minute for light to reach Planet Mirth from Earth, and 1 minute to make the return journey. You are going to send a message, carried by light, to the explorers on Mirth.

1. Decide who is going to stay on the Earth, and who is going to Planet Mirth. (Your teacher will tell you where Mirth is.)
2. Decide who is going to be the light, carrying messages between you.
3. Now the astronauts must go off to Mirth.
4. Decide what message is going to be sent from the Earth.
5. Send the message. The light messenger will take 1 minute to reach Planet Mirth.
6. When the people on Mirth receive the message, they must send the light messenger back with a reply.
7. When the reply arrives on the Earth, send a reply back to Mirth.
8. Discuss these questions:
   - **a** Why does it take a long time for messages to reach Mirth?
   - **b** What difference would it make to the time delay if Mirth had travelled to the other side of the Sun from the Earth?
   - **c** Why is there not a time delay if we are sending messages using light on the Earth?

Mirror, mirror

You can use a mirror to make a reflection. What you see depends on the angle of the mirror.

With the mirror like this, you see a straight line …

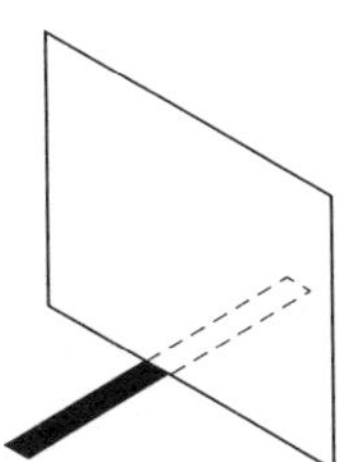

… but like this, you see a bent line.

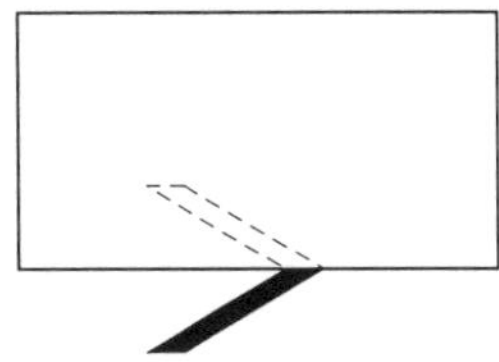

1 **a** Copy this line onto a piece of paper.

b Try putting your mirror on the line so that you can see a straight line.

c Draw a line on the paper along the back of the mirror.

d Measure the angle between the mirror and the line.

2 Where must you put your mirror on this diagram to see a square? Where must you put it to see a triangle?

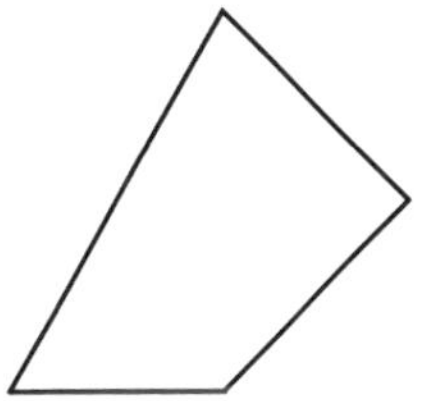

3 **a** Put your mirror on the dotted line along the middle of these words. What do you notice?

b Can you make some other words like this?

CARBON DIOXIDE

4 Use two mirrors to see yourself in an unusual way. Stand them up so that they make a right angle. Look at your reflection in this mirror. Wink your left eye. What do you notice?

5 **a** Use two mirrors to make a kaleidoscope. Put them on this diagram. What shapes can you see? (Try looking for a circle, a hexagon, a star.)

b Measure the angle between the two mirrors. What is the angle that allows you to see everything six times over?

c Now make your own kaleidoscope pattern. Use colours and interesting shapes to give an attractive pattern.

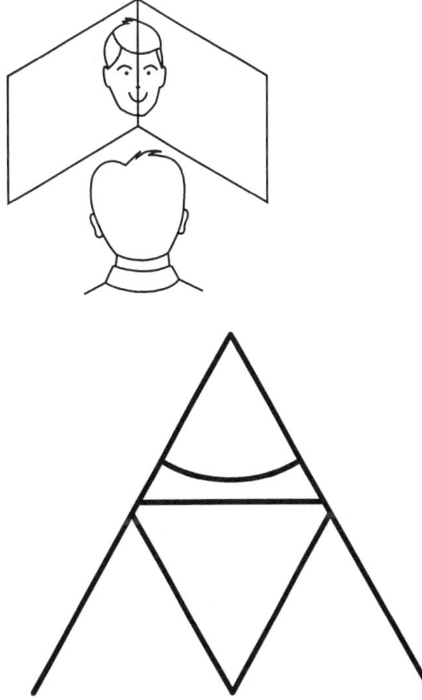

From ice to steam

How does the temperature of water change as the water changes from ice to steam? You can find out by following these instructions.

Take care

Boiling water will scald you badly if you spill it on yourself.

If you do get burnt, cool the skin at once with plenty of running water, and keep it in the cold water for 10 minutes.

Be very careful when taking temperature readings.

1 Set up the apparatus as shown in the diagram.

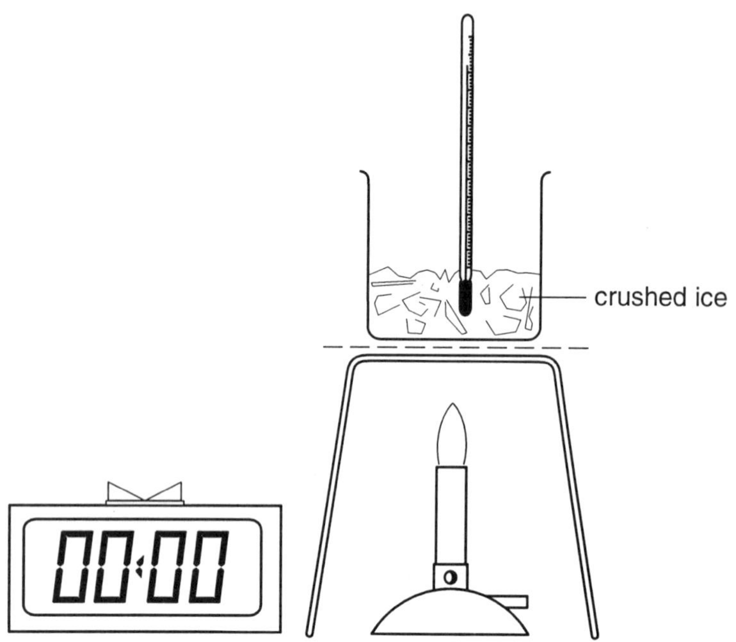

2 Draw up a suitable table to record the temperature every minute.

3 Take the temperature at the start. Then light the Bunsen burner. Take the temperature every minute.

4 When the water has been boiling for 5 minutes, turn off the gas and leave the water to cool down.

5 Draw a suitable graph of your results.

6 Describe the graph in words.

7 At what temperature did the ice start to melt?

8 What happened to the temperature while the ice was melting?

9 At what temperature did the water start to boil?

10 What happened to the temperature while the water was boiling?

11 What do you think would have happened to the temperature if you had kept the water boiling for another 5 minutes?

From ice to steam

How does the temperature of water change as the water changes from ice to steam? You can find out by following these instructions.

Take care

Boiling water will scald you badly if you spill it on yourself.

If you do get burnt, cool the skin at once with plenty of running water, and keep it in the cold water for 10 minutes.

Be very careful when taking temperature readings.

Wear eye protection

1. Make sure that you know how to read a thermometer correctly.
2. Draw a table like this:

Time (min)	Temperature (°C)

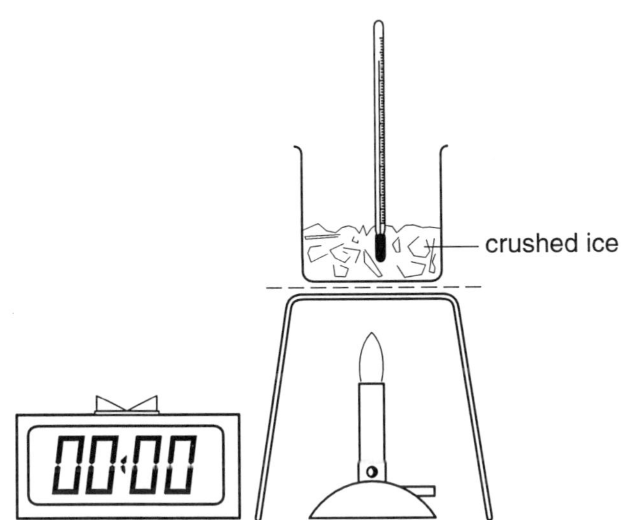

3. Set the apparatus up as shown in the diagram.
4. Make sure that the beaker is stable. Wipe any water from the sides of the beaker using a paper towel.
5. Light the Bunsen burner and set it to a medium flame with the gas tap half on.
6. Take the temperature of the ice and start the stopclock.
7. Take temperature readings every minute and put them in your table.
8. When the water has been boiling for 5 minutes, carefully turn off the gas and leave the water to cool down.
9. Look at your table.
   - When did the ice start to melt?
   - When did the water start to boil?

From ice to steam

Data logging

Have you ever wondered what happens to water in an ice tray when you put it in the freezer? If you open the door occasionally and have a look, you find that ice crystals start to grow quite quickly, but it takes ages for the water to completely freeze solid. But if you keep opening the freezer to read a thermometer, you let warm air in. How can you follow the temperature of the water as it freezes?

Help is at hand using a computer. A **temperature sensor** is a kind of electronic thermometer. You can put it into the water in the ice tray. Its wires run outside and are connected up to a computer through a special box.

You can use a program on the computer to take readings as often as you like, for as long as you like. Most data-logging programs can collect about 1000 readings. The computer could take a temperature reading every 15 seconds for over 4 hours, without getting bored! The computer will then print out the graph for you.

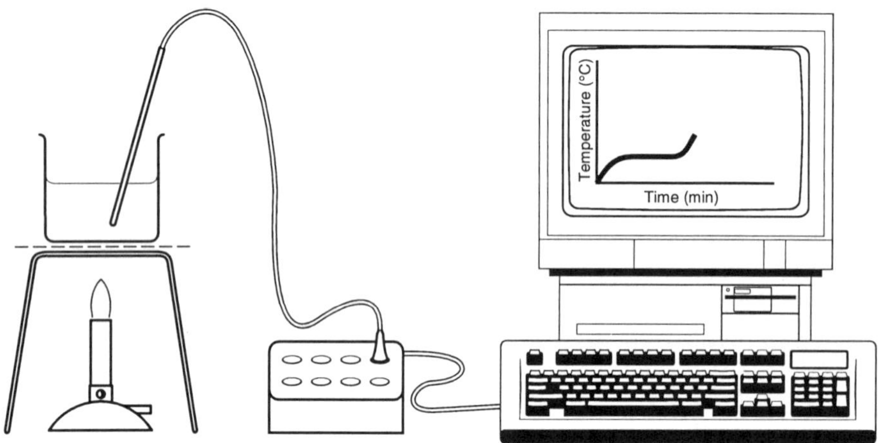

Try it yourself

You could use data-logging equipment for the activity 'From ice to steam'. You will have to read the instructions for your equipment carefully before you start.

Some clever tricks with data logging

Here are some other ideas to try. They are all things that would be difficult to do using an ordinary thermometer.

- Plot the temperature graph of a block of ice as it melts and then boils in an oven.
- Check the temperature of a gas oven at different settings.
- With two sensors, plot the temperature of a frozen meal compared with the oven temperature as it cooks.

Take care

Make sure your plans are approved before you try any practical work.

Working with wax

In ancient times, letters were sealed with wax. A special ring was pressed into the wax to show who it had come from. Wax is easy to melt, but it soon sets to a solid again. You can use wax to make casts of things like shells or fossils.

> ⚠ **Take care**
>
> If you drop molten wax on your hands, it will burn you.
>
> If you do get burnt, cool the skin at once with plenty of running water, and keep it in the cold water for 10 minutes.

Sealing a letter

1 Fold a piece of paper over to make a simple letter.

2 Drip some wax from a lighted candle onto the join. Press your seal into it (a coin will do) while it is still liquid.

3 Leave to cool and you have a sealed letter.

A shell cast

4 Make an impression of a shell in a piece of Plasticine.

5 Drip molten wax from a candle into the Plasticine until the mould is full up.

6 When it is cool, press out your wax cast.

The melting point of wax

A test tube full of wax will melt if you stand it in a beaker of boiling water. If you put a thermometer into this molten wax, you could record the temperature of the wax as it cooled down and solidified. This would help you to find its melting point.

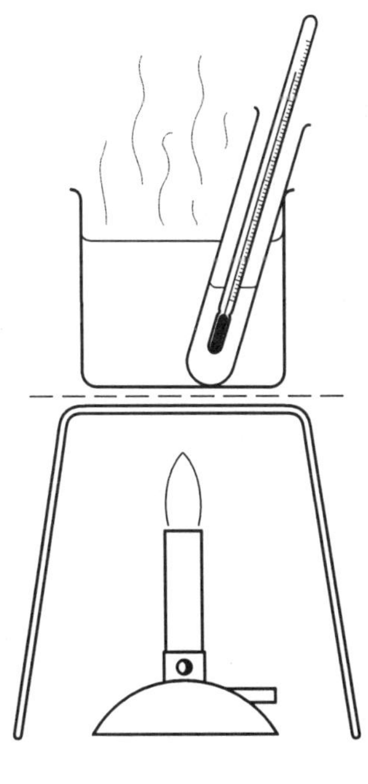

7 Plan an experiment like the one suggested above.

8 **Once your plan has been approved**, carry it out and record your results.

9 Plot a graph of your results.

10 What is the melting point of the wax, according to your graph? Explain how you know this.

11 Is your wax a pure substance? Explain your answer.

Working with wax

The melting point of wax

Take care

If you drop molten wax on your hands, it will burn you.

If you do get burnt, cool the skin at once with plenty of running water, and keep it in the cold water for 10 minutes.

1. Half-fill a test tube with powdered wax, and put a thermometer in it.
2. Put this test tube in a beaker half-full of water.
3. Heat the water over a Bunsen burner.
4. As soon as the water starts to bubble and boil, turn off the gas.
5. Check the test tube. The wax should have melted and the temperature should be above 90 °C.
6. Take the temperature and start the stopclock.
7. Draw a table like this. Record the temperature every minute, noting any changes that you see happening to the wax.

Time (min)	Temperature	What is it like?
0		clear liquid

8. When the temperature falls below 40 °C and all the wax is solid again, stop taking readings. Do not try to take the thermometer out of the wax or it might break.
9. At what temperature did the liquid start to turn to a solid?
10. What is the melting point of your wax?

Fudge

If you have ever tried to make toffee or fudge, you will know how tricky it can be. You have a sugary solution and you boil the water off until it will set when cooled. To find this out, you have to keep testing drops of the mixture in cold water. It is all rather hit and miss. What you are trying to find is the exact concentration of sugar solution when the fudge will set.

As a solution becomes more concentrated, its boiling point goes up. You can use this fact to find the right concentration for the sugar solution. If you follow the temperature of the solution as it boils, you will know exactly when to stop boiling without any fuss.

Take care

The temperature may rise to 150 °C, so you need to check the range of the thermometer you use.

Do not get hot sugar solution on your skin as it will burn you very badly.

If you do get burnt, cool the skin with plenty of running water, and keep it in the cold water for 10 minutes.

Do not eat food cooked in a laboratory. It may be contaminated.

Heat 500 g of sugar and 300 cm^3 of milk gently in a pan until all the sugar has dissolved.

Add 50 g of butter.

Bring to the boil and boil steadily until the temperature reaches 115 °C.

Remove from the heat and leave to cool for a minute.

Beat with a wooden spoon until the mixture is thick and creamy.

Pour into a well-greased shallow tin and allow to cool and set.

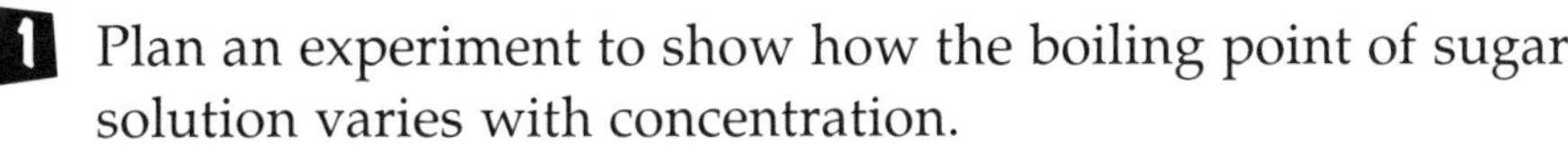

1 Plan an experiment to show how the boiling point of sugar solution varies with concentration.

- You could test the boiling point of known concentrations of sugar solution.
- You could follow the rise in temperature as the water boils off, reweighing the solution regularly to check how much water has been lost.

Wear eye protection

2 **Once your plan has been approved**, try it out.

Looking at expansion

Here are some experiments. For each one:

1. Describe what happens.
2. Explain why this has happened.
3. Does the change reverse as things cool down?

Take care

Hot pieces of metal stay hot for a long time. They can still burn you badly several minutes after you have stopped heating them. Use tongs to handle hot metal.

Ball and ring

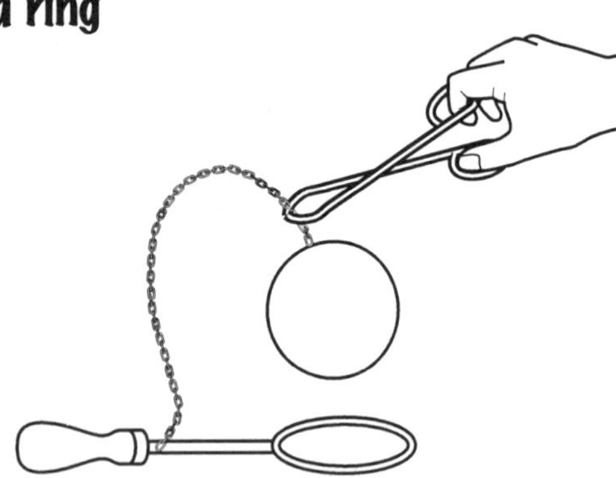

- Does the ball fit through the ring?
- Heat the ball and try again!

Flask of air

- Gently heat the flask with your hands.
- Now place a cool, damp cloth over the flask.

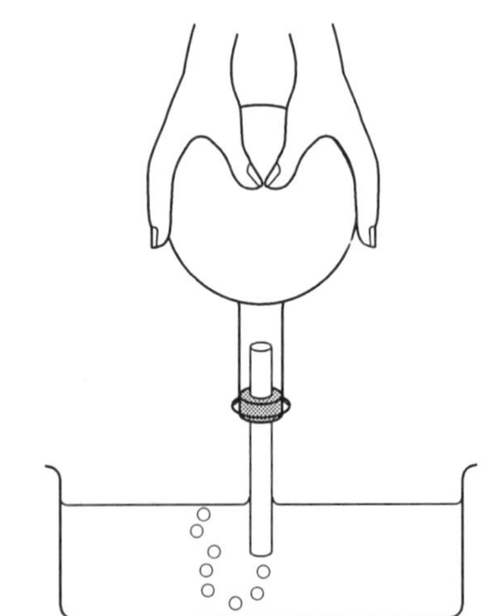

What happens to liquids?

- Mark the levels on all the bottles.
- Pour hot water into the trough. Mark the new levels.
- Measure how high each liquid rose above the first mark.
- Did each liquid rise by the same amount?

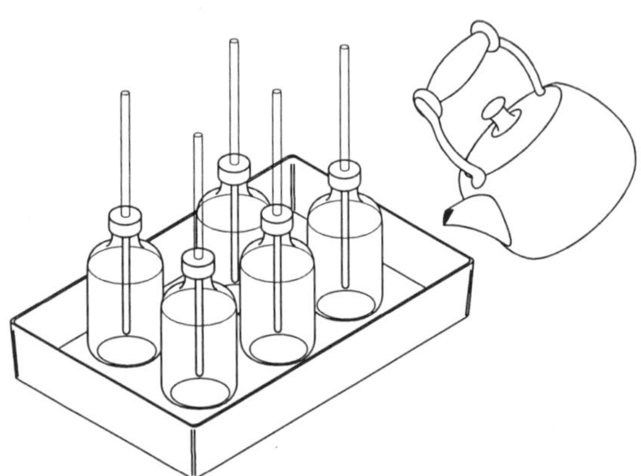

Are all solids the same?

This **bimetallic strip** is made of two different metals riveted together. The two metals are brass and invar.

- Heat the strip.
- What would happen if you put this bimetallic strip in the fridge?

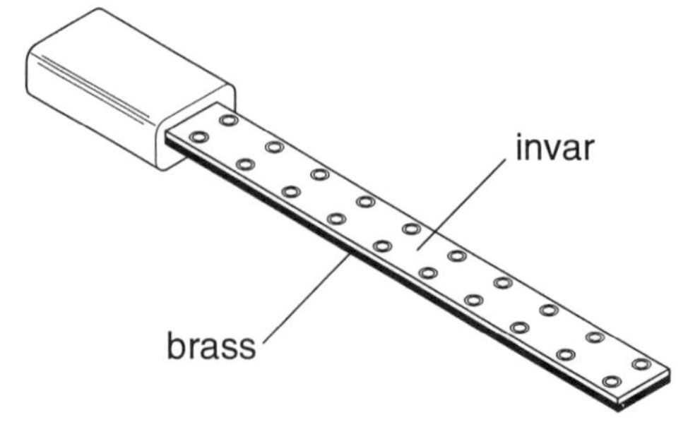

Looking at expansion

All these bottles are the same size. They are all filled to the same mark. Bottles 1–4 are made of glass. Bottle 5 is made of polythene.

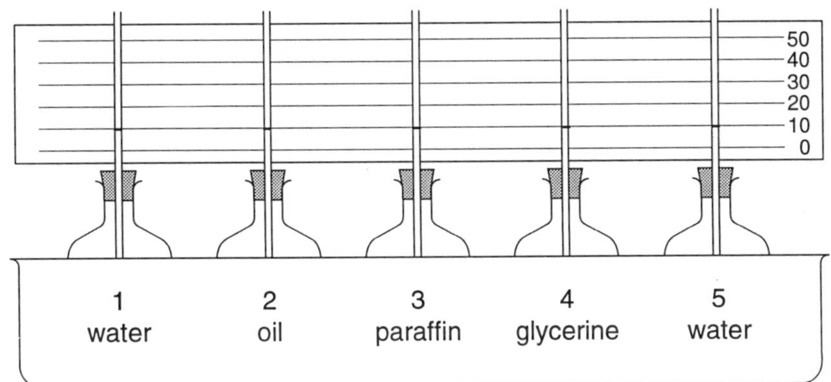

1 When hot water is added to the trough, the liquid levels in the tubes change. For bottles 1–4, use the scale shown to measure how much each liquid rose. Record this in a table.

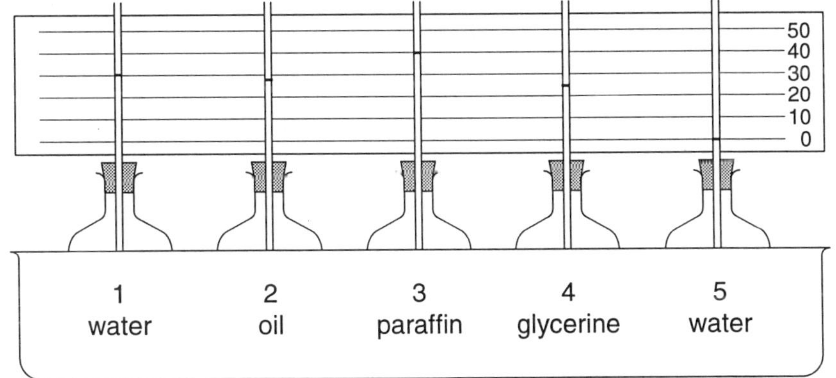

2 List the four liquids in order of increasing amount of expansion. Draw a bar chart of your results in this order.

Bottles 1 and 5 both contain water.

3 What has happened to the water level in bottle 5?

4 Try to explain this effect, compared with the result from bottle 1.

Make a water thermometer

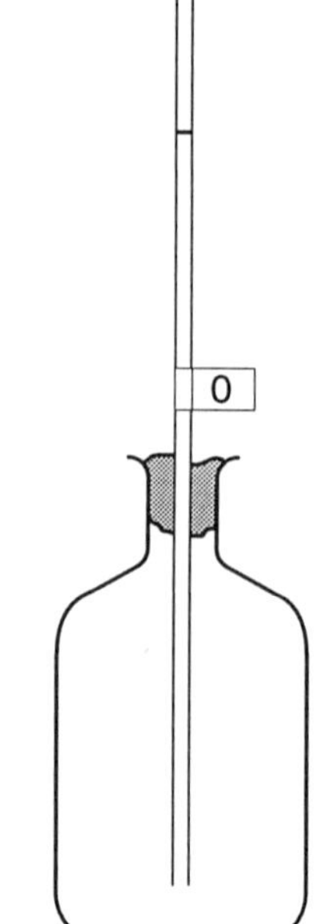

1. Fill a bottle with water. Fix a long, narrow tube into the top with Plasticine, so that the water level is half-way up the tube.

2. Put the bottle in a mixture of ice and water. Mark the lowest level that the water in the tube reaches. This is the 0 °C mark.

3. Now put the bottle in a trough of hot water. Adjust the water temperature in the trough so that it reads 50 °C on an ordinary thermometer. Mark the highest point that the water reaches in the tube.

4. Measure the distance between the 0 °C and 50 °C marks. Divide this distance by 10. Make new marks along the tube to represent 5, 10, 15, 20, 25, 30, 35, 40 and 45 °C.

5. You have now calibrated your water thermometer. You can use it to find the temperature in different places, for example:
   - in a warm room
   - outside on a cold day
   - on top of an oven or radiator
   - in a pond.

6. Describe how you made your water thermometer.

7. How accurate do you think your water thermometer is?

8. How quickly does it reach the new temperature when you move it to a different place?

9. Why would it become less accurate if you kept it for a few days?

10. What are the advantages of the liquid-filled thermometers that you use at school?

Physical or chemical?

It is not always easy to tell the difference between a physical change and a chemical change. You have to watch carefully for clues. Remember that chemical changes:

- make completely new substances
- are not easily reversed.

Here are some pairs of experiments to try. In each pair, one shows a physical change, the other a chemical change.
For each experiment:

a Describe what the substances look like at the start.

b Describe what happens during the experiment.

c Describe what the substances look like at the end.

d Decide whether any changes are physical or chemical.

e Give reasons for your answers.

f Plan how to record your results.

1 Heat a spatula measure of these substances one at a time in a test tube:

i zinc oxide

ii copper carbonate.

2 Heat a spatula measure of these substances one at a time in a test tube:

i powdered wax

ii sugar.

3 Drop a crystal of sodium carbonate into test tubes half-full of:

i cold dilute hydrochloric acid

ii cold water.

4 Pour the following into a test tube, cork it and shake:

i oil and salt solution

ii lead nitrate solution and salt solution.

Take care

Lead nitrate solution is toxic.

Wash your hands after using chemicals.

Physical or chemical?

Take care
Wash your hands after using chemicals.

Experiment/ substances	What is it like at the start?	What happens?	What is it like at the end?	Physical or chemical?
1 i heating zinc oxide				
1 ii heating copper carbonate				
2 i heating powdered wax				
2 ii heating sugar				
3 i sodium carbonate in acid				
3 ii sodium carbonate in water				
4 i shaking oil and salt solution				
4 ii shaking lead nitrate solution and salt solution				

Burning fuels in air

Activity 8e Core

Candles need air

When a candle burns, the wax is reacting with the oxygen in the air.

1 Place a jar upside down over a burning candle.

2 Describe what happens.

3 Explain why this happens.

How much air does a candle need?

4 Roughly how long did it take for the candle to go out?

5 How long do you think it might have taken if the jar was twice as big?

6 Plan an experiment to see whether the size of the jar does affect how long it takes for the candle to go out.

How will you measure the size of the jar?

7 **Once your plan has been approved**, try it out.

8 Plot a graph of your results and explain your conclusions.

Burning fuels in air

How much air does a candle need?

If you cover a lighted candle with a jar, the flame goes out after a little while. Why does this happen? And what do you think would happen if the jar was twice as big? Here's an experiment to try out your ideas.

You will need:

- a candle
- a stopclock
- a measuring cylinder
- a selection of different sized jars.

1 Find the volume of each of your jars by filling each jar with water from the measuring cylinder.

2 Write down the volumes of your jars in the table.

Volume of jar (cm^3)	Time taken for candle to go out

3 Light the candle, put the first jar over it and start the stopclock.

4 Stop the stopclock when the candle goes out. Write the time down in the table.

5 Repeat steps **3** and **4** with each jar.

6 Can you see any pattern in your results? Write down what you think it is.

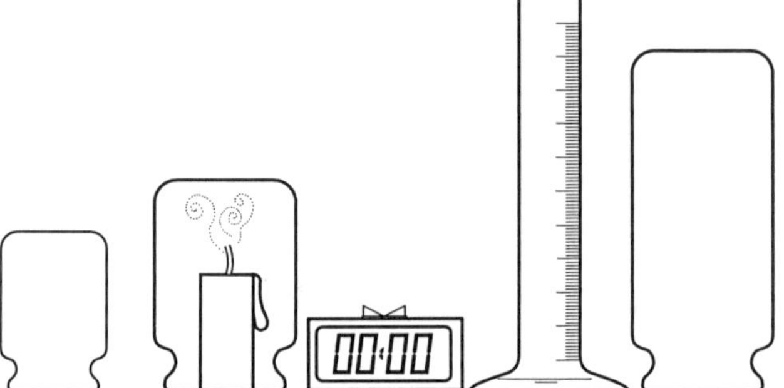

Burning fuels in air

How much oxygen is there in air?

1 Float a burning night-light candle in water in a trough.

2 Lower a tall jar over it.

3 Describe what happens.

4 Roughly how much of the air is oxygen?
Explain your answer.

5 Burning a candle makes a new gas, carbon dioxide. Why doesn't this just take the place of the oxygen in the jar? (If you can't answer this, go on to question **6**.)

6 Limewater is a clear liquid which turns milky with carbon dioxide. Replace the water in the trough with fresh water mixed with some limewater. Repeat steps **1** and **2**.

7 What happens to the limewater as the candle goes out?

8 Where has the carbon dioxide gone?

(Now go back and answer question **5**, if you couldn't answer it before!)

The secrets of a flower

Plants don't have sex organs all the time. They make flowers when they need to reproduce. You are going to look closely at some flowers to see how they are arranged, and to find where the male and female sex cells are.

1 Where would you expect to find:

a the male sex cells

b the female sex cells?

2 Collect a flower from the jar of brightly coloured flowers. You will also need a white tile, a mounted needle and a hand lens.

3 Look at your whole flower from the side and from above. Make two drawings, and label them to give as much information about the flower as you can.

4 Remove the petals from the flower. Count them and arrange them on the white tile or on a piece of white or black paper.

a How many petals are there?

b Are there any markings on the petals?

c If so, what do you think they are for?

5 Draw one petal in detail to show its size and shape, and any markings it has. If the petals differ a lot in size and shape, draw them all (without colouring them).

6 Carefully take all the stamens off the flower and arrange them on the white tile.

a How many stamens are there?

b What colour is the pollen?

c Can you see individual pollen grains?

d Could you measure or count them?

7 Have a good look at one of the stamens with your hand lens and draw what you see.

Now you are left with just the female parts of the flower, the stigma and the ovary joined together by the style.

8 Look carefully at the stigma with your hand lens and draw it. Then ask your teacher to cut the female parts in half. Look inside the ovary with your hand lens.

a What do the ovules look like?

b How big are they?

c Approximately how many ovules are there?

9 How do you think your flower is pollinated?

Activity 9b Extension

The secrets of a flower

1 Collect a flower from the jar of greenish brown flowers. You are going to compare this with the brightly coloured flower you have already looked at. You will find it helpful to look at spread 9b in your textbook.

2 Plan what you are going to do to compare the flowers. You need to look for differences in:

a the petals

b the stamens

c the female flower parts.

3 Discuss your plan with your teacher. **Once it has been approved**, carry out your investigation. Remember to draw and make a note of everything you see.

4 How do you think this flower is pollinated?

5 Make a table to show the differences between the petals, stamens and female parts of these flowers and brightly coloured flowers.

Make your own flying fruit

Plants often spread their seeds using the wind. You will be given a seed. You have 15 minutes to make a 'fruit' to carry it.

1. Design and make a fruit that will float as far as possible when it is dropped from a height. The fruit must carry your seed. Use only paper or card, Sellotape and Blu-Tack in your design.
2. Try out your flying fruit and see how far it will fly.
3. See if you can improve its performance so that it flies further. (Remember, you have 15 minutes to produce your finished flying fruit.)
4. Test the flying fruits from all the groups to see which will travel the furthest.

   How can you make sure that the test is a fair comparison?
5. What features help to make sure that your seed is carried as far as possible?
6. Can you think of any real flying fruits which are similar to your designs?
7. Write about what you have done. Make a drawing of your design. If you improved your first effort, explain how and why you made improvements.

The male reproductive system

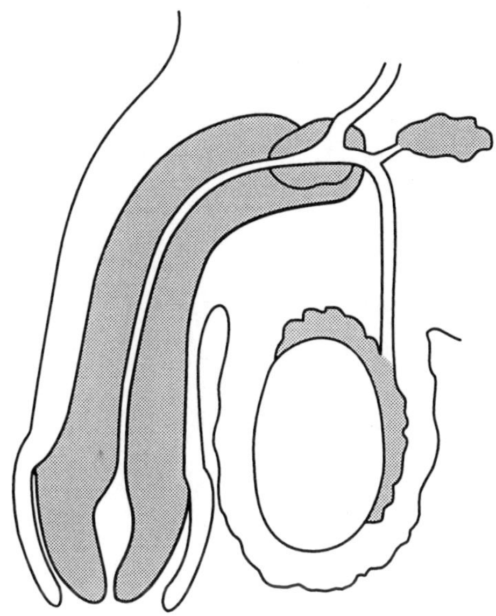

Here is a diagram of the male reproductive system like the one in your textbook.

1 Label these parts:

a penis

b urethra

c sperm duct

d testis

e scrotum.

2 Copy and complete this paragraph. Use the words below to fill the gaps.

sperm duct	**penis**	**scrotum**	**sperm**	**urethra**	**testes**	**semen**

The main male reproductive organs are the ____________ which are held in a bag of skin called the ____________. The testes make ____________ which are the male sex cells. When the sperm leave the testes they are carried along the ____________ ____________ past a gland. This adds a liquid to produce ____________. The semen is then squeezed out through the ____________ along the ____________, which at other times also carries urine away from the body.

The male reproductive system

In some books, the male reproductive system is shown from the front like this, instead of from the side.

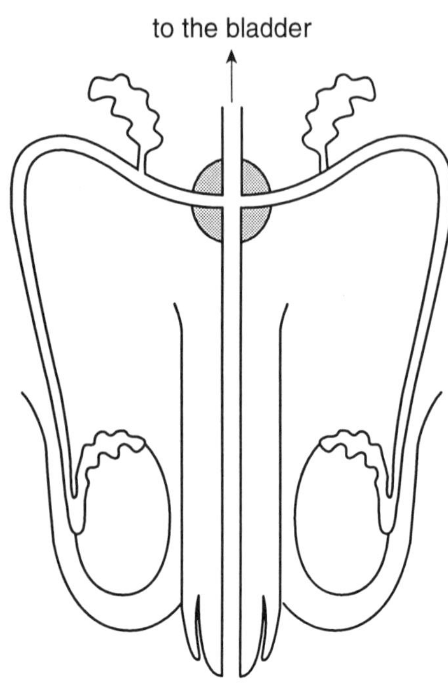

1 Look at this diagram of the male reproductive system and label all the parts.

2 Is there anything which you think is clearer on this view? Do you think it makes anything more difficult to understand?

The female reproductive system

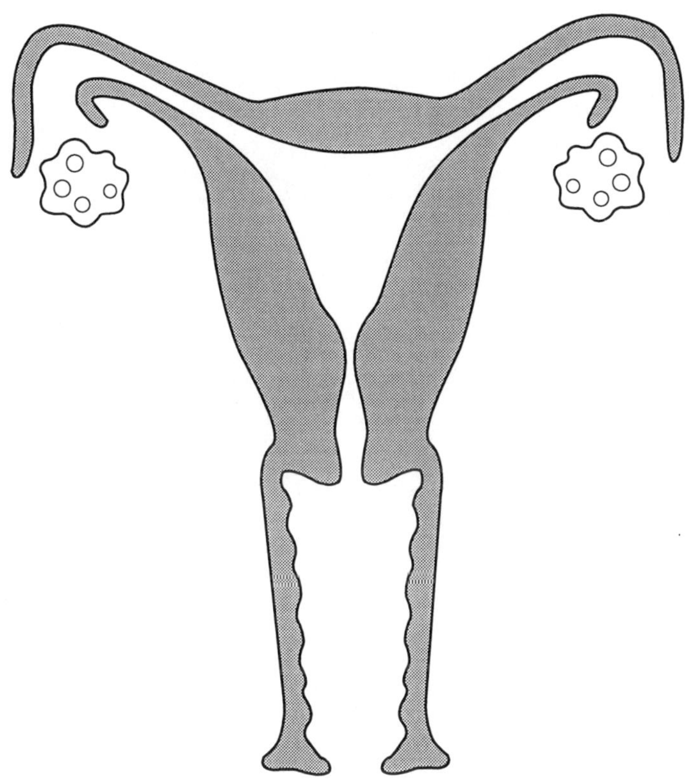

Here is a diagram of the female reproductive system like the one in your textbook.

1 Label these parts:

a ovary

b Fallopian tube

c uterus

d vagina.

2 Copy and complete this paragraph. Use the words below to fill the gaps.

ovum	**ovaries**	**uterus**	**Fallopian tube**

The main female reproductive organs are the ___________.

They contain the female sex cells, and after puberty they produce an ___________ every month. This travels down the ___________ ___________ to the ___________.

This is where a baby will grow and develop if the ovum is fertilised.

The female reproductive system

In some books, the female reproductive system is shown from the side like this, instead of from the front.

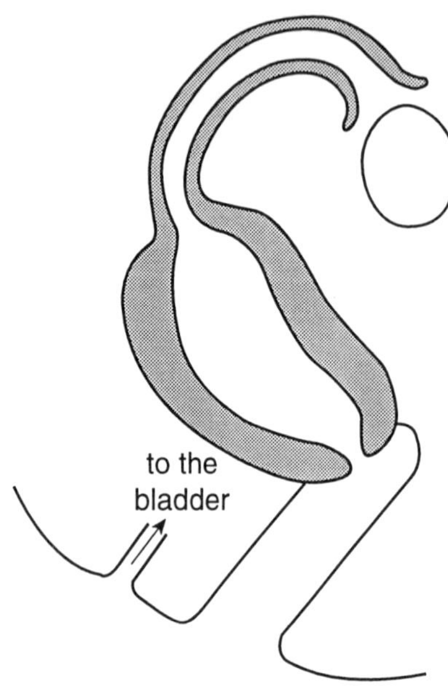

1. Label this diagram of the female reproductive system.
2. What do you think are the advantages and disadvantages of the two different views?

A new body, a new you?

You have lived with your body all your life, and it has always been pretty much the same. It has just got bigger over the years. At puberty you have to get used to a body which is quite different.

1. In your group, make a list of the changes you would expect to occur at puberty in a boy.
2. Decide which you think would be the most difficult to get used to, and give reasons for your choices.
3. Make a list of the changes you would expect to occur at puberty in a girl.
4. Decide which you think would be the most difficult to get used to, and give reasons for your choices.
5. With the help of your teacher, compare your list with those of other groups in the class. Make a bar chart to show the changes your class predicts will be the most difficult to get used to for each of the sexes.
6. Produce a leaflet which could be handed out in youth clubs to help young people cope with worries about their bodies at puberty.

The human life support machine

Here is a diagram like the one in your textbook of a fetus developing in the uterus.

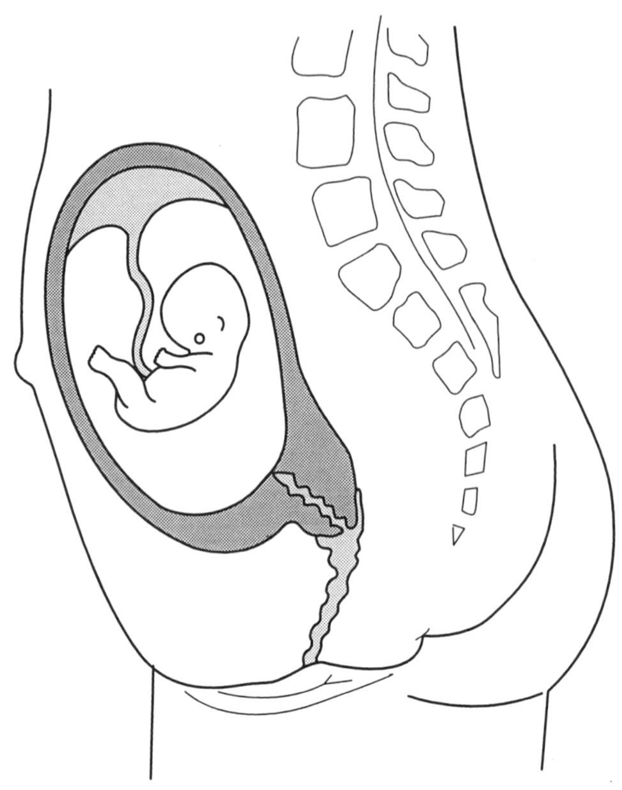

1 Label these parts:

a fetus

b uterus

c placenta

d umbilical cord

e bag of fluid.

2 Make a table like this to show what each part does.

Name of part	Function (What does it do?)

A fetus can't say no!

The placenta provides a growing fetus with all that it needs, and takes away its waste. The placenta also protects the fetus from many diseases and harmful substances, but not all!

Some things, like alcohol, can cross the placenta. They can reach the fetus and damage it. Alcohol is bad for the developing brain of the fetus. Babies have been born who are already alcoholics, addicted to alcohol because their mothers drank so much while they were pregnant.

Smoking when pregnant also affects the growing baby. The drug nicotine gets into the blood of the fetus. Even worse, the mother's blood doesn't supply as much oxygen to the growing baby. Smokers' babies are often smaller than they would have been if the mother hadn't smoked. The babies are also more likely to die before or soon after birth. Cot deaths in the first year of life are much more common in the babies of smokers. In fact, doctors think that more than 1000 babies die each year just because their parents smoke cigarettes.

1 Design a leaflet or poster to help midwives and doctors explain to pregnant women why they should try to stop drinking and smoking.

Burglars beware!

When this burglar steps on the rug, the alarm will go off. You are going to design and make a switch to go under the rug.

1 Here is a list of some possible materials that you might use to make your switch:

cardboard	copper sheet	aluminium foil
plastic	connecting wire	wood

a Which of these are good conductors of electricity?

b Which are electrical insulators?

2 Decide whether your alarm will make a buzzer sound, or a lamp light up.

3 Your switch must be part of a circuit. What other things will you need in your circuit?

4 Now make your switch and connect up your circuit.

5 How will you test your circuit?

6 Draw a picture to show your finished burglar alarm.

7 Write a sentence or two to explain how your switch works. What happens when the burglar steps on the rug?

8 Does your buzzer keep buzzing when the burglar steps off the rug? Is there any way you could improve your burglar alarm design?

Burglars beware!

Here is one idea of how to make a switch for a burglar alarm:

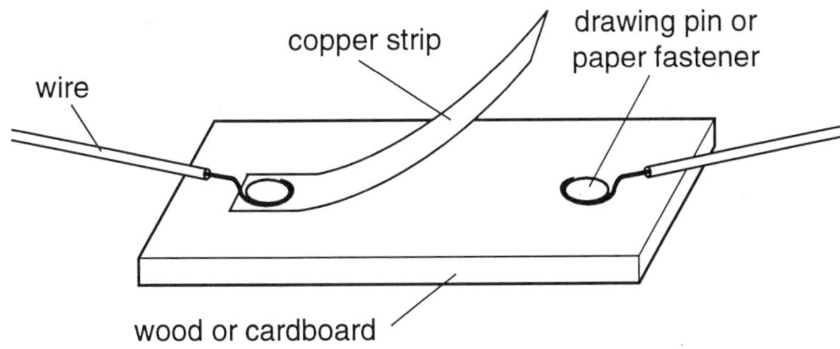

Here is one way to connect up your circuit:

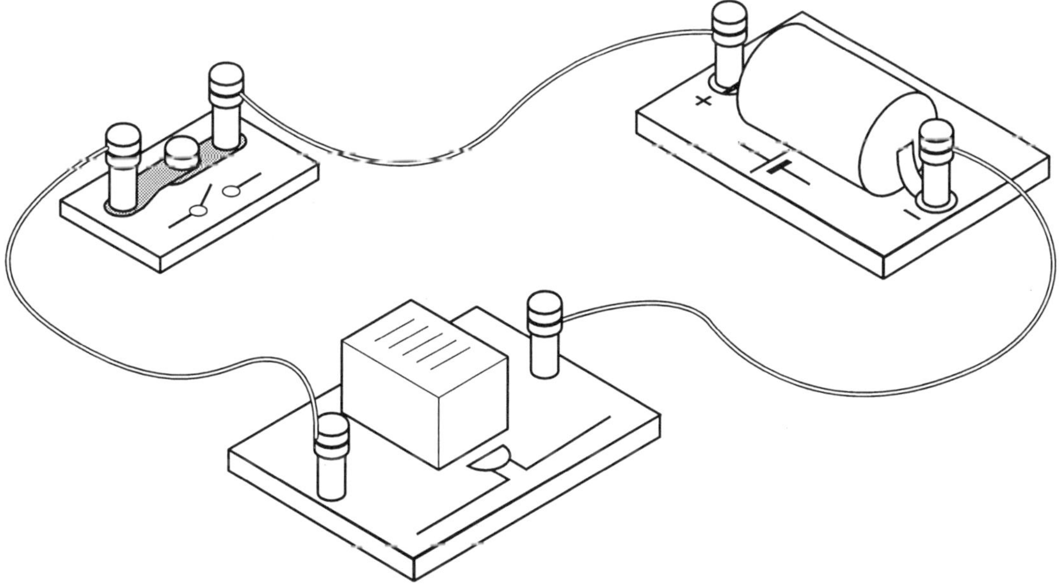

- Do you know the name of each thing in the circuit?
- Can you connect up a circuit like this?

If you are not sure, check with your teacher.

Circuits and symbols

1. Set up each of these circuits in turn.
2. Draw a circuit diagram for each circuit.
3. Write a sentence to say what happens when you close the switch or switches.

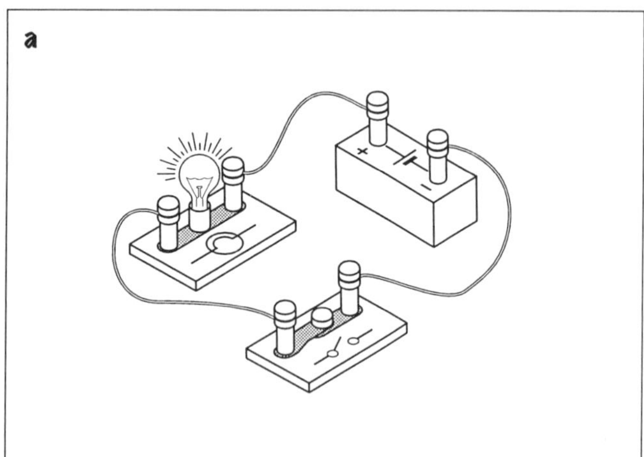

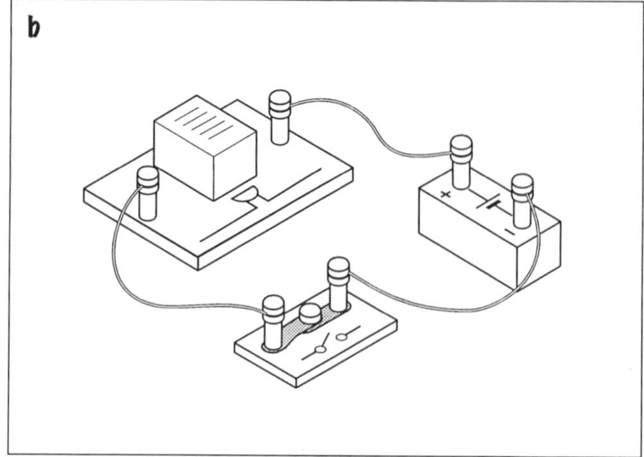

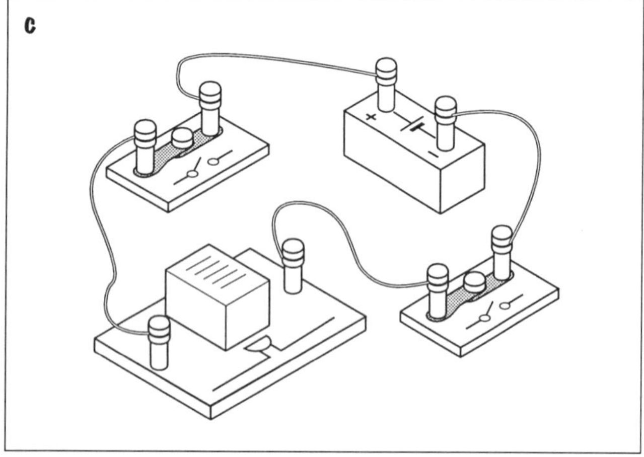

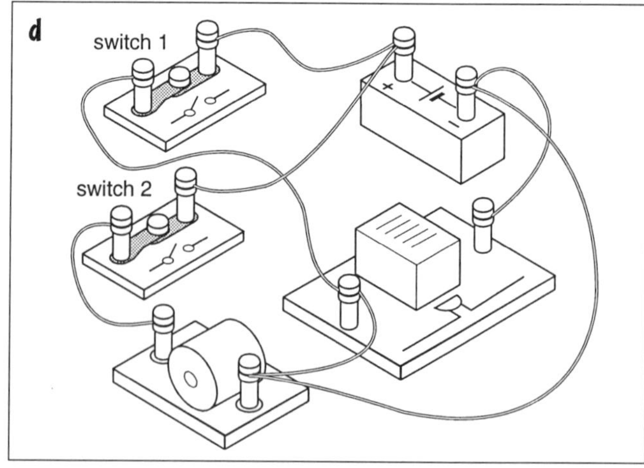

Circuits and symbols

1 Set up each of these circuits in turn.

2 Complete the sentence to say what happens when you close the switch.

a

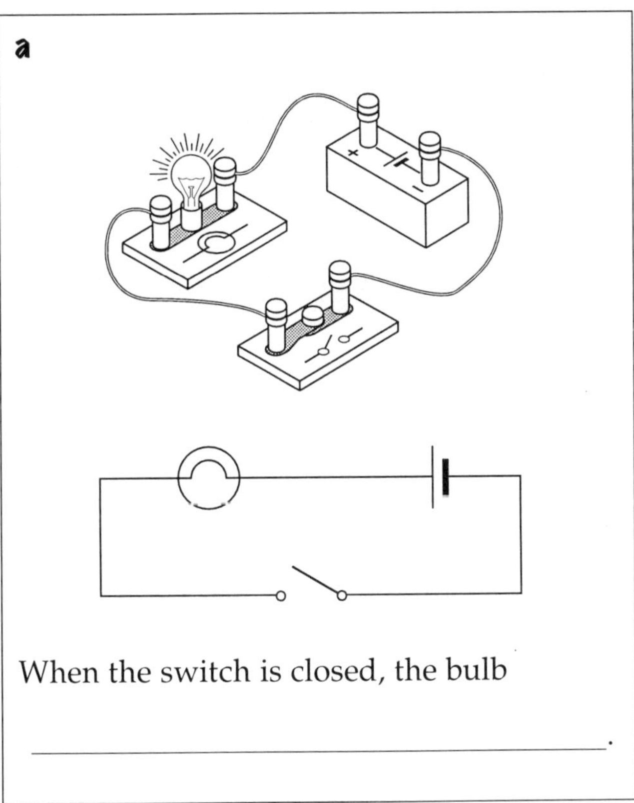

When the switch is closed, the bulb

______________________________.

b

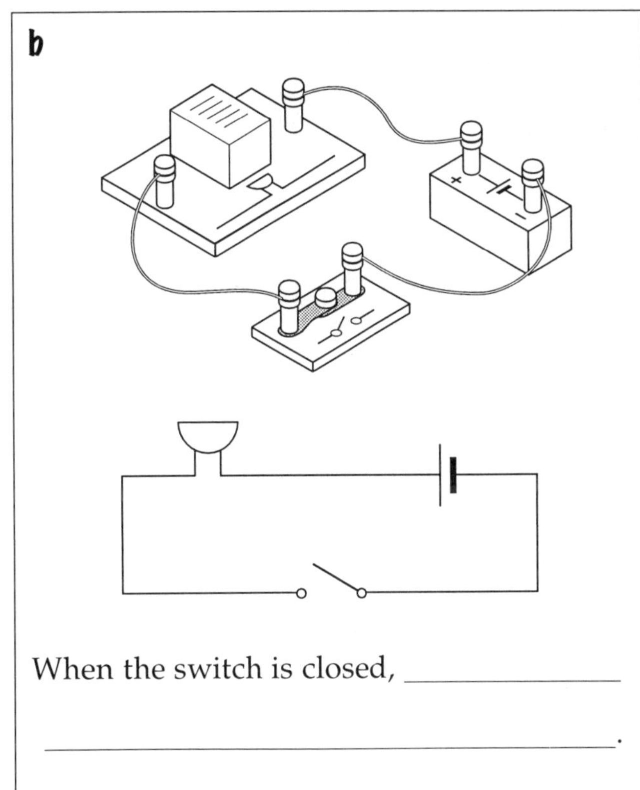

When the switch is closed, ______________

______________________________.

c

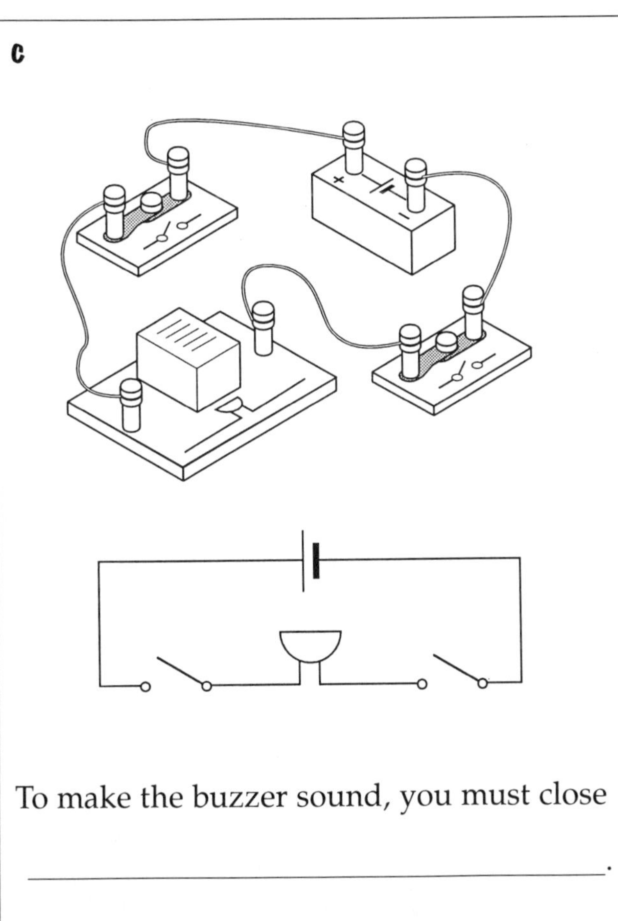

To make the buzzer sound, you must close

______________________________.

d

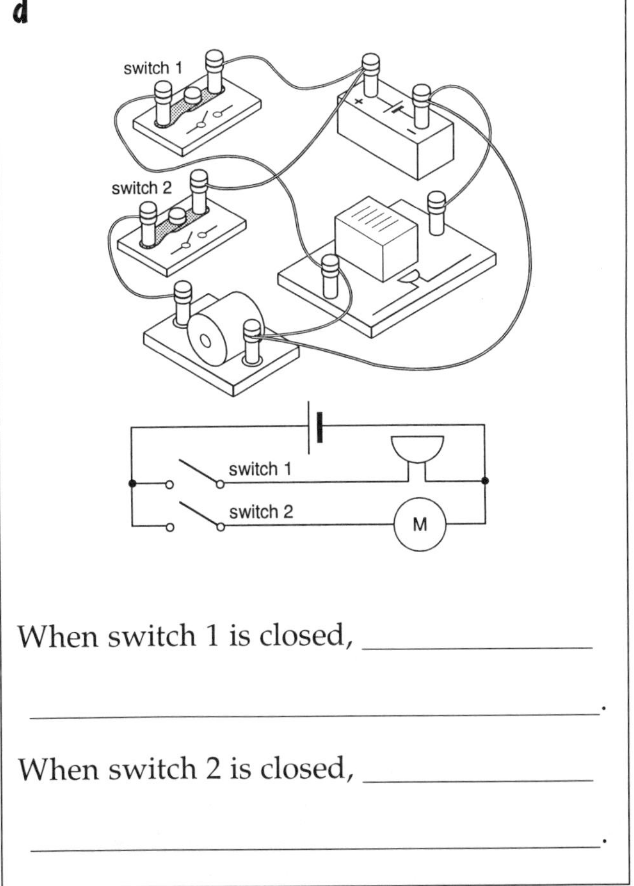

When switch 1 is closed, ______________

______________________________.

When switch 2 is closed, ______________

______________________________.

Circuits and symbols

1 Set up each of these circuits in turn.

2 Draw a circuit diagram for each circuit.

3 Write a sentence for each to say what happens when you close the switch.

a

A battery and a switch are connected in a simple circuit to a bulb.

b

A battery and a switch are connected in a simple circuit to a buzzer.

c

A buzzer is connected to two switches, one on each side. The switches are connected to a battery.

d

A buzzer is connected to a switch. A motor is connected to another switch. Both work from the same battery.

e

Two bulbs are connected to a switch and a battery.

f

How many different circuits can you make with the following components?

- 2 bulbs
- 2 switches
- a buzzer
- a battery
- connecting wires

Puzzlectric

Can you remember how two bulbs must be arranged if they are connected **in series**? And if they are connected **in parallel**? Test yourself with these problems.

In each case, before you complete the circuit, make a prediction to say what you think will happen.

1 Connect one bulb up to a battery, to see how brightly it shines. Now change the circuit, so that two bulbs are connected in series.

a How bright will they be?

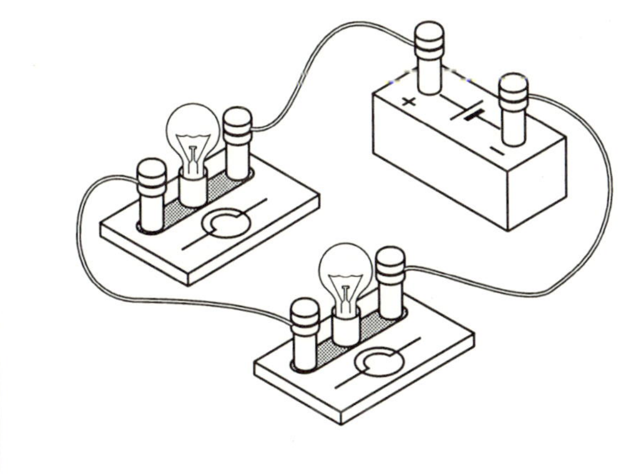

2 Connect two bulbs in parallel.

a How bright will they be, compared with the first bulb?

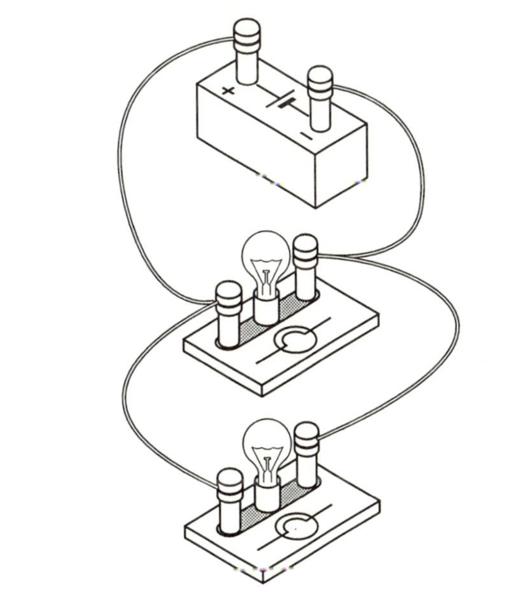

3 Connect three bulbs in parallel.

a What do you predict?

4 Make a puzzle box. Have two bulbs sticking out of the box, but keep the wiring hidden inside. Challenge a friend to work out whether they are connected in series or in parallel.

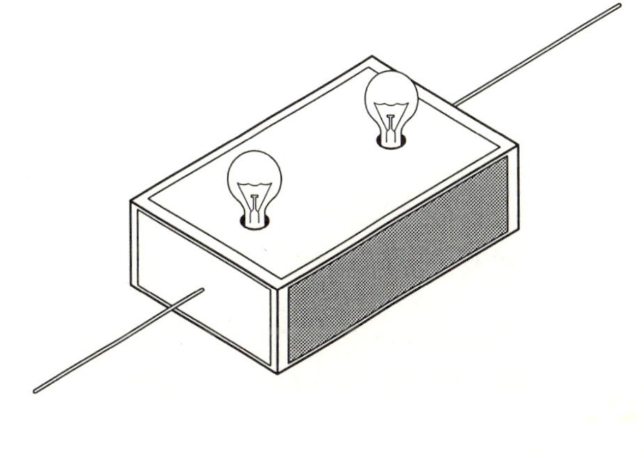

Ammeters at work

You can't see electric current. You need an ammeter to measure how much current is flowing in a circuit.

1 **a** Set up this circuit.

b How much current flows around it?

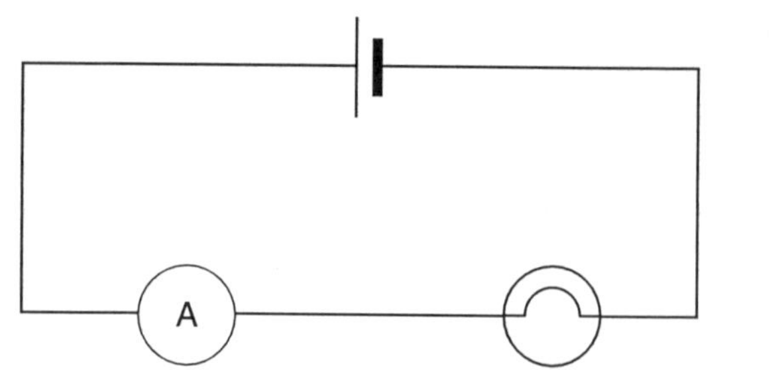

2 **a** If you move the ammeter to this position, do you expect the reading to change?

b Set up the circuit and find out.

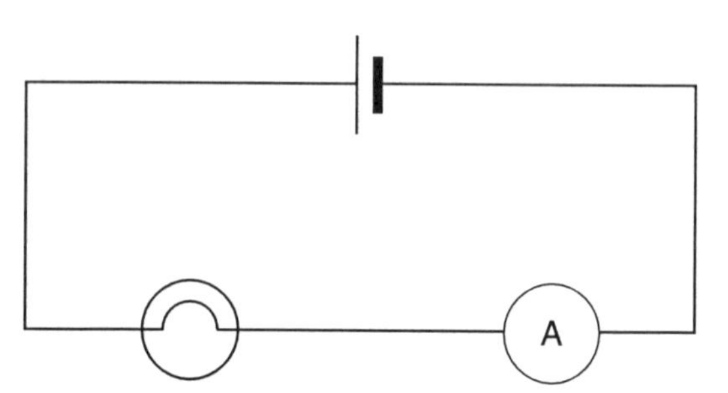

3 **a** Do you expect the current to be bigger or smaller with an extra bulb in the circuit? Explain your answer.

b Set up the circuit and find out.

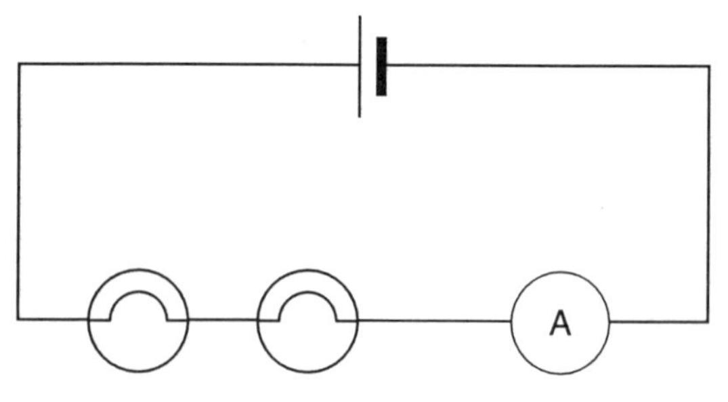

4 **a** Set up a circuit with a variable resistor controlling the brightness of a bulb. Include an ammeter to measure the current.

b Move the slider on the resistor gradually from one end to the other. Draw a table to record:

- the position of the slider
- the current flowing
- the brightness of the bulb.

c How does the bulb's brightness depend on the current flowing through it?

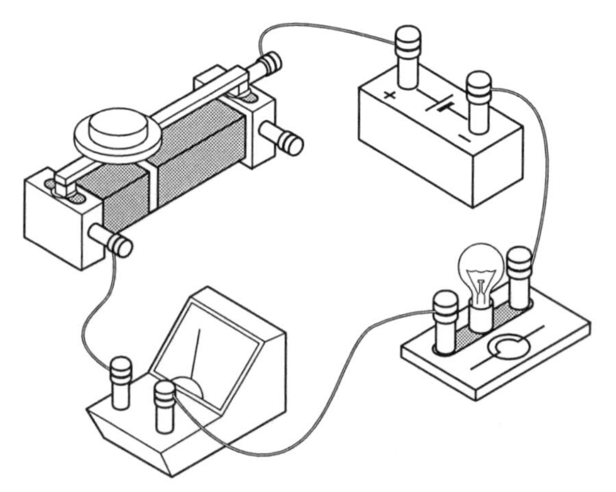

Ammeters at work

You can't see electric current. You need an ammeter to measure how much current is flowing in a circuit.

1 **a** Set up this circuit.

b How much current flows around it?

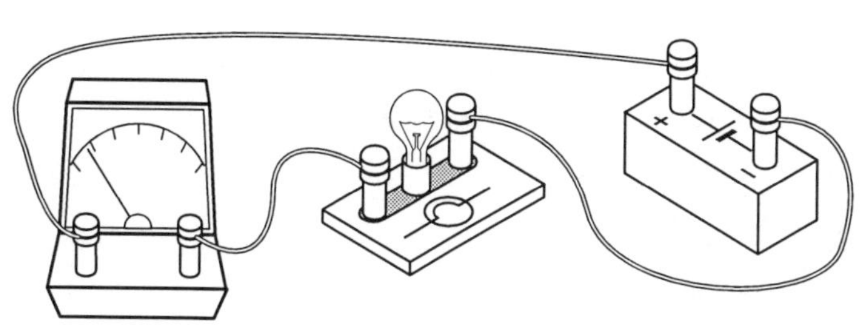

2 **a** If you move the ammeter to this position, do you expect the reading to change?

b Set up the circuit and find out.

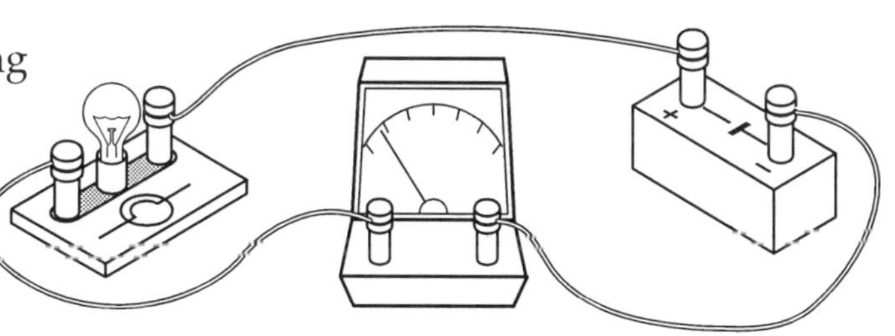

3 **a** Do you expect the current to be bigger or smaller with an extra bulb in the circuit? Explain your answer.

b Set up the circuit and find out.

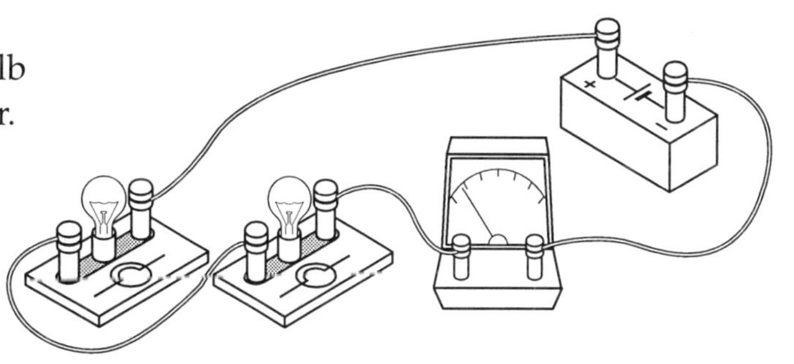

4 **a** Set up a circuit with a variable resistor controlling the brightness of a bulb. Include an ammeter to measure the current.

b Move the slider on the resistor gradually from one end to the other. Copy and complete the table below.

Position of slider	Current flowing	Brightness of bulb

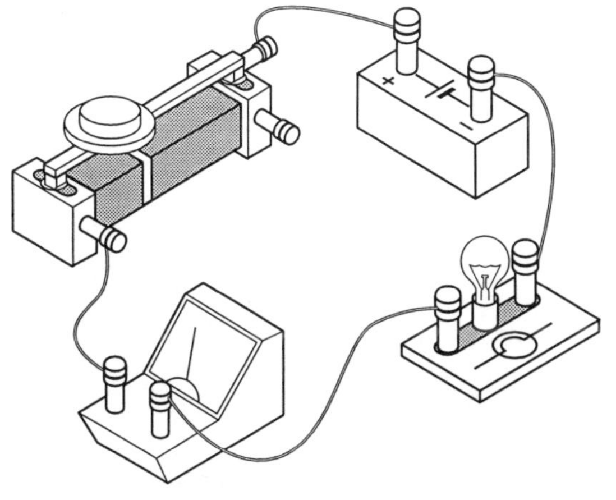

c How does the bulb's brightness depend on the current flowing through it?

Activity 11a Core

Igneous rocks

Crystal size

Igneous rocks form when crystals grow from molten rock as it cools.

1 Look at some samples of igneous rocks. Divide them up into those with large crystals (bigger than 2mm across) and those with tiny crystals (or no visible crystals at all).

2 Why is there this difference in crystal size?

Crystals from molten sulphur

Your teacher will grow some crystals from molten sulphur for you to see. When the sulphur cools in the funnel, the outside cools rapidly, but the inside stays hotter for longer.

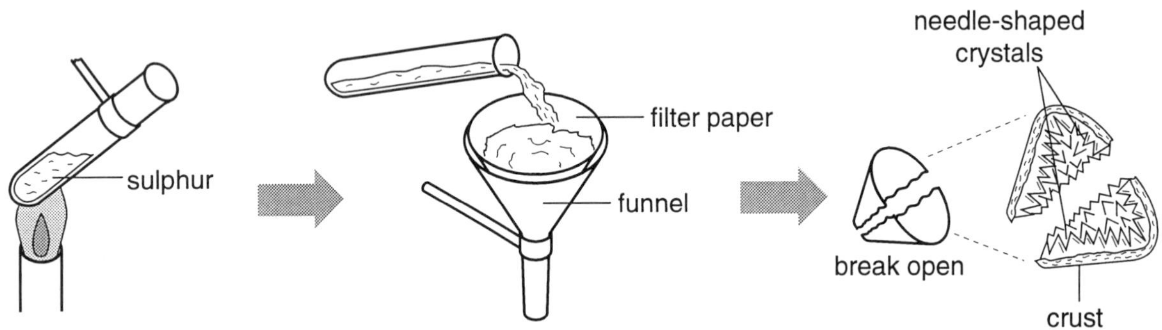

3 Look at the crystals. What is the difference between the crystals that grew in the centre and those in the outer crust?

4 Why are the crystals different like this?

5 Can you think of a reason why some igneous rocks have larger crystals than others?

Test out your idea

You will need:

- some warm microscope slides from an oven
- some cold microscope slides from the fridge
- some molten salol.

6 Pour a drop of molten salol onto the warm slide, and another drop onto a cold slide.

7 Let them cool down and watch the salol crystallise.

8 What is the difference between the crystals on the warm slide and the cold slide? What has caused this difference?

9 Does this confirm your ideas from question **5**?

Igneous rocks

Identifying igneous rocks

1 Use this key to identify some igneous rocks.

2 For each rock, try to work out whether it cooled rapidly on the surface as lava, or deep underground like granite or gabbro.

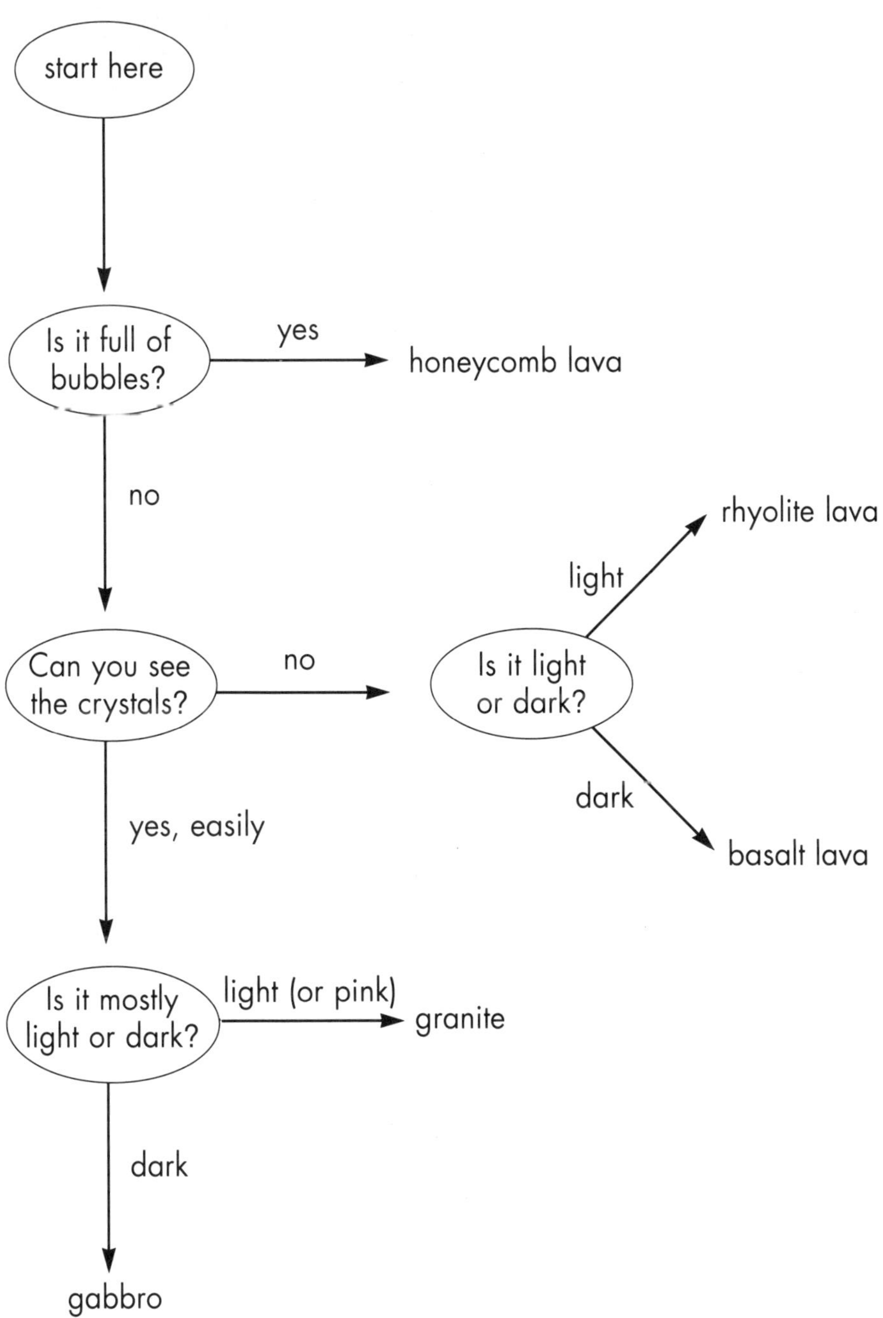

Weathering

Most rocks seem hard enough to last forever, yet in time they are worn or weathered away by physical or chemical attack.

Wear eye protection

Smash them up

1 Heat the end of a glass rod in a Bunsen burner until it glows red. Holding it at arm's length, plunge it into a beaker of cold water. Watch what happens.

2 Explain the connection between this experiment and rock fragments found in the desert.

3 Look at the bottle full of water that was left in the freezer overnight.

4 Explain the connection between this experiment and rock fragments found at the bottoms of cliffs in the mountains.

5 Are these physical or chemical changes?

Here comes the rain

Rain affects rocks very slowly. Rainwater is slightly acidic, so you can speed up the process by using dilute hydrochloric acid.

Take care

Wash your hands after using chemicals.

6 Collect some samples of different types of rock.

7 Put each rock sample in turn onto a watch-glass. Add a few drops of acid and record what happens in a table.

8 Which rocks are most likely to be affected by the rain?

9 For rocks that fizz, is this a physical or chemical change?

How hard are the rocks?

How easy would it be to wear the rocks away by rubbing? Here's one way to find out.

10 Design an experiment to test the hardness of the rock samples. You might work out a fair test using sandpaper, for example:

- Weigh the rock sample.
- Rub it with sandpaper 20 times.
- Reweigh the rock sample.

Weathering

Activity 11b Help

Smash them up

Complete these sentences.

1 When the glass rod is put in the flame of the Bunsen burner it gets very __________. This makes it __________ (get bigger). When it is put in the water it __________ very quickly. This makes it __________. The sudden change makes the glass __________.

2 The rocks in the desert get very __________ during the day and expand. They get very cold at __________ and contract. This makes them __________ into pieces.

3 Water is unusual because it __________ when it freezes. This has __________ the glass bottle.

4 __________ gets into cracks in the rocks in the mountains. At night, this freezes and turns to __________. This expands and pushes the crack open, __________ the rock.

Here comes the rain

5 Write the results from your experiment in this table:

Rock type	What happens with acid?

6 Which rock types were affected by the acid?

____

____

7 Which rock types were not affected by the acid?

____

____

Rivers in action

You can model a lot of the features of a river using a tap, a plank of wood and some home-made soil.

1. Cover the plank with a layer of home-made soil: sand, a little gravel and some powdered chalk for clay.
2. Prop the board at an angle, with its base in a large tray.
3. Connect some rubber tubing to the tap. Clamp the free end of the tubing at the top of the plank, in the centre.

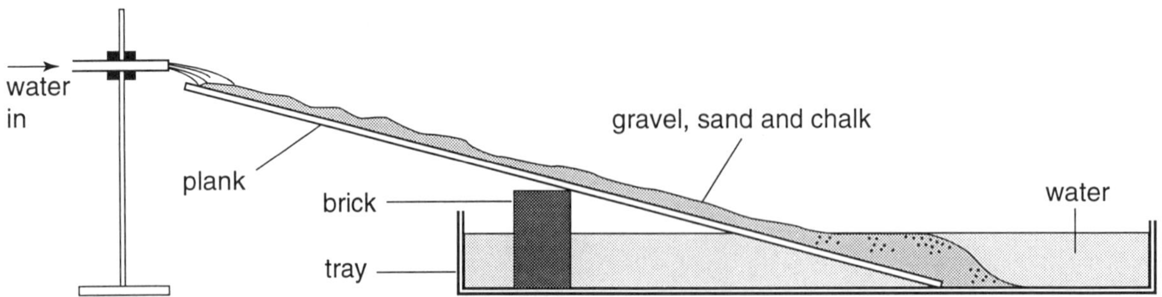

4. Turn the tap on slowly, until a stream of water flows gently down the plank.
5. Watch carefully and record what happens. In particular:
   - **a** Look at the water flow. Does it keep to a straight line?
   - **b** Look at the way material is carried along in the flow.
   - **c** What happens in the tray, as a pool of water starts to collect?
   - **d** How does the appearance of your river change with time?
6. Discuss your observations with your group. Try to compare the things you see in your model with what happens in a real river.
7. Write a report on your experiment. Describe the different river features you have seen in your model.
8. Draw a picture of your model river and label as many interesting features as you can. Add a few words to explain what is happening in each case.
9. Try varying factors such as the rate of flow or the angle of slope of your model river. Note what happens.

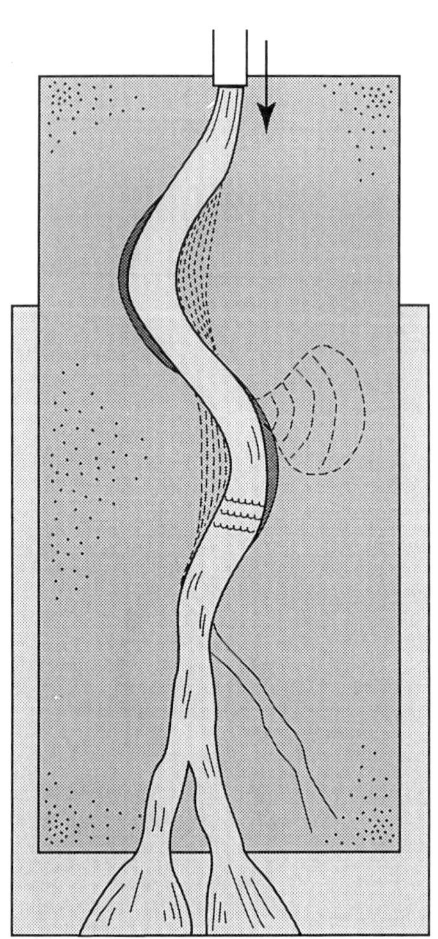

How does your river compare with this one?

Sedimentary rocks

Sedimentary rocks are made from broken fragments of other rocks, or the shells of sea creatures.

1 Look closely at some pieces of sedimentary rock. How can you tell they are sedimentary? Can you see any broken fragments or fossils?

2 Use this key to name the different rocks. Use your answers to the key questions to write a description of the rocks you have named, for example:

- A is chalk. It fizzes with acid and is pure white.

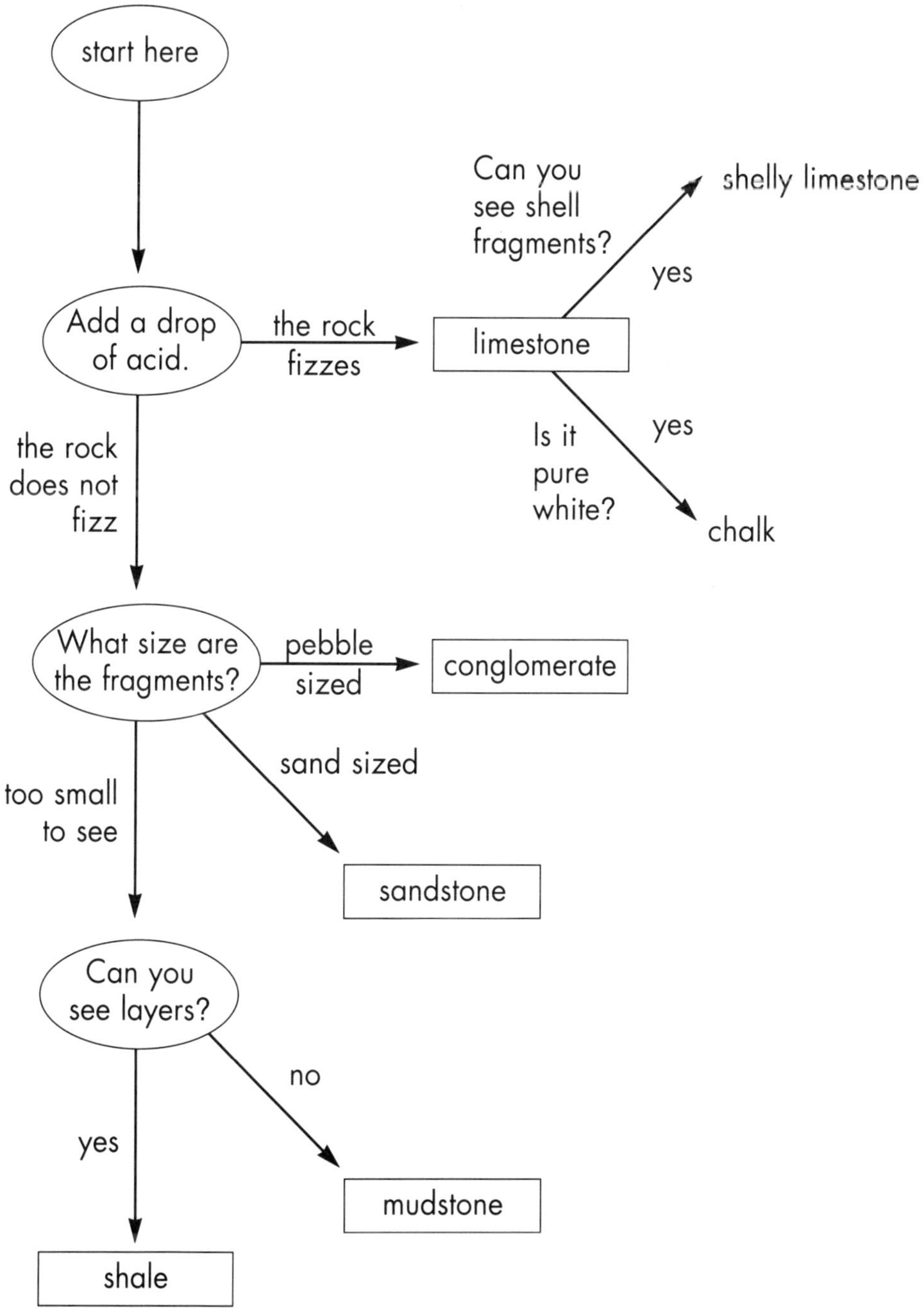

start here

Add a drop of acid.

the rock fizzes → limestone

Can you see shell fragments? yes → shelly limestone

Is it pure white? yes → chalk

the rock does not fizz

What size are the fragments?

pebble sized → conglomerate

sand sized → sandstone

too small to see

Can you see layers?

yes → shale

no → mudstone

3 Did any of the rocks have fossil sea shells? If so, where do you think that rock was formed? Explain your answer.

Sedimentary rocks

1 Choose a rock sample. Look at it and test it with acid. Underline the words in the box that describe it.

2 Cross out any words or phrases that you have not underlined.

3 Now look at the key and find the words you underlined in the box. They will lead you to the name of your rock.

4 Put the name of the rock in the box. Choose another rock sample and start again at step **1**.

Rock A

The rock fizzed / did not fizz with acid.

You can see pieces of shell in it.

You cannot see shells and it is pure white.

The rock is made from fragments that are pebble sized / sand sized / too small to see.

The rock shows layers.

It is called ____________________.

Rock B

The rock fizzed / did not fizz with acid.

You can see pieces of shell in it.

You cannot see shells and it is pure white.

The rock is made from fragments that are pebble sized / sand sized / too small to see.

The rock shows layers.

It is called ____________________.

Rock C

The rock fizzed / did not fizz with acid.

You can see pieces of shell in it.

You cannot see shells and it is pure white.

The rock is made from fragments that are pebble sized / sand sized / too small to see.

The rock shows layers.

It is called ____________________.

Rock D

The rock fizzed / did not fizz with acid.

You can see pieces of shell in it.

You cannot see shells and it is pure white.

The rock is made from fragments that are pebble sized / sand sized / too small to see.

The rock shows layers.

It is called ____________________.

Rock E

The rock fizzed / did not fizz with acid.

You can see pieces of shell in it.

You cannot see shells and it is pure white.

The rock is made from fragments that are pebble sized / sand sized / too small to see.

The rock shows layers.

It is called ____________________.

Types of rocks

The rock cycle

Igneous, sedimentary and metamorphic rocks are all related in the rock cycle. How can you tell them apart? This activity will show you.

1 Use this simple key to name some rock samples as igneous, sedimentary or metamorphic.

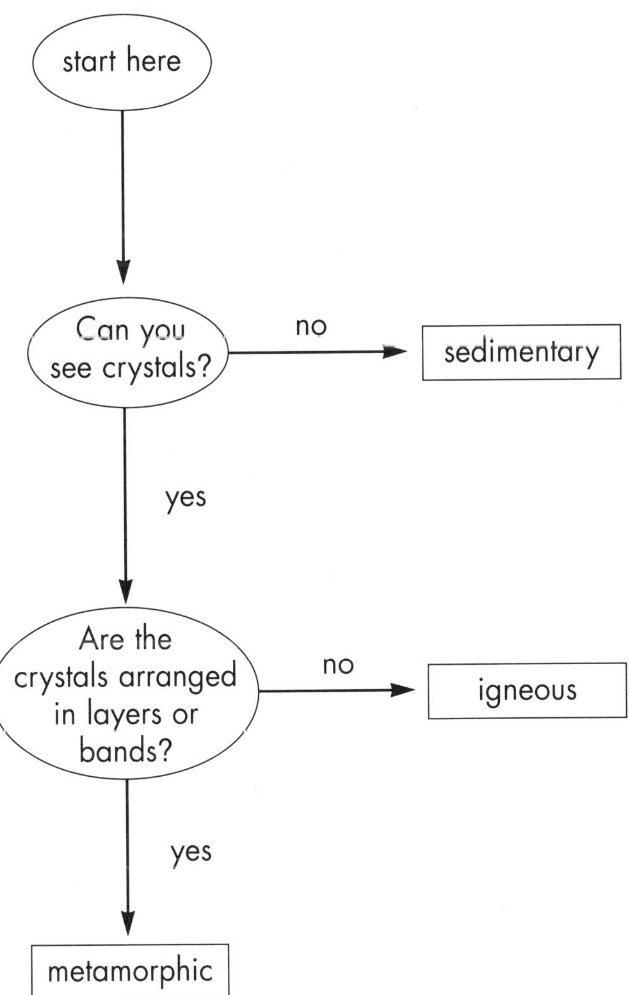

2 Look at samples of the rocks described below. They do not fit this simple pattern.

- **Slate** has crystals in layers. That's why the rock splits into layers easily. But the crystals are too small to see. Slate is metamorphic.
- **Lava** has crystals that are too small to see. The crystals are not in layers. Lava is igneous.
- **Marble** may have crystals that are not in layers. Marble often looks like sugar. But marble is metamorphic.

Types of rocks

A local survey

If you walk down any high street, you will see lots of rock types:

- Igneous rocks are used as kerbstones or cobblestones.
- Sedimentary rocks are used as paving stones, road chippings or building stone.
- Metamorphic rocks such as marble may be used for statues. You see marble more often as a facing stone to smarten up shop-fronts. Slate is used for roofs.

Churches and graveyards may also show a wide range of rocks used as building stone or gravestones.

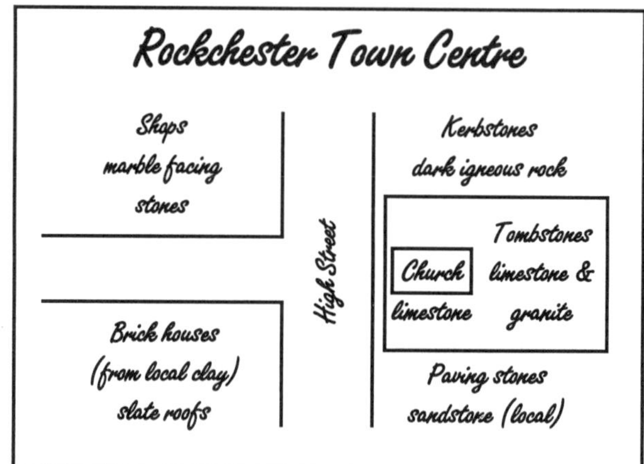

Helen's survey

1 Draw a simple large-scale map of your local high street and its surroundings.

2 Walk around the area, trying to identify as many different types of rock as you can. If you can't name them exactly, then igneous, sedimentary or metamorphic will do.

3 Mark on your map where you find the different rock types.

4 You could have a class competition. See who can draw up the longest list of different rocks used in your area.

5 Try to find out if any rocks are dug out of the ground in your area. Mark on your map the rocks that have been quarried locally.

6 See if you can find out where some of the other rocks have come from.

7 If you have an old graveyard nearby, you will find gravestones over 100 years old made from different rock types. They might show signs of weathering, as the words become harder to read. Which rock types are weathering faster, igneous or sedimentary?

Measuring volumes

Skills sheet 1

Liquids

You can easily measure the volume of a liquid using a measuring cylinder.

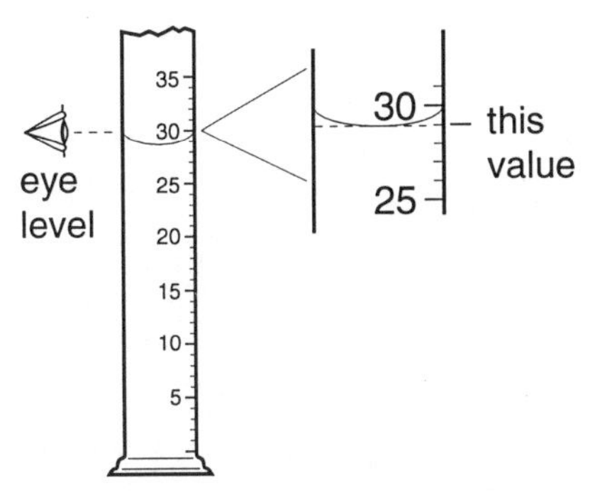

1. Pour the liquid into the cylinder.
2. Look at the scale, making sure your eye is level with the top of the liquid.
3. Water has a curved upper surface (called the **meniscus**). Find the point on the scale level with the bottom of this curve. This is the volume of water in the cylinder.

   Note: Some cylinders have a scale marked in cubic centimetres (cm^3), others in millilitres (ml).

1 ml is the same volume as 1 cm^3.

Small solids

When you get in the bath, the water level rises as your body pushes water out of the way. You can use this method to find the volumes of small solid objects.

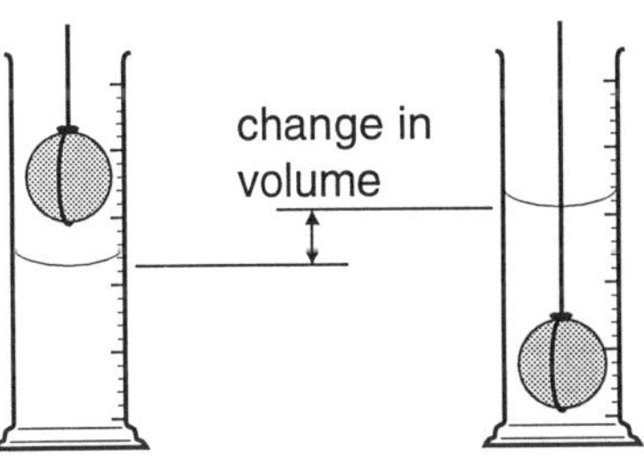

1. Half-fill a measuring cylinder with water and take the reading.
2. Completely submerge the solid in the water and take the new reading.
3. The volume of the solid is the **difference** between the two readings.

You can use this method to find the volume of an irregularly shaped solid object.

Skills sheet 2

Plotting graphs

Organising your results

Look at your table of results.

Mass (g)	Extension of spring (cm)
10	1.9
20	4.1
30	6.0
40	7.8
50	10.1
60	12.2

1. Which factor did you change in your investigation? In this example, pupils changed the mass on the end of a spring.
2. Write the values of this factor at equal spaces on the horizontal (x) axis. This is a **scale**. Make sure the scale goes up at least as far as the highest value for the factor. The highest value is 60g.
3. Write a label which says what the units are. Here the units are grams.
4. Which factor is changing as a result of what you do? In this example it is the length (extension) of the spring.
5. Make a scale for the values of this factor on the vertical (y) axis. Remember that the scale must go at least as far up as the highest value for the factor.
6. Write a label saying what the units are.

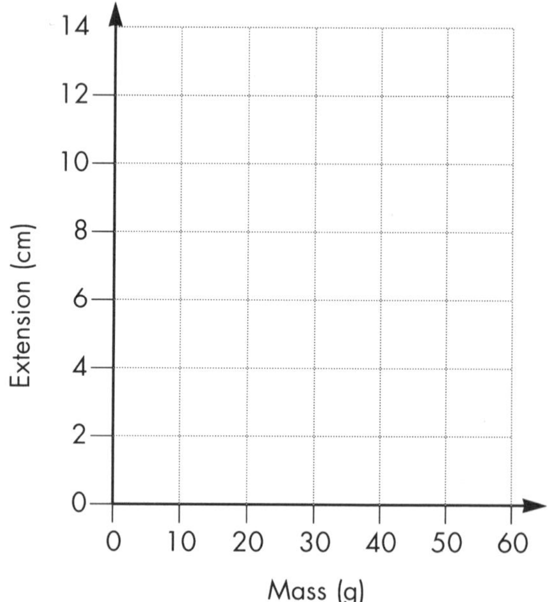

Plotting the points

Make a cross for each pair of readings from your table.

It might help if you:

- draw a faint line up from the x axis
- draw a faint line across from the y axis
- mark a cross where these lines cross.

Drawing the line

Look at your points. Do they lie in a slightly 'fuzzy' straight line?
If so, use your ruler to draw a straight line which goes as near to as many points as possible, and has as many points above the line as below it.
This is called a **best fit** line.

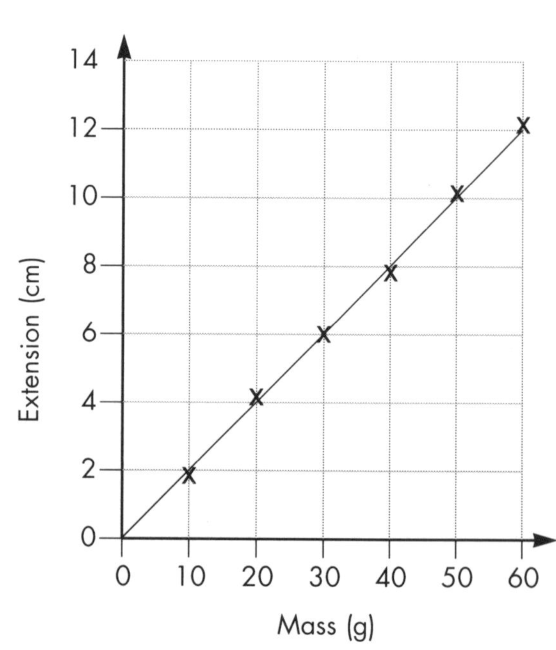

Skills sheet 3

Report writing

Use these ideas to help you when you write a report on an experiment.

You will need to include:	Ideas to think about:
1 A title and date	
2 What you were trying to find out	What was the point of the experiment? What factor did you change? What changed when you did this? What factors did you keep the same (control)? What result did you expect to find (your best guess)? What scientific idea made you expect this result?
3 What you did	Give a step-by-step account, with a diagram if that helps. Were there any special safety precautions you had to take? Did you have to use any special or unusual apparatus? How did you take any measurements you made? How did you control any other factors to make sure that it was a fair test?
4 Your results	If you recorded your results in a table as you did the experiment, show it in your report. You could present your results as a bar chart or line graph.
5 What you found out	Did you find any pattern in your results? Did your experiment give you the results you expected? If so, say how this agrees with the scientific idea you were using. If not, try to explain your results. Was the idea you were using wrong? Is a different scientific idea at work?
6 Was it a fair test?	Did you manage to control all the other factors to make it a fair test? Or were there some factors that affected the result that you did not expect?
7 Could the experiment be improved?	How reliable do you think your results were? Did other groups get the same sort of results as you? If you did the experiment again, what changes, if any, would you make?

Using a Bunsen burner 1

Lighting the Bunsen burner

1. Make sure that the air-hole is completely closed.
2. Light a match or splint and hold it to one side of the Bunsen burner, 2 or 3 centimetres above the top of the barrel.
3. Turn on the gas tap and bring the match or splint sideways into the stream of gas, moving your hand away quickly when the gas catches fire.
4. Adjust the collar over the air-hole to give a suitable flame as shown.

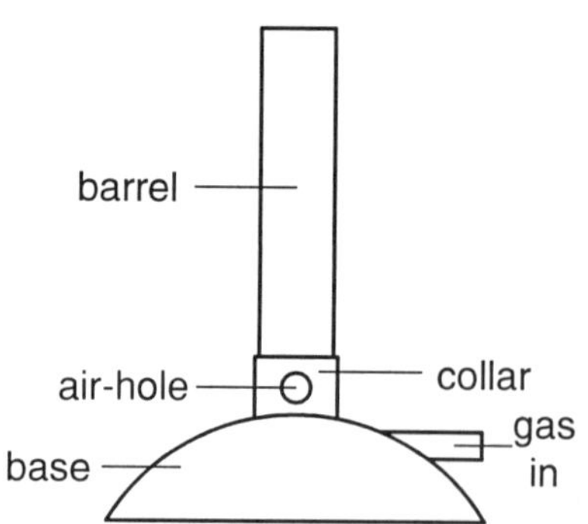

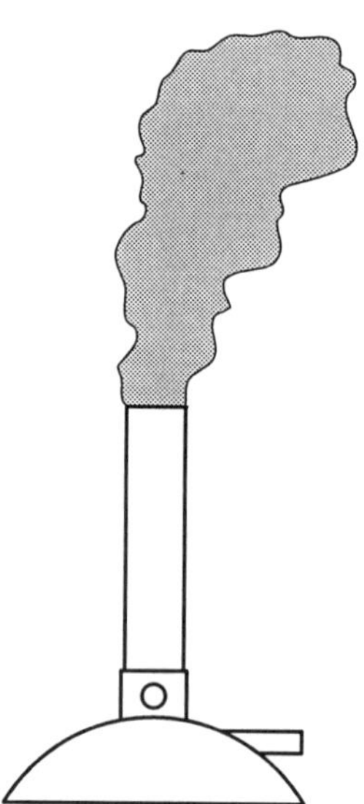

Air-hole closed: the luminous flame
This gives a yellow flame like a candle. This flame is no good for heating things. But if you leave a Bunsen burner on for any reason, always turn it back to this flame so that it is easy to see.

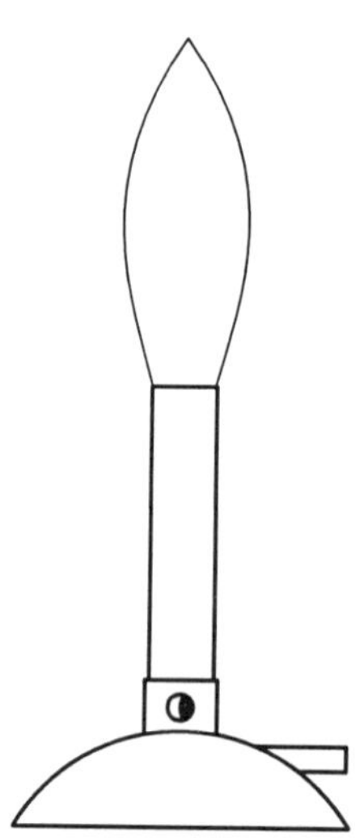

Air-hole half open: the medium flame
This gives a pale mauve, almost colourless flame. This is the best flame for heating most things. You can turn the gas tap down to adjust its height and power.

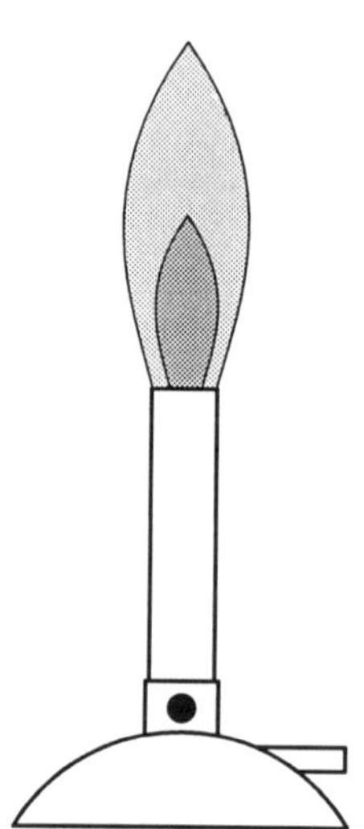

Air-hole fully open: the roaring flame
This is the hottest flame. You use it when you need to heat things as strongly as possible. The hottest part of the flame is just above the blue cone.

Using a Bunsen burner 2

Heating water using a Bunsen burner

1. Half-fill a beaker with water.
2. Put a gauze on a tripod stand over a heat-proof mat as shown. Stand the beaker on this, making sure it is steady.
3. Light a Bunsen burner. Adjust the gas tap to give a medium flame.
4. Carefully push the Bunsen burner under the beaker. Turn up the gas if necessary.

Heating a solid in a test tube

1. Light a Bunsen burner. Adjust the gas tap to give a medium flame.
2. Hold the test tube in a test tube holder, near to the top. Angle it away from the flame, so that the top is not pointing towards you or anybody else.
3. Put the bottom of the tube into the flame for a short time. If nothing happens, put it in again and again, leaving it a little longer each time.
4. If the solid does not change, you may need to turn the gas up. If you are told to heat really strongly, then open the air-hole fully. Hold the base of the tube just above the blue cone.

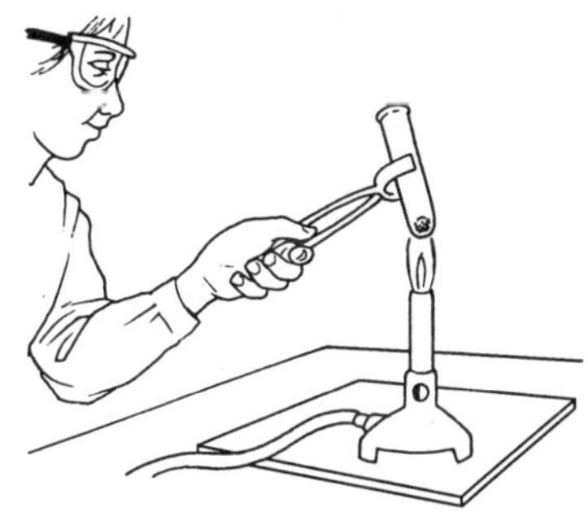

Heating liquids in a test tube

You can heat liquids the same way as solids, but you must take extra care. If the liquid boils suddenly it may squirt out.

⚠ Take care

1. Never have the tube more than one-third full of liquid.
2. Always heat very gently at first.
3. Make doubly sure that the tube is not pointing at anyone.
4. You can heat water solutions like this if you are careful. But **never** heat flammable liquids such as ethanol using a Bunsen burner.

What is a microscope?

A microscope makes small things look bigger, rather like a complicated magnifying glass. It can make things look 4 times, 10 times, 100 times or even 400 times bigger than they really are.

We can use a microscope to see bits of living organisms, or non-living things like crystals, which we could never see just using our eyes. We call the material we are looking at the **specimen**.

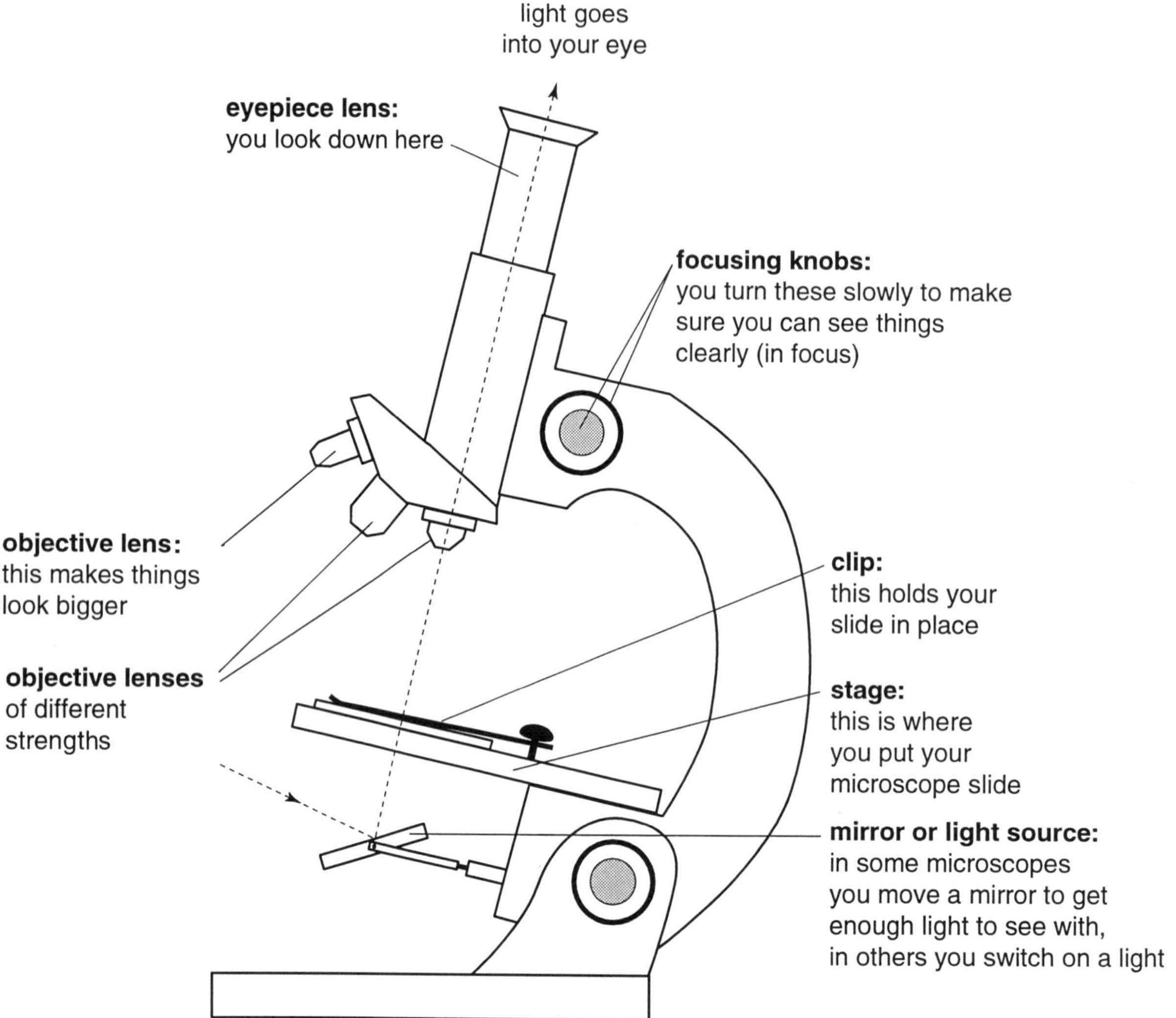

How the microscope works

Light from the light source goes through your specimen on the microscope slide.

The light travels up through the objective lens and the eyepiece lens.

Then it reaches your eye, giving you a clear, magnified picture of your specimen many times bigger than it really is.

How to use the microscope

Take care

a **Always be very careful.** Microscopes are easy to break and expensive to mend!

b **Never** look down a microscope without a slide or piece of paper on the stage. The light could hurt your eyes.

c When you are learning to use the microscope, practise using a piece of graph paper instead of a slide. Graph paper doesn't break, and you can easily tell when it is in focus because you can see the lines clearly.

d **Do not use the largest objective lens.** It is difficult to focus and you can easily break your slide or scratch the lens when using it.

1. Switch on the light source.
2. Make sure that the **smallest** objective lens is pointing down to the stage.
3. Put a slide or a piece of graph paper on the stage. Make sure it is held in place by the clips.
4. Look down the eyepiece lens with one eye. Don't screw the other eye up too tightly or you may end up with a headache!
5. Use the large focusing knobs to get the picture clearly in focus. Turn the knobs very slowly and gently. If the picture does not get clearer when you turn them one way, slowly turn them back the other way until the picture is in focus.

   Remember that if someone else looks down the microscope, they will probably need to change the focus, because everyone's eyes are different.
6. Once you can see clearly with the smallest objective lens, move the middle-sized lens into position. This time, turn the smaller focusing knobs to get a clear picture.

1a Getting started

1 Draw diagrams to show the following forces. Remember to label the arrows.

a the push of a hammer on a nail

b the pull of a hand on a door handle

2 In this rugby scrum, the Stripes team is stronger than the Spots team.

a Draw a diagram to show the pushing forces of the two teams.

b Are the forces balanced?

c What will happen?

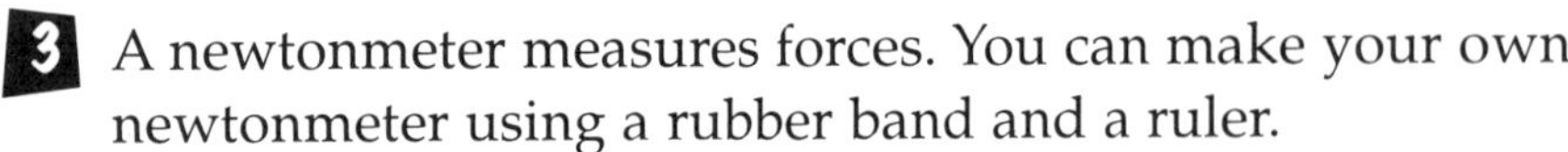

3 A newtonmeter measures forces. You can make your own newtonmeter using a rubber band and a ruler.

Draw a design for a newtonmeter. What other materials would you need?

1b The pull of gravity

4 Who will win the race to push their rock over the cliff first?

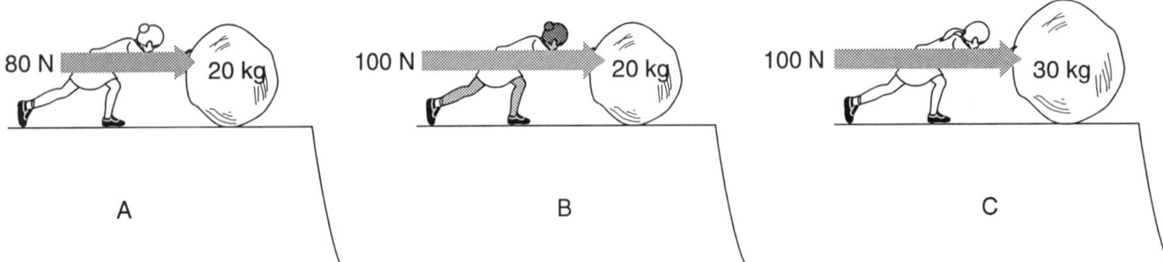

5 Copy these sentences.
Replace the underlined words with the correct words.

"Hi! I'm Pat the space hero. My <u>weight</u> is 70 kilograms. That means that the <u>push</u> of the Earth's gravity on me is 700 <u>kilograms</u>. When I went to the Moon, my <u>mass</u> was much less than on the Earth."

6 The Earth's gravity pulls on everything on the Earth's surface. Its pull is about 10 N for every 1 kg of mass.
a What do N and kg stand for?
b What do we call the pull of the Earth's gravity on something?
c Work out the pulling force of the Earth's gravity on a 20 kg child.

1c Moving along

7 Look at this list.

- a skater sliding easily over the ice
- using the brakes on a bicycle
- putting oil into a car engine
- hammering a nail hard to make it go into a piece of wood

Which of these is an example of:
a using friction to stop something moving
b using lubrication to reduce friction
c friction making it hard to move something
d keeping moving when there is little friction?

8 Use the idea of friction to explain why:
a you can slide a long way on a well polished floor
b if you push a toy car on the carpet, it won't go very far
c if you slip on a hard tennis court, you may get a nasty graze.

9 Fran is trying to push this heavy box along.
a Are the forces balanced? How do you know?
b Will she succeed?
c How could she make it easier to push the box along?

1d Moving in water and air

10 A shark is a large fish which travels fast under water. It has a streamlined shape. It needs to escape from its enemies and catch its prey. Some fish jump out of the water when they see a shark coming, and travel above the surface of the water.

a How does a shark's shape help it to escape from enemies and to catch its prey?
b Why do some fish jump out of the water when they see a shark coming? How does this help them to escape?
c Which is bigger, the friction in water or the friction in air?
d Why is there no friction in space?

11 **a** The gannet is diving into the sea. Why does it make this shape?
b The puffin is landing by its nest. Why is it holding its wings out like this?

puffin

gannet

12 These seeds need to travel through the air, away from their parent plant.
a Why do they fall slowly to the ground?
b Which do you think will travel furthest? Explain your answer.

thistledown

dandelion seeds

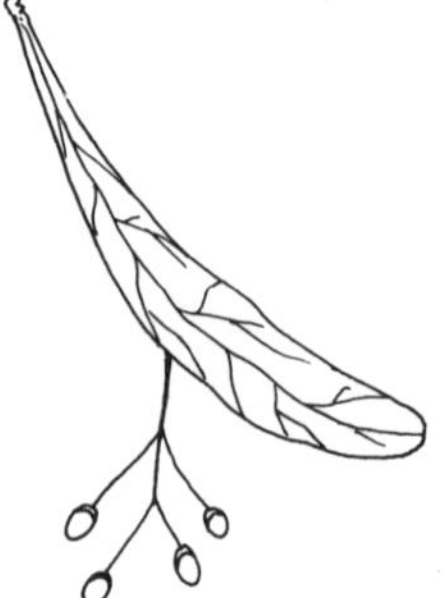
lime seeds

1e With a little help

13 Copy the diagram of the lever.

a Label the pivot.

b Draw an arrow to show the weight of the rock.

c Where should you press down to make it easiest to lift the rock? Draw an arrow to show your answer.

14 Is this see-saw balanced? What will happen when the man lifts his feet off the ground?

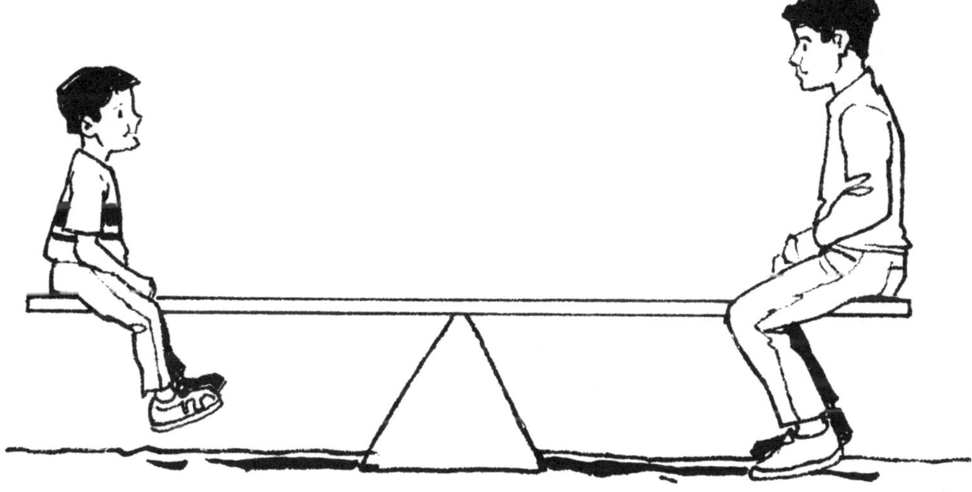

15 If you dig with a spade in the garden or on the beach, you are using a kind of lever.

Explain why a spade is like a lever.

2a Alive or . . . ?

1 Here is a list of all sorts of different things.
Make a table and sort them into two groups, living things and non-living things.

watch	**holly tree**	**hamster**	**bicycle**	**computer**	
worm	**oyster**	**popcorn**	**potato**	**snake**	**aeroplane**

2 The name MRS GREN contains the first letter of the features of all living things. Copy and complete this list of all seven.

M________________________

Respiration

Sensitivity

G________________________

R________________________

Excretion

N________________________

3 One of these fish is alive, one is dead and one is non-living.
What are the differences between them?

2b Moving matters

4 **a** Think of two animals and two plants that you might eat.
b Think of two animals and two plants which might share your home or garden.

5 **a** Why do we divide living organisms into plants and animals?
b You are working in a team of scientists exploring a remote desert. You have found several mysterious specimens which you have never seen before. How will you decide if the specimens are plants or animals?

6 **a** Why does moving matter?
b Make a list of all the different types of movement in this description of a hedgerow at dawn.
(The features of all living things will give you some clues.)

Dawn breaks over the old hedgerow. As the light gets stronger, night creatures like the badger and the owl go home to sleep during the day. The flowers slowly open their petals and turn their leaves to the sun to catch as much light as they can.

In their nests, baby birds begin to move restlessly as they feel hungry. Their parents sing in the hedgerow before flying off to collect food.

Insects begin to crawl and fly as it gets warmer. Blackbirds hunt for worms, which try hard to avoid being pulled from the ground. In the undergrowth, mice feed on new grass before curling up to sleep, their sides rising and falling as they breathe.

The hedgerow fills with the noise and bustle of the animals that feed during the day.

2c The Lego of life

7 Here are some of the parts of cells:

nucleus **cell membrane** **chloroplasts**
cell wall **cytoplasm**

a Which of these parts would you expect to find in an animal cell?
b Which would you expect to find in a plant cell?

8 **a** Plant cells have two important parts that animal cells don't have. What are they?
b Why are these two things important to plants?

9 Design a clear, colourful poster for display in the science area of your school to show and explain the main differences between animal and plant cells.

2d What is an animal?

10 Animals are divided into two main groups, the vertebrates and the invertebrates.
What is the main difference between them?

11 If everyone in your class made a list of 20 animals and divided them into vertebrates and invertebrates, you will have a record of the choices of animals you all made.
You can use those results for this question.
If not, you can use these results from a class at another school.

Number of invertebrates in the list of 20 animals	Number of pupils making this choice
1	2
2	4
3	5
4	9
5	6
6	3
7	1

a Make a bar chart to show these results.
b Did the pupils choose more vertebrates or more invertebrates in their lists of 20 animals?
c Why do you think they made these choices?

12 Invertebrate animals come in a much wider variety of shapes and sizes than vertebrates. Think about the main differences between them, and say why you think this is so.

2e Creepy crawlies

13 Here are some drawings of animals which belong to the main groups of invertebrates.
Which group does each animal belong to?
(The most difficult ones have been done for you!)

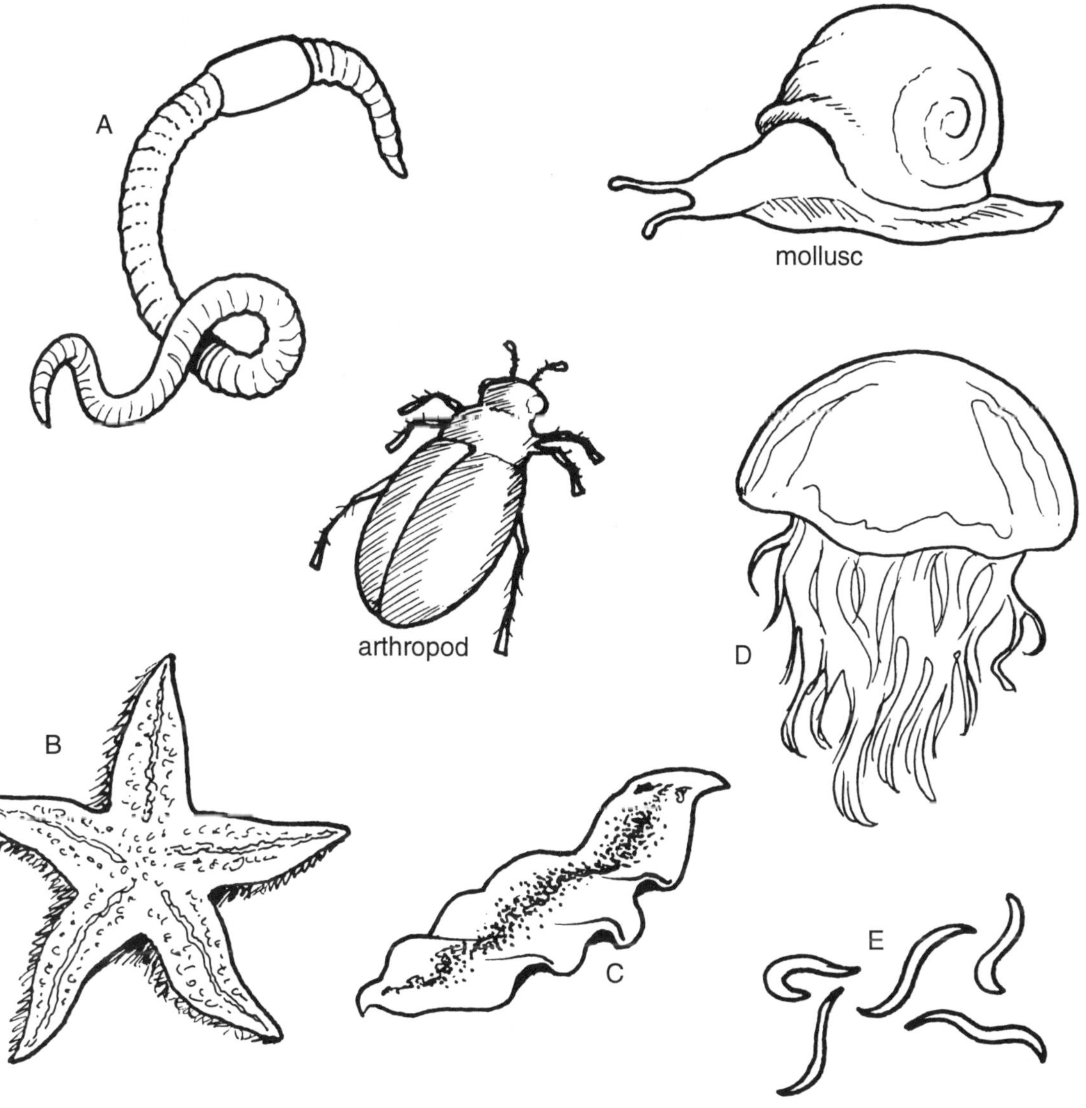

14 Design a way of showing all the main groups of invertebrates that someone could take out in the garden, the playground or the countryside to help them identify any creepy crawlies they might find. Think about:

a how you can describe the types of invertebrate as clearly as possible

b how it can be used outdoors and stay clean and dry.

15 Using your own general knowledge and books from home, from school or the library, find out as much as you can about one particular type of invertebrate. Write about it in no more than 300 words, and draw pictures if you want to. Think about how your invertebrate moves, what it feeds on and where it lives.

2f Where do we belong?

16 **a** Name the five main groups of vertebrates.
b Write down at least one main feature of each group.

17 Australia has many interesting animals that are found nowhere else in the world. Here are two examples:

Kangaroo	Duck-billed platypus
 Kangaroos are warm-blooded vertebrates known as marsupials. Their babies are very tiny when they are born. They have to pull themselves through their mother's fur until they reach her pouch. Once in the pouch, they feed on milk produced by their mother until they are big enough to come out of the pouch and begin to look after themselves.	 The duck-billed platypus is known as a monotreme. They are furry creatures with webbed feet and a strange, duck-like bill. The young hatch out of eggs laid by the mother. The babies are then fed from milk produced by special glands in the fur on the mother's stomach.

a What tells you that kangaroos are a type of mammal?
b What suggests that a platypus is also a type of mammal?
c What features about a platypus suggest that it might not be a mammal?

18 Using your own general knowledge and books from home, from school or the library, find out as much as you can about one particular type of vertebrate. Write about it in no more than 300 words, and draw pictures if you want to. Think about how your vertebrate moves, what it feeds on and where it lives.

2g Pick a plant, any plant

19 Here are drawings of four different types of plant. Which plant groups do they belong to?

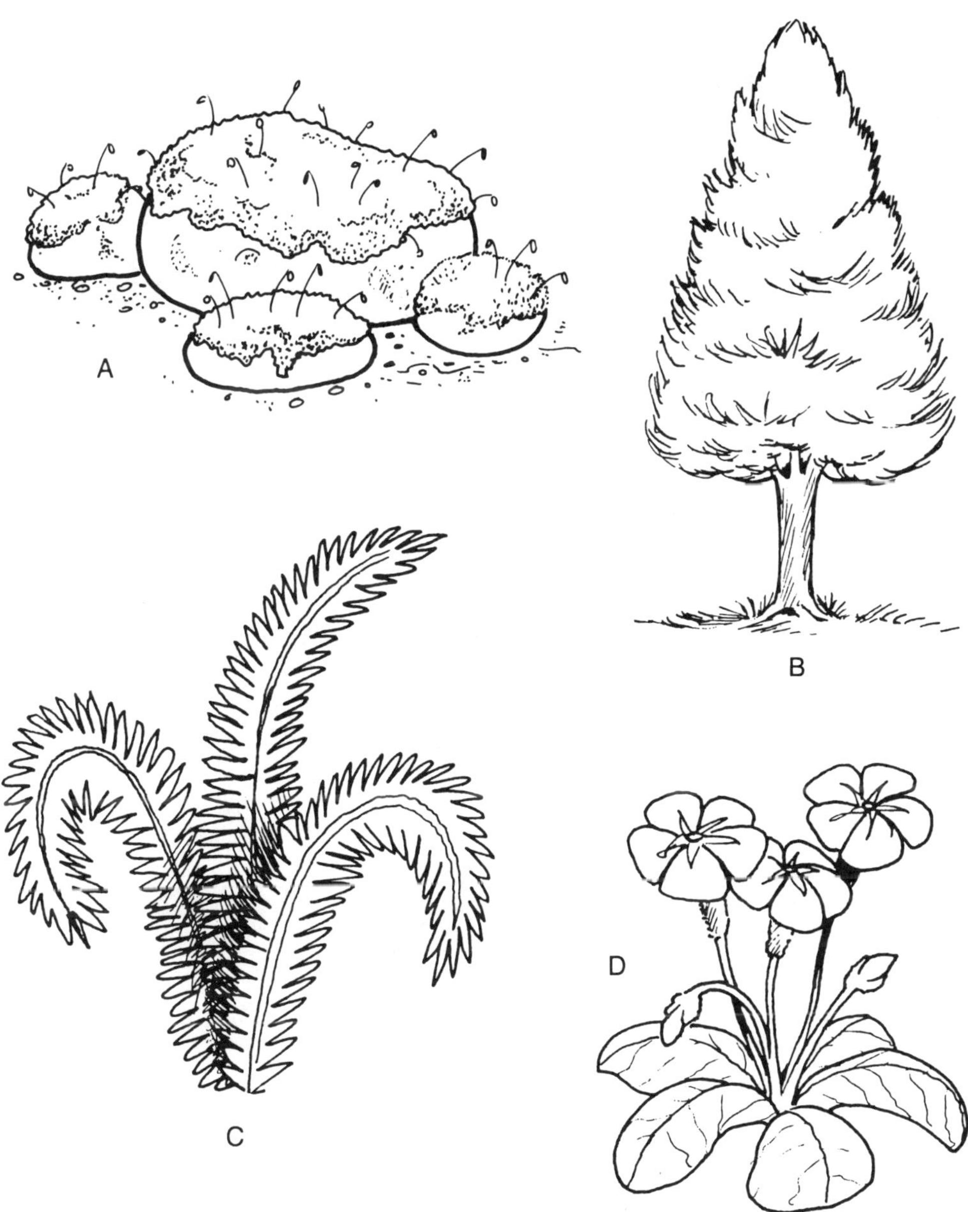

20 Name:

a a plant that produces edible fruit
b a plant with needle-like leaves and cones
c a plant that produces green flowers
d a plant that may live for several hundred years.

21 Your local supermarket has started to sell arrangements of plants. Some of the arrangements are made of ferns and mosses, others are made of flowering plants.

Design a label for each type of arrangement. The label needs to give customers information about the types of plants, and also advice on how to look after them.

2h Organise your organisms

22 You have been given a group of plants to name. Look at these leaves taken from the plants, and use the key below to find out which plant is which.

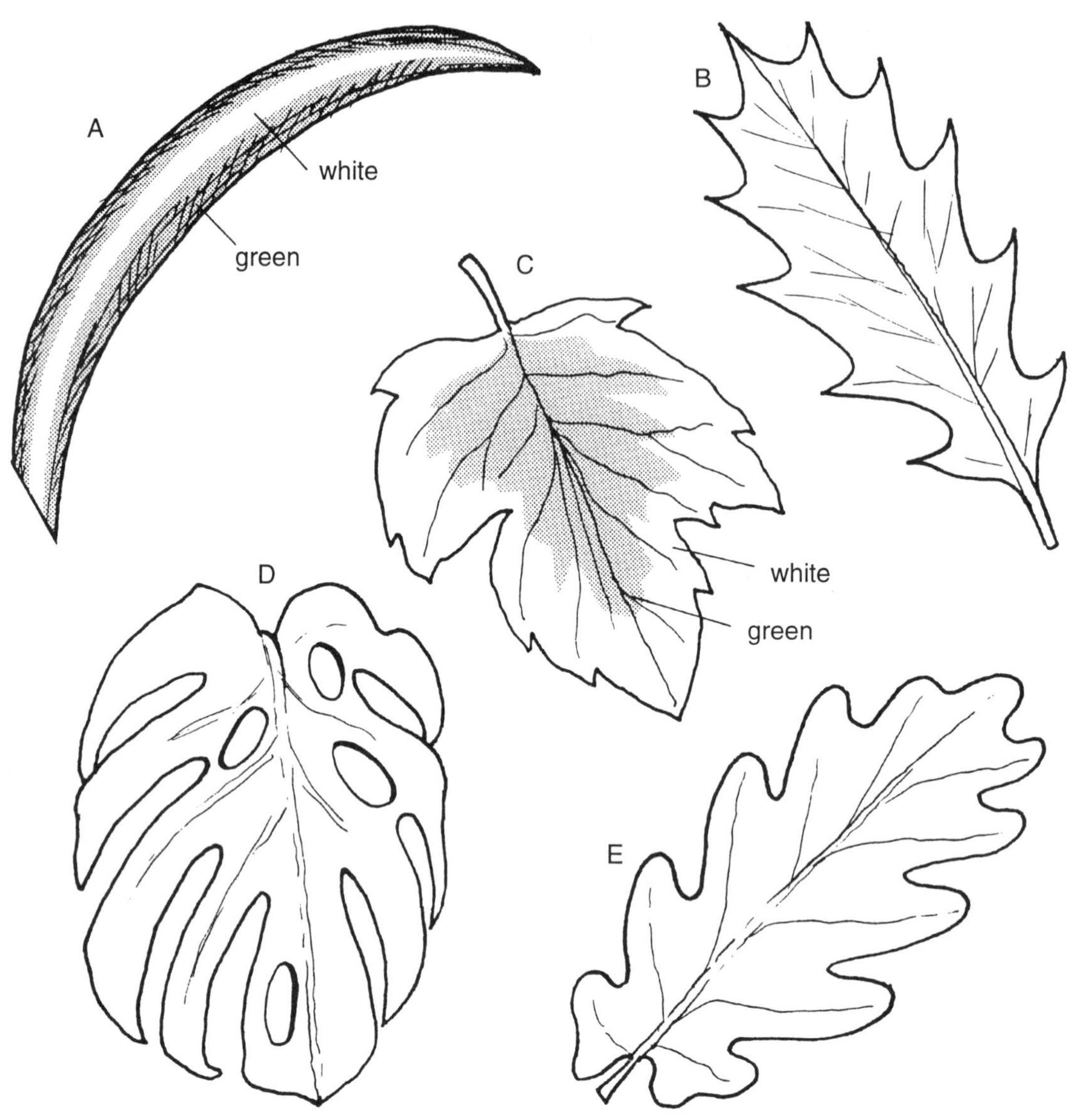

1 Are there two colours on the leaf?	Yes: go to question **2**	No: go to question **3**
2 Is the leaf long and thin?	Yes: spider plant	No: variegated ivy
3 Are there holes in the leaf?	Yes: cheese plant	No: go to question **4**
4 Is the leaf prickly?	Yes: holly	No: oak

23 Choose six animals. Either draw them or stick pictures of them onto a sheet of paper. On the same sheet, make a key to identify the animals you have chosen. Try out your key on a friend or a member of your family to see if it works.

3a Solids, liquids and gases

1 On Earth, Jane invited four friends to her party. She had a large cake to eat and a 1-litre (1000 cm^3) bottle full of cola to drink. As she is a very fair person, she wanted to share out these treats in five equal portions.

a Explain how she could do this for the cake.
What property of cake lets her do this?

b Explain how she could do this for the cola.
What property of cola lets her do this?

c On the planet Splgrnge, Frrgoo invited four friends to her party. She had a nice 1-litre bottle of brown gas to share with her friends as their party treat.
Why will Frrgoo have great trouble sharing this out equally?

2 You need bread and water to live. The water comes through pipes to your home. You have to go out to buy the bread. Why can't bread be piped in, like water?

3 Washing up can be very messy if people forget to scrape the waste beans or carrots off the plate! When you tip the washing-up water away, why is it that the water can go down the drain, but the carrots and beans get stuck in the plughole?

4 Your heart is a pump which works by squeezing the blood inside it, forcing it out and around your body. If bubbles of air get into the bloodstream and reach the heart, this pumping action cannot work properly. Why not?

3b Solid properties

5 Which would you rather carry around in your pocket, 50 pound coins or a £50 note? Why?

6 If you have a brace on your teeth, it will be stuck on with glue. This glue is not very strong, so you can sometimes accidentally knock parts of the brace off when eating. Superglue is very strong.
Why don't dentists use strong glue like that for braces?

7 Bob the glazier likes to show off. He wears lots of jewellery, including a huge diamond ring. But he has lost his glass cutter. How can he cut new glass to fix the broken window?

8 Shoes with leather soles wear out quite fast. If you take them to the cobblers, they are often resoled with a hard kind of rubber.

a Why is this used instead of leather?

b Iron is even more hardwearing than rubber. Can you think of two reasons why we don't wear iron-soled shoes?

3c Useful metals

9 **a** Make a list of five things you use that are made from metal.
b For each, try to think why a metal has been used instead of a non-metal. (Think about the properties that are needed.)

10 **a** Esme the plumber uses some pipes made of copper and others made of plastic. Pipes A are light and easy to carry about, but are not very strong, so they have to be supported with clamps. If the pipe has to change direction, it has to be cut and a special joint fitted.

Pipes B are heavier but they are also stronger, so they need fewer clamps to support them. They can also be bent to go around corners.

Which type is copper and which is plastic? Give reasons for your answer.
b Mains electrical wiring has to be **earthed**, with a special wire leading to a metal pole hammered into the earth. If there is a fault, the dangerous electricity is carried safely to the earth. Copper water pipes pass down into the ground, so some electricians fix the earth wiring on to the water pipes. This used to be fine. But nowadays some copper water pipes are being replaced by plastic ones. Why is this a problem?

11 The cables that carry electricity over the pylons of the National Grid have to be very good conductors of electricity. They also have to be very strong, so that they do not snap. These cables are made from a mixture of aluminium and steel.

Use this table of properties to explain why steel or aluminium are not used on their own.

Metal	Electrical conductivity	Strength
steel	poor	very strong
aluminium	very good	weaker

3d 'Heaviness'

This table shows density data in grams per cubic centimetre. Use it to answer questions **12** to **15**.

dry sand	1.8	wet sand	2.1
iron and steel	7.8	aluminium	2.7
dried beans	0.75	tinned beans	3.9
dried milk	0.4	long-life milk	1

12 Tony's Tippers runs tipper lorries full of sand from the quarry face to the processing plant. The manager thought he'd save some money by buying cheap, poor quality tyres. They were fine throughout the dry summer. But when it rained heavily in the autumn, the tyres started to burst. What might have caused this sudden failure?

13 Aeroplanes use vast amounts of fuel, which is very expensive. The heavier the planes are when they take off, the more fuel they use and the more expensive the trip.

a Why are aeroplanes built from aluminium rather than steel?

b You are putting together a plane-load of relief food supplies to send to earthquake victims in Armoulia. Armouli Airways have said they will provide the planes if you pay for the fuel.
You want to send milk and beans. Do you send:

- dried beans or tinned beans
- long-life milk or dried milk?

Explain your choices.

14 The density of water is 1 gram per cubic centimetre.

a What is the mass of 20 cm^3 of water?

b What volume would 10 g of water take up?

15 The Earth has a density puzzle. The Earth as a whole has an average density of 5.5 grams per cubic centimetre. But the rocks at the surface of the Earth have an average density of only 2.7 grams per cubic centimetre.

Scientists think that the Earth has a core of molten iron. Explain how this helps to solve the puzzle.

3e A model for materials

16 Use this information to design a poster for your class, to show how small the particles that make up solids, liquids and gases really are.

> How many grains of sand are there in a mountain?
> A silly question perhaps, but try to scale it up.
>
> If a grain of sand was 1 mm across, there would be:
>
> - 1 thousand grains in a 1 cm cube of sandstone
> - 1 million grains in a 10 cm cube of sandstone
> - 1 billion grains in a 1 m cube of sandstone
> - 1 billion billion grains in a 1 km cube of sandstone.
>
> That would make a small hill. For a mountain about the size of Ben Nevis, you would need nearer 30 billion billion grains.
>
> So how many of these tiny particles would there be in one grain of sand? About the same as the number of grains of sand in Ben Nevis!

17 **a** Liquid water can turn into a gas (steam). One litre of water weighs 1000 g, but one litre of steam weighs less than 1 g! Why does the steam weigh so much less than the liquid water?

b 18 cm^3 of water weigh the same as about 24 litres of steam. What does that tell you about the number of particles in 18 cm^3 of water and 24 litres of steam?

18 When you squash a gas, you push the particles a little closer together. What would happen to a gas eventually if you kept squashing it more and more?

4a Getting moving

1 There are many different ways of transferring movement energy to something. Make a table like the one shown here listing at least six different ways.

Transferring movement energy by . . .	To . . .
kicking	a ball

2 Here is a list of objects. Put them in order, starting with the one with most kinetic energy (movement energy), and ending with the one with least kinetic energy.

a a 40-tonne truck travelling at 100 kilometres per hour
b an eagle soaring at 50 kilometres per hour
c a pupil sitting still at her desk
d a butterfly fluttering by at 2 kilometres per hour
e a high-speed train travelling at top speed

3 Kinetic energy is the scientific name for movement energy. Why do you think the cinema was originally called the kinema?

4b Getting warm

4 Fuels are stores of energy. When fuels burn, they give us heat energy. There are many different kinds of fuel. The table shows some of them, together with some of their uses. Unfortunately, the table is all mixed up. Sort it out, so that each type of fuel is next to its correct use.

Type of fuel	Use of fuel
petrol	bonfire
charcoal	candle
wax	steam train
wood	car
coal	barbecue

5 Copy and complete these sentences, adding the name of a type of energy.

a Something that is moving has ________________ energy.

b Something that is hot has ________________ energy.

c A fuel is a store of ________________ energy.

6 Draw energy transfer diagrams to show these transfers.

a When petrol is burned in the engine of a car, the car moves.

b Wood is burned on a bonfire to give heat and light.

4c Body fuel

7 We need food so that we can move around, and so that our bodies can keep warm. Use words from this list to answer the following questions.

kinetic energy	**chemical energy**	**heat energy**

a What kind of energy is stored in food?

b What two forms of energy do we get from our food?

8 The table shows the energy values of some drinks.

Drink	Energy value of 100 cm^3
orange juice	150 kJ
cocoa	1300 kJ
soda water	0 kJ
orange squash	90 kJ
cola	200 kJ

a Which drink stores most energy?

b Which stores more energy, orange juice or cola?

c Which drink stores no energy at all?

9 Look at the drinks in the table in question **8**. Which drink would you take with you if you were going for a long hike in the mountains? Give a reason for your answer.

4d Electrical mysteries

10 A television set changes electrical energy into light energy (the picture) and sound energy (the sounds you hear). Name a device that can change electrical energy into:

a heat energy
b light energy
c kinetic energy and heat energy.

11 Copy and complete the following table. Use the words below to fill the gaps. You may need to use some of the words more than once.

batteries **dynamo** **oil** **generator** **power station** **coal**

stores of chemical energy	
fuels used at power stations	
they spin round to generate electricity	
where electricity is generated	

12 The wheel of a bicycle turns the dynamo to generate electricity for the lights.
Draw an energy transfer diagram to show these changes.

4e Hidden energy

13 Copy the sentences below. Choose the correct word from each pair.

a A battery stores **chemical/electrical** energy.
b A stretched rubber band stores **gravitational/elastic** energy.
c Water behind a dam stores **kinetic/gravitational** energy.

14 Some toy cars work like this:
You pull the car back to wind up a spring. Then you let go, and the car speeds off.

Draw an energy transfer diagram to show what happens when you release the car.

15 Explain which type of stored energy is important for these people:

a a downhill skiing champion
b a disabled person in a battery-powered wheelchair
c Robin Hood and William Tell.

5a Sun for supper

1 Plants can make their own food by photosynthesis.
To do this they need:
a light
b water
c carbon dioxide.
Where do they get these things from?

2 Caterpillars eat leaves. If a plant is covered with caterpillars, it will die.
a Why will the plant die?
b Why are the leaves of a plant so important?

3 Here are three pieces of information.

a Carbon dioxide gas can build up in the atmosphere. If it does, many scientists think that it acts as a 'greenhouse', making the surface of the Earth get warmer.
b Oxygen is needed for almost all living things to respire.
c More and more forests are being destroyed all over the world, and many people are getting worried about the loss of plants.

Use what you know about photosynthesis to write a paragraph linking these three bits of information.

5b The root of it all

4 The roots of a young plant spread out and grow much more quickly than the shoots.
a Give two reasons why plants need roots.
b Plant roots are covered in root hairs. How do these help the plant to take in water?
c What do plant roots do for the soil?
d Why was there so much concern when many trees were uprooted during great storms at the end of the 1980s?

5 Plants use their roots to get substances from the soil. These substances are called **minerals**.
a Why do plants need minerals from the soil?
b Which minerals are particularly important to plants?
c When crops are grown on a field, the plants quickly use up all the minerals from the soil.
How do farmers make sure that their crops do not go short of the minerals they need?

5c Animals need food too

6 Make a table like this:

Carbohydrate	Protein	Fat	Vitamins and minerals

In each column, list six examples of food containing that substance. You may find that some foods fit into more than one box.

7 Keep a careful record of your own diet for a week, including snacks! Which sorts of foods do you eat a lot of? Are there any types of food you eat very little of? Does it seem a fairly balanced diet?

8 Write a paragraph to explain why each food substance (carbohydrate, protein, fat and vitamins and minerals) is important in your diet.

5d Healthy eating

9 Why do we need:
a calcium
b iron
in our diets?

10 Some people grow up with their legs bending outwards in the middle. This is called rickets.
a Why do their legs bend like this as they grow?
b Children get rickets if they do not get enough of one particular vitamin as they grow up. Which one?

11 Branno is a new high-fibre breakfast cereal. You are in charge of meals at your local hospital. Write a message to all the patients, explaining why you would like them to change from their normal cereal to Branno.
(You may be able to find out more about fibre on cereal packets, or in books at home.)

5e Food for thought

12 Pine martens are fierce carnivores that live in the pine forests of Scotland. They eat squirrels. Many grey squirrels live in the forests, feeding on pine cones.

a Make a food chain using these animals.

b Animals such as pine martens and squirrels are called consumers. Plants like the pine trees are called producers. Explain why.

13 A children's television programme is producing a special show about the environment. You have been given two minutes to describe a food chain to viewers.

Write a plan for your presentation. You can have living members of your food chain with you in the studio, so choose your example carefully!

5f The web of life

14 Think of the simple food chain:

grass → rabbits → foxes

This chain suggests that if something terrible happened to wipe out the rabbits, the foxes would die out as well, and the grass would get very long and take over the countryside. Yet when most rabbits were wiped out by disease early in the twentieth century, none of these things happened. Explain this.

15 Think about a garden habitat which you know, either at home, at school or at a local park. Here are some of the organisms it will contain. You may be able to think of some more.

plants worms greenfly ladybirds flies
slugs spiders birds snails

Make a food web to show the connections between the different organisms living in your habitat. Start by looking for the food chains and then link them together.

6a Separating mixtures

1 When water is pumped from a reservoir, it often has a lot of mud and sand mixed with it. These must be removed before the water reaches your tap. To do this, the water is allowed to stand in a tank for a while. It is then run off and trickled down through a bed of clean fine sand.

a Why is the water left to stand for a while first?
b Why is it then passed through a bed of fine sand?
c Every so often, the sand in this filter bed has to be replaced. Why do you think this is?

2 You make 'real' coffee by pouring boiling water onto ground-up coffee beans. Then you separate the liquid coffee from the ground-up beans before drinking it.

Look at these two different coffee-makers and explain how each type does this.

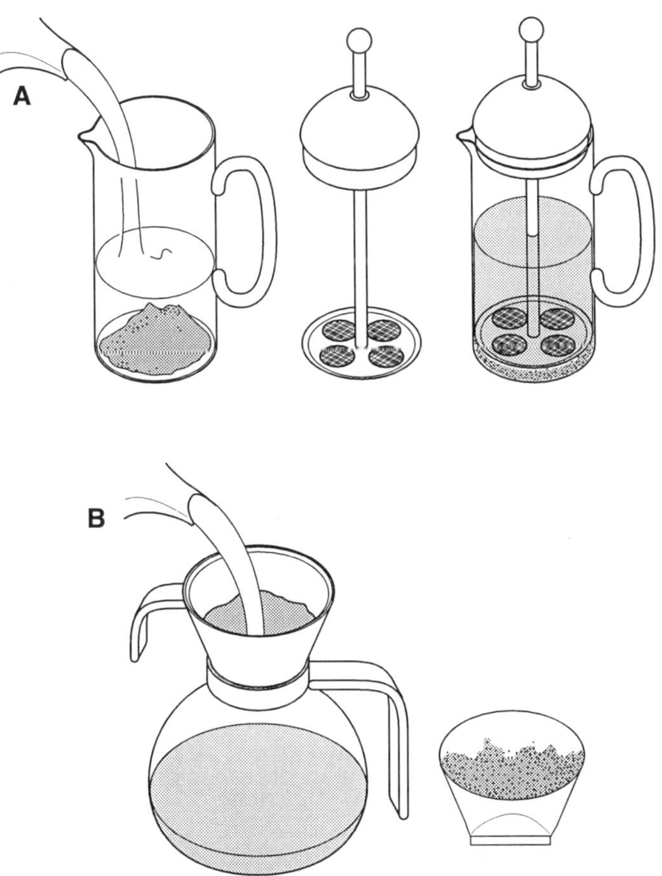

3 Wheat is used to make flour and bread. First the wheat grains have to be knocked out of their seedheads. Then they have to be separated from the broken straw pieces (the chaff). In the old days, this was done by throwing the mixture of grains and chaff up into the air on a windy day. Explain how this method works.

6b Solids in liquids

4 **a** What happens to salt when you sprinkle it onto your soup?
b How do you know that the salt is still there, even though you cannot see it any more?

5 To make jam, you need to dissolve a lot of granulated sugar in fruit juice or water.
a Suggest two ways to speed this process up.
b Caster sugar would dissolve faster than granulated sugar. Explain why.
c Suggest a reason why caster sugar is not generally used for jam-making.

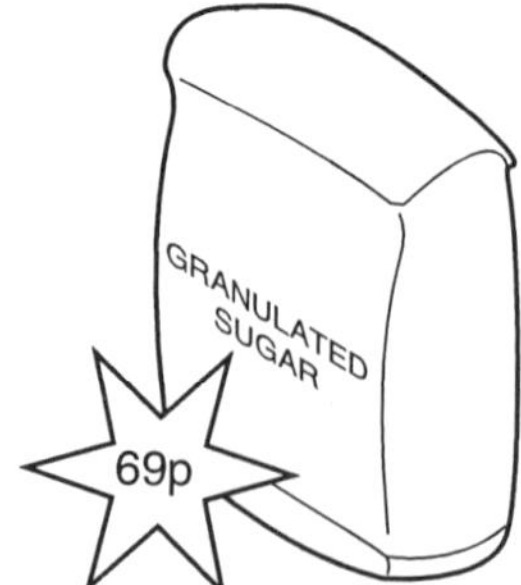

6 Daniel is very messy. When he drinks, he slops some of the cola out of his glass, so it leaves a sticky ring behind on the table. When he is varnishing his models, he drips varnish down the side of the jar, and that leaves a sticky ring too. His table is covered in sticky rings!

When his dad tried to clean the table with a wet cloth, half the rings cleaned off, but the other half didn't.
Explain why.

6c Separating solutions

7 Josie and Judy went swimming. Afterwards, Josie dried her hair thoroughly with a towel, while Judy ran about on the beach. An hour later, they both had dry hair.
How did Judy's hair dry?

8 You may have seen 'air plants' for sale. These strange spiky plants have no roots, yet they grow well, especially in bathrooms. Where do they get their water from?

9 The Great Salt Lake in Utah, USA is very salty. Rivers carry small amounts of salt into the lake, but no rivers run out from it. The surrounding area is a hot desert.
a Explain why the Great Salt Lake is so salty.
(Hint: what happens to the water that comes in … and what happens to the salt?)
b The lake is surrounded by a huge flat area, thickly covered with salt crystals.
i How did the salt crystals get there?
ii What must be happening to the Great Salt Lake?

6d Getting the water back

10 Patrick's house had a cellar that he wanted to use as a workshop. It was very cold, but the house was very warm as the central heating was on. Patrick decided to leave the cellar door open for a week, to let the warm air in. When he went back, the cellar was indeed warmer, but the ceiling and walls were covered with water droplets. Explain what had happened.

11 The air is always warmer at sea level than it is at the top of a mountain. If the wind blows towards a mountain, the air is forced to rise. Explain why it is often cloudy above mountains.

12 If you look closely at a boiling kettle, you will see that the clouds of 'steam' start to form a few centimetres away from the spout.

a What invisible gas is in this gap?
b What makes up the clouds that you can see near the spout?
c Ahmed says that the clouds are not really steam at all. Is he right? Explain your answer.

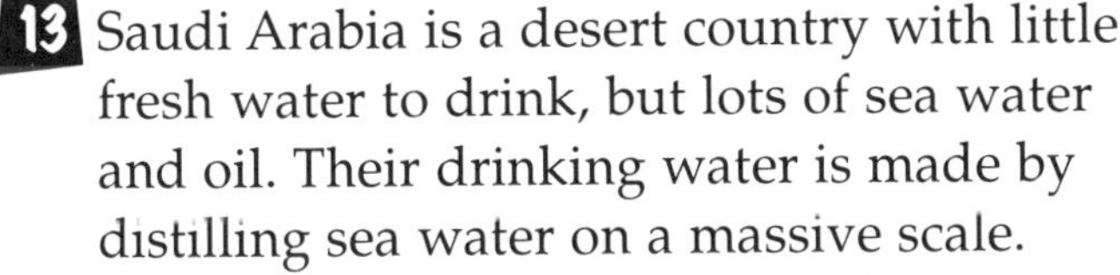

13 Saudi Arabia is a desert country with little fresh water to drink, but lots of sea water and oil. Their drinking water is made by distilling sea water on a massive scale.
a Look at the diagram and work out how the distilled water is produced.
b Copy the diagram and write a few words on it to show how it works.
c Why is this method not used to make drinking water in Great Britain?

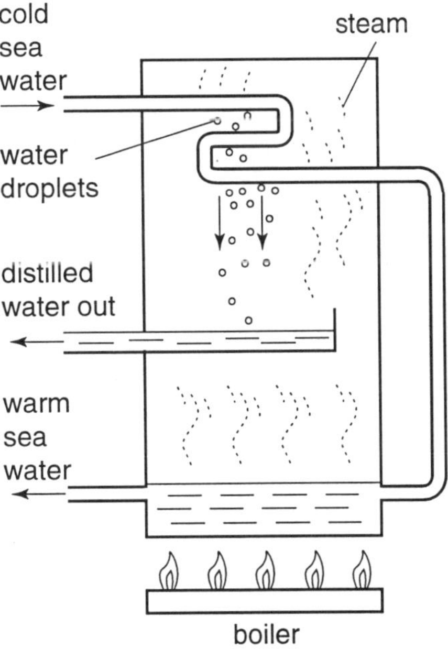

6e What's in that mixture?

14 You can only separate a dye by chromatography using water if the dye dissolves in water.
Fortunately the process works just the same with any solvent.
These diagrams show how the green substance found in plants can be tested to see if it is really a mixture.

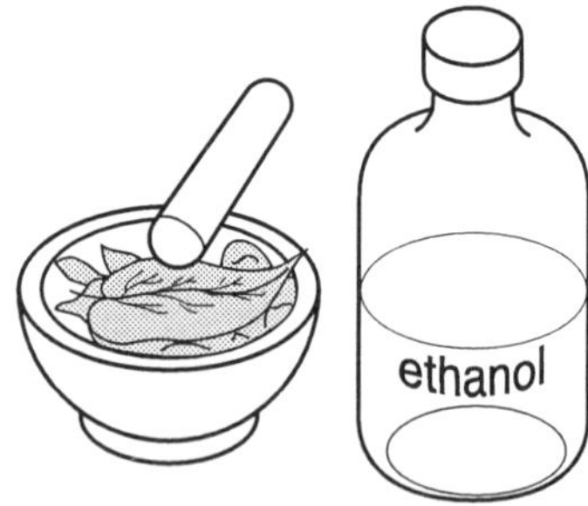

a Some leaves are crushed in a mortar with ethanol.

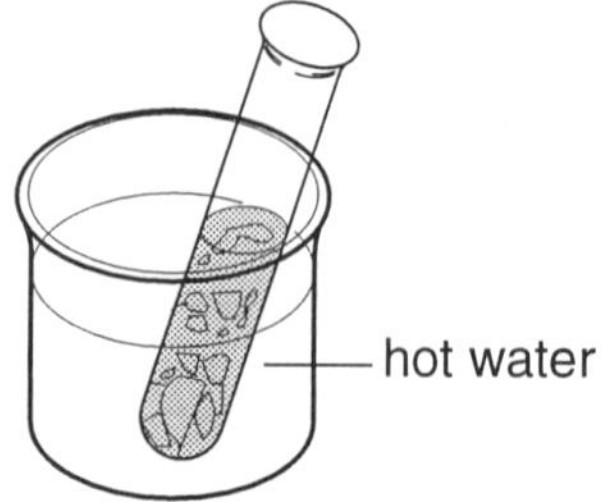

b The mixture is warmed in a water bath.

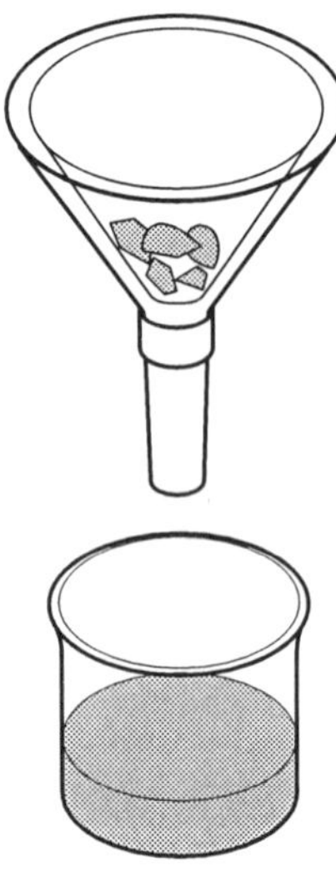

c The mixture is filtered.

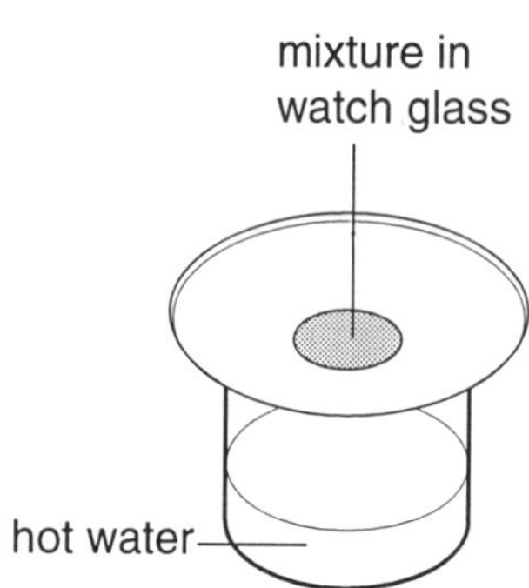

d Some of the ethanol is evaporated from the filtrate.

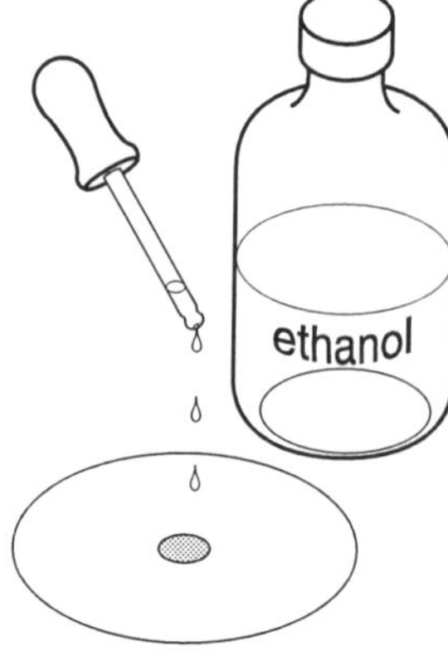

e Chromatography is used to see if the colour is a mixture or not.

Look at the diagrams for steps **a** to **e**.
For each step, describe the process, and explain why it is being used.

7a What causes noise?

1 Copy and complete the following table.
Use the words below to fill the gaps.

noises	vibrations	ear protectors	eardrums	vocal cords

These vibrate when a sound enters our ears.	
These are worn to stop loud sounds reaching our ears.	
These are any sounds that we don't like.	
These vibrate when we speak or sing.	
These travel through the air when we make a sound.	

2 Explain the following. For each, give some examples of animals (or birds) like this.
a Some animals have very big ears.
b Some animals can move their ears around.

3 **a** The three small bones in your middle ear may be damaged by loud sounds. Why does this make you deaf?
b Some people become deaf because they work in noisy places. Give three examples of very noisy workplaces.
c How can people who work in noisy places avoid damage to their hearing?

7b Less noise please!

4 In a house, some things are good absorbers of sounds.
Other things are good reflectors of sounds.
Put the following things in two lists, good absorbers and good reflectors.

curtains	windows	tiles	brick walls	carpets	bedclothes

5 Copy and complete the following sentences. Use the words below to fill the gaps.

higher	**louder**	**lower**	**softer**

a A young person can hear ______________ sounds than an older person.
b A bat can hear ______________ sounds than a human.
c An owl can hear ______________ sounds than a human.
d The ______________ a sound, the more it may damage your hearing.

6 Imagine that you are sitting on the beach, near the cliffs, watching a game of beach cricket. You notice that, every time someone hits the ball, you hear the sound twice.

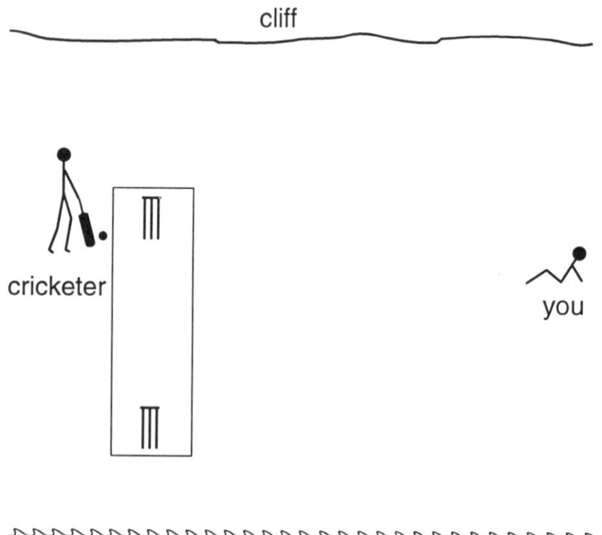

a Copy this sketch plan of the scene.
b Draw a line to show how the sound from the bat travels directly to your ear.
c Draw a second line to show how the sound also travels to the cliff, and then reflects off to your ear.
d What do we call a reflected sound like this?

7c Travelling light

7 At the cinema, you may have noticed that the light from the projector shows up as a beam. You can see the beam because of dust or smoke in the air.

Draw a diagram to show the rays of light from the projector as they go towards the screen.

8 Copy and complete the following table.
Use the words below to fill the gaps.

darkness	**sundial**	**ray**	**shadow**

uses a shadow to tell the time	
where no light reaches	
light travelling in a straight line	
what it's like with no light	

9 **a** Copy this picture of the flag on a golf course.

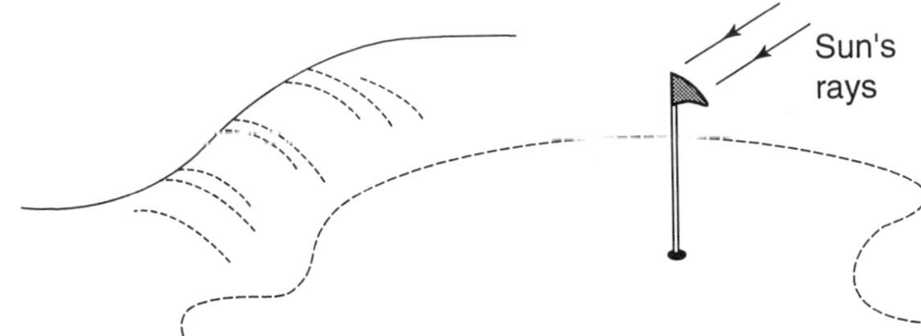

b Draw a ray of light from the Sun to the top of the flag and on to the ground.
c Mark on the ground where the shadow of the flag will be.

7d Long time coming

10 Look at the things in this list:

a ray of light	**a cyclist**	**a sound**	**a toy train**

Put them in order, from slowest to fastest.

11 **a** If the Sun stopped shining, we wouldn't know about it for 8 minutes. Explain why this is so.
b Why does it take less time for light from the Moon to reach us?

12 Imagine setting off a sky-rocket on bonfire night. Rearrange the following sentences to describe what you might observe.

You hear the rocket explode.
You light the blue touchpaper and stand well back.
You hear the dead rocket land on the greenhouse next door.
You see the rocket rush up into the sky.
You see the rocket explode into many stars.

7e Bouncing light

13 Copy these sentences, choosing the correct word from each pair.

a A smooth surface is a **good/bad** reflector of light.
b A **stick/flame** is a source of light.
c Light is scattered by a **smooth/rough** surface.
d The Moon is a **source/reflector** of light.

14 Why might mirrors be useful for these people?

a a bus driver
b an actor
c a shopkeeper
d someone shipwrecked on a desert island

15 This drawing shows someone's idea about how a lamp helps her to see her work. She has got a wrong idea! Draw a correct diagram to show the right idea.

8a More about water

1 Helena was out camping on a snowy mountainside in midwinter as part of her survival course. She had brought all her cooking gear, but had forgotten to bring any water!

a Will she have to go thirsty?

b If not, why not?

2 John wanted the coffee in his flask to be really hot at lunchtime, so he let the kettle boil for 10 minutes before he filled the flask up.

a Did this make any difference?

b If not, why not?

3 On a cold night after a warm day, water often condenses out of the air as dew. What happens to this water if the temperature drops below 0 °C?

4 Paper starts to burn if it is heated to nearly 200 °C. Sherina says she once made a cup of coffee at a barbecue by boiling water in a paper cup over the fire. Could she be telling the truth? Explain your answer.

The flames may reach 1000 °C, but the paper cup cannot get hotter than the water it contains!

8b Changing state

Use this table to answer question **5** on the next page.

Metal	**Melting point (°C)**
mercury	–39
solder	183
tungsten	3400
iron	1540
copper	1080
Wood's metal	70

5 **a** A light bulb filament gets white-hot, over 2000 °C!

i Which metal would you use for a light bulb filament?

ii Why not use iron?

b **i** Which metal is used in a thermometer?

ii What is unusual about this metal?

c You can buy trick spoons made from Wood's metal. What happens if you stir your freshly poured coffee with one of them?

d A roaring Bunsen burner flame can reach just over 1100 °C.

i What happens to thin copper wire in this flame?

ii What happens to thin iron wire?

e A soldering iron can reach over 200 °C. Explain how solder can be used to join electrical wires together.

f Brass is made by mixing zinc into liquid copper. Zinc boils at 907 °C. What is the problem with making brass?

6 Plants are damaged and often killed if their sap freezes in winter. Pure water freezes at 0 °C. If chemicals such as salt or sugar are dissolved in the water, the melting point (freezing point) is lowered.

The sap in some soft-stemmed plants gets more and more sugary as the weather gets colder. How does this protect them in the winter?

8c Expansion and contraction

7 Railway lines suffer from expansion and contraction. Old tracks had to be laid with a gap between each section of rail, to leave room for expansion. As the train ran along the track it made a rhythmic 'clickety-click' sound, as the wheels jumped over the gaps.

Why is the 'clickety-click' of railway lines louder in winter than in summer?

8 **a** Central heating pipes often creak and groan as hot water starts to be pumped through them. What is happening to the pipes to cause this?

b Long metal pipes which carry steam in factories often have loops or bends in them, as shown. Why aren't they straight and just clamped tightly at each end?

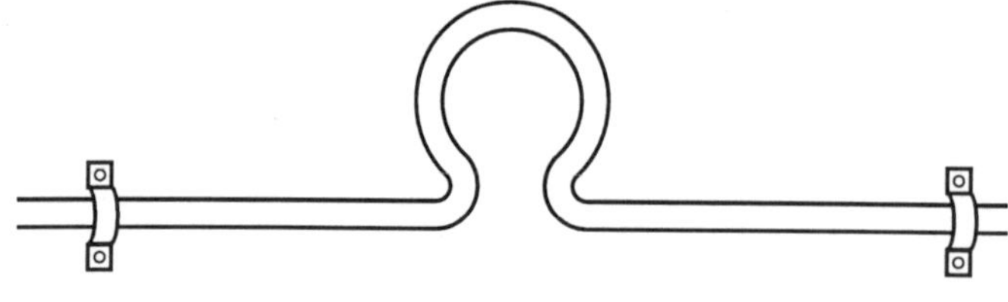

9 Bridges are often fixed firmly at one end, but supported on rollers at the other end. Why is this?

10 In light bulbs, the filament is supported by wires embedded in the glass.
Look at the table. Which metal would you use for these supporting wires? Explain your answer.

Material	Expansion number*
glass	9
invar	1
platinum alloy	9
iron	11
brass	19

*The bigger the number, the more it expands.

8d Physical or chemical?

11 There are three clues that tell you if a change is a physical or a chemical one. What are they?

12 Are the following changes physical or chemical?
Give reasons for your answers.

a cleaning a greasy mark from a plastic tablecloth with washing-up liquid
b getting rid of limescale from the toilet using Harpic (it fizzes)
c a jelly cooling and setting in a mould
d the white of an egg solidifying in a frying pan
e fat splashed from roasted food forming a layer of carbon on an oven wall
f the oily layer on meat stock setting to form dripping as it cools

13 Sodium hydroxide solution is a colourless liquid. Hydrochloric acid is a colourless liquid. When they are mixed, nothing seems to happen, and you are left with another colourless liquid. The mixture does get warm, however.

Has any kind of change occurred here?
Explain your answer.

14 Copper(II) oxide is a black powder. Sulphuric acid is a colourless liquid. If you warm the two together, the black powder dissolves and the liquid turns blue. If you let the water evaporate from the blue liquid, deep blue 'squashed diamond'-shaped crystals grow.

a What are these blue crystals? (You have seen them before.)

b What kind of change has occurred here? Explain your answer.

8e More chemical changes

15 **a** When you burn a fuel, you get new chemicals. But it is not the chemicals you are interested in. What else do you get?

b **i** What other chemical is needed if a fuel is to burn?

ii Where does this usually come from?

iii How much of this chemical is there in the air?

16 Hydrogen burns to give water, and carbon burns to give carbon dioxide. Oil contains both carbon and hydrogen. What two new chemicals are formed when oil burns in air?

17 **a** Jet engines burn a kind of light oil called kerosene. They carry this with them in their fuel tanks. Where do they get the oxygen from?

b Why can't jet engines work in space?

c The Space Shuttle carries oxygen for its crew to breathe. Why else does it need to carry oxygen?

18 Look at this word equation:

iron + water + oxygen ⟶ rust

Why doesn't iron rust:

a if it is covered in paint or grease

b in space

c in the Sahara Desert

d at the South Pole?

9a More of the same

1 **a** What are the male and female sex cells in a plant called?
b What are the male and female sex cells in an animal called?
c What happens to the male and female sex cells in sexual reproduction?

2 Here are some organisms that use sexual reproduction to make more of themselves. They use very different ways of making sure the sex cells meet.

deer **sea anemone** **stickleback** **grass** **peacock** **hazel**

a Choose two organisms where the male displays or fights to attract a female.
b Choose two organisms that rely on wind or water to carry their sex cells.

3 The number of female sex cells produced by living things varies a lot, depending on how the cells are going to meet. This table shows the numbers of eggs produced at a time by different animals.

Animal	Where the eggs are fertilised	Number of eggs produced at a time
plaice	in the sea	500 000
pike	in rivers	20 000
stickleback	in streams	500
bluetit	in the body	10
crow	in the body	4
elephant	in the body	1

a Think about where each animal lives. What are the dangers for the eggs after they are produced but before they are fertilised?
b Why do you think each animal produces the number of eggs it does?

9b Flowers

4 Why are some flowers big and brightly coloured, while others are small and green?

5 Design your own super-flower. Make it **either** insect pollinated **or** wind pollinated. Make sure it has all the things it will need to be good at its job. Label these things, explaining why they are there.

6 What is the difference between pollination and fertilisation?

9c Scattering the seeds

7 Why do plants produce fruits?

8 Do a survey of your home to find out in how many different places fruits or substances from fruits are used. Look in the fruit bowl, in the fridge, in tins, and also in the bathroom, on the dressing table and in the cupboard under the sink! Make a table to show the different types of fruit products that you find.

9 Find a fruit. First check with an adult, and then cut it in half.
Draw what you can see. Label the different parts of the fruit.

9d From boy to man

10 Write a paragraph to describe how you think it would feel to be one of the first boys to enter puberty in your year at school. Would you be pleased or not? What might worry you? What might make you feel good about it?

9e From girl to woman

11 Do you think it would be more difficult to be the first or the last girl in your year to start your periods?
Give reasons for your answer.

12 In the early years of secondary school, the changes of puberty and the time when they happen seem very important. By the time most people are 18, it no longer matters very much. Why not?

9g What's happening inside?

13 This circle shows the average number of days in a menstrual cycle. Explain what is happening at A, B and C.

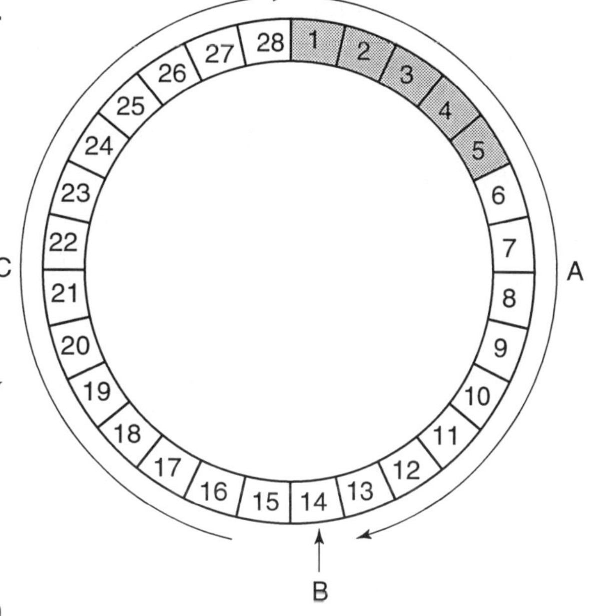

14 Nagesswar and Sangita are twins. Sangita was born 15 minutes before Nagesswar and she never lets him forget this!

Hannah and Charlotte are also twins. They look so similar that only their very best friends can tell them apart.

No one will believe James and Edward are twins. James is tall and fair haired, but Edward is short with dark curls.

a Explain why one twin is always older than the other.
b Which of these sets of twins are identical? How can you tell?
c Nagesswar and Sangita are not the same sex. James and Edward are, but they look very different. They are both the same type of twins. How do twins like these come about?

9h The end and the beginning

15 A human fetus grows and develops inside its mother.

a How does the fetus get food and oxygen?
b How does it get rid of waste?
c How is it protected from knocks and bumps?
d How is the fetus kept warm?

16 How does a baby get out of the uterus when the 40 weeks of pregnancy are over?

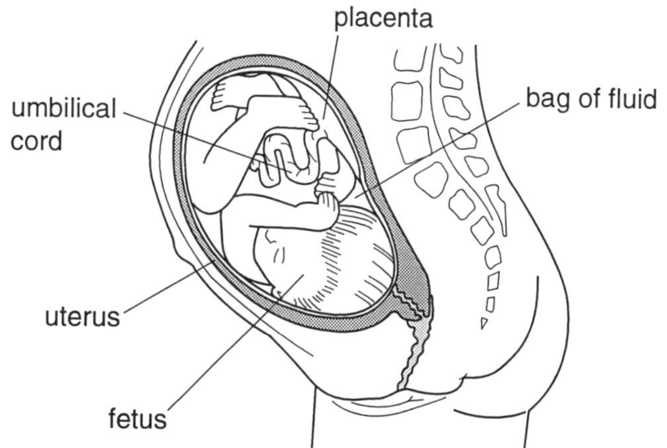

10a Switching on

1 Copy and complete the following table.
Use the words below to fill the gaps.

switch **circuit** **current** **electrical conductor**

	flows round a complete circuit
	close this to complete a circuit
	metal is an example
	must be complete for a current to flow

2 These pupils have set up a circuit to test whether certain things are electrical conductors or insulators. Can you predict their results for the following experiments? Give a reason for each answer.

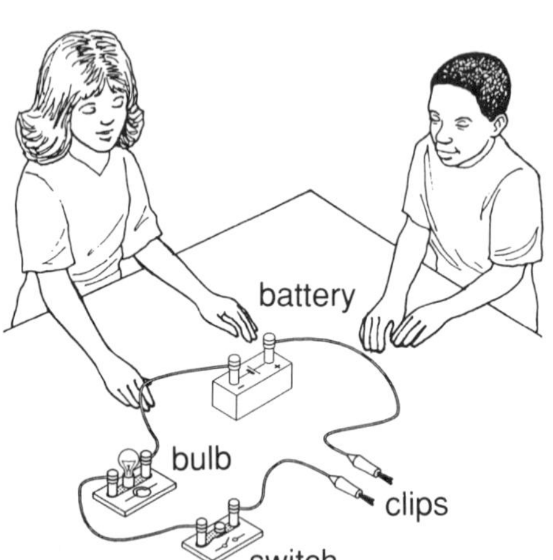

a The switch is open and a metal teaspoon is connected in the circuit. Will the bulb light up?

b The switch is closed and a plastic teaspoon is connected in the circuit. Will the bulb light up?

c The switch is closed and a piece of aluminium foil is connected in the circuit. Will the bulb light up?

3 **a** Draw a picture of an electrical circuit made up of a battery, a switch and a bulb.

b How many connecting wires are there in your circuit?

c Mark a plus (+) at one end of the battery.

d Draw arrows on your drawing to show how the electric current flows when the switch is closed.

10b Wiring up

4 **a** What will happen if you just close switch 1?

b What will happen if you just close switch 2?

c What will happen if you close both switches?

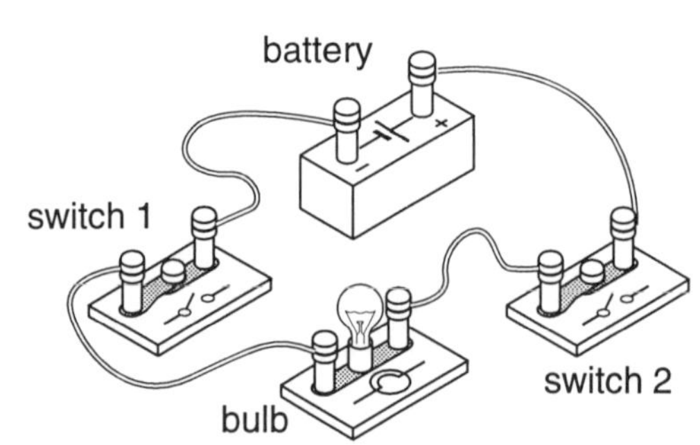

5 **a** This is a very untidy circuit diagram. Draw a neater version of it.
b Beside each component, write its name.
c Say what will happen when the switch is closed.

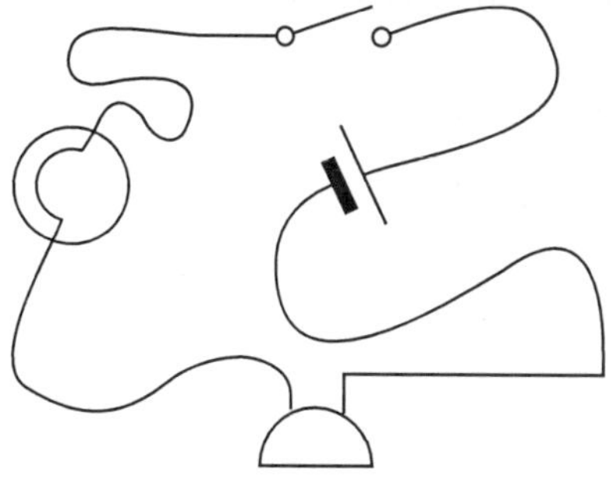

6 Electricity is very useful. The table shows some useful electrical appliances.
a Copy and complete the table. The first row has been done for you.

Appliance	Used for	Could this be done without electricity?	How?
electric kettle	boiling water	yes	use a gas cooker
electric fire			
radio			
torch			
calculator			

b Which things can you do more easily with electricity?

10c Seeing the light

7 Draw a circuit diagram to show three batteries connected end to end (in series), together with a motor.

8 **a** Which circuit, A or B, shows two bulbs connected in series?
b Which circuit shows two bulbs connected in parallel?
c In which circuit will the bulbs be brighter?

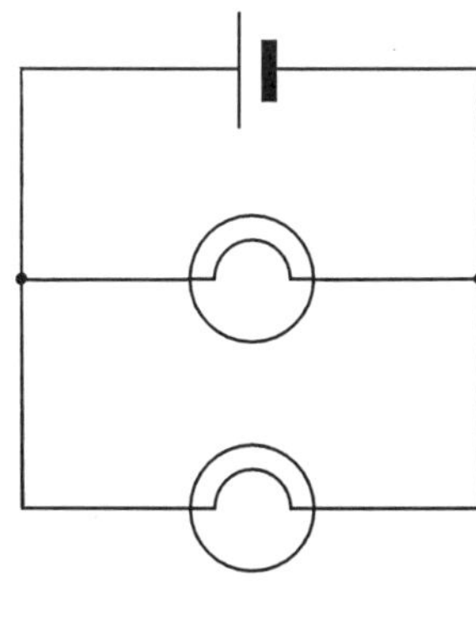

A

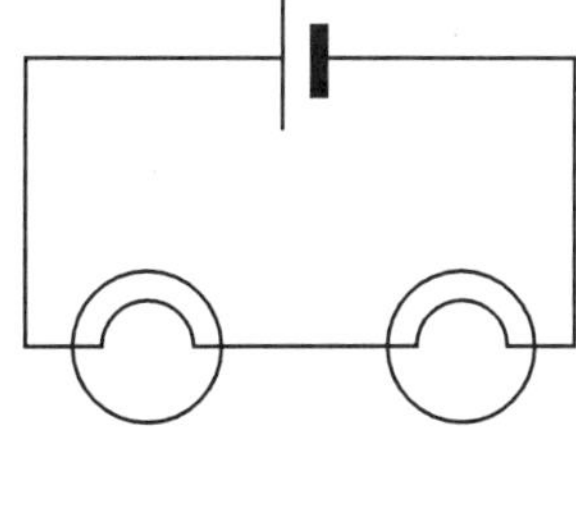

B

9 This girl is disappointed that the lights for her dolls' house are not very bright. Suggest **two** ways she could make them shine more brightly.

10d Resistance to change

10 Copy and complete the following sentences, explaining how an electric cooker works. Use the words below to fill the gaps.

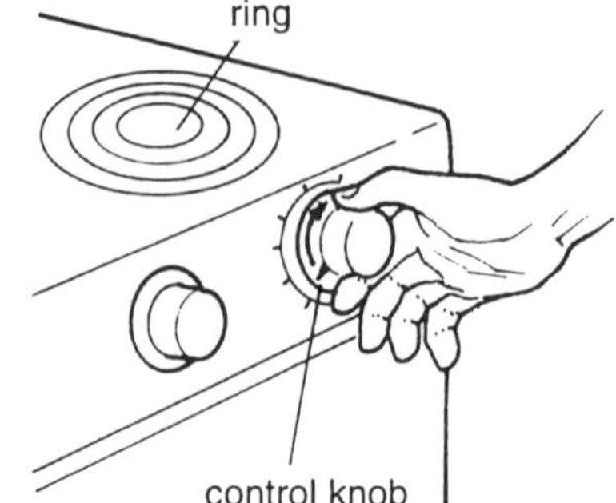

hotter	variable resistor	current	resistance

When you turn the knob, you are turning a __________ __________. There is less __________ in the circuit, so a bigger __________ flows, and the ring gets __________.

11 An electric cooker uses variable resistors to control the flow of electric current, so that the rings are at the right temperature.
Name two other places where variable resistors are used.
What are they used for?

12 It can help us to understand the flow of electric current if we think about other things that flow. Imagine blowing air through these different tubes.

a Which tube is easiest to blow through? Give a reason.
b Which tube is hardest to blow through? Give a reason.
c Which tube has most resistance to the flow of air?
d Draw two pieces of wire, A which has high resistance, and B which has low resistance. Explain why your wire A has higher resistance than your wire B.

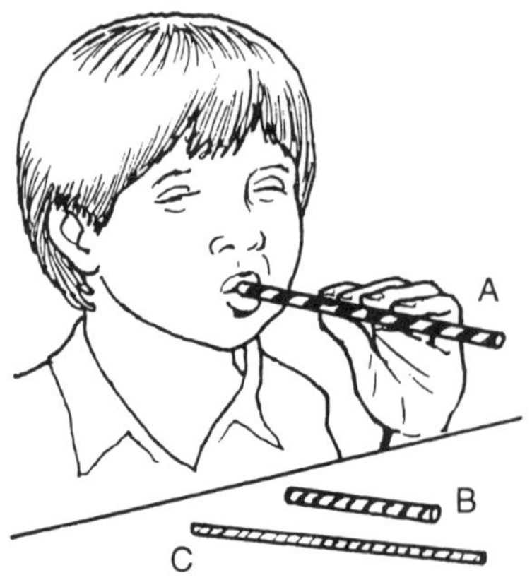

10e How much current?

13 These ammeters are measuring electric currents. How many amps is each one measuring?

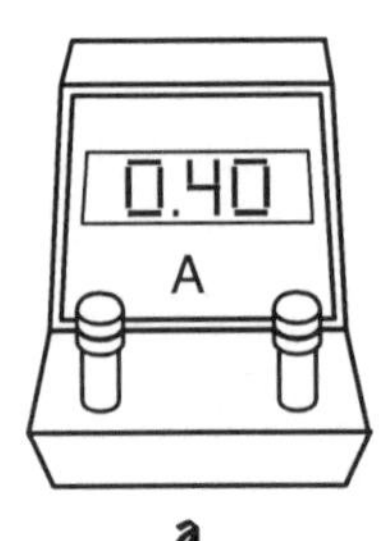

a

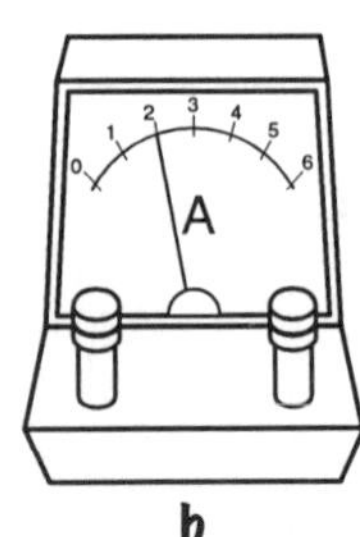

b

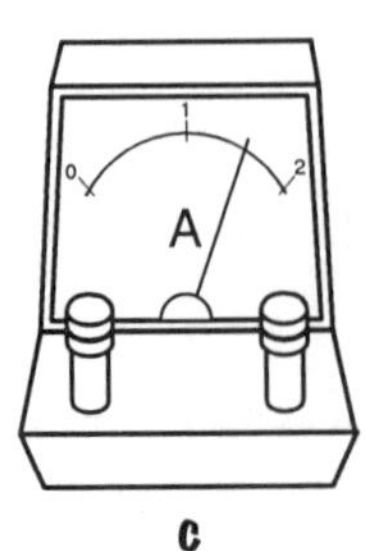

c

14 Electric current does not get used up as it flows around an electric circuit. What will ammeter 1 read in this circuit?

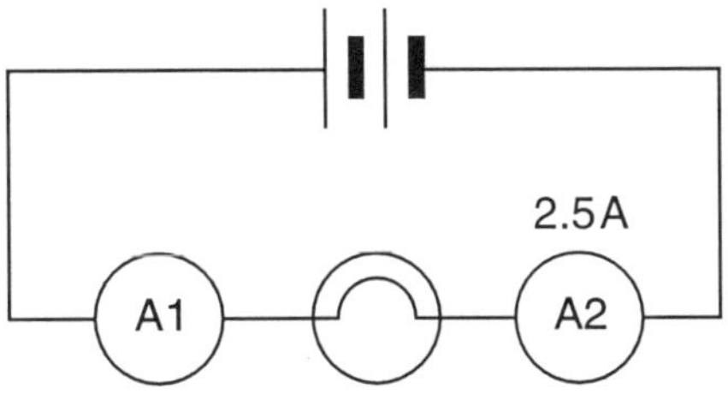

15 Copy and complete this table to show the symbols for different components in circuit diagrams.

Component	**Symbol**
ammeter	
motor	
	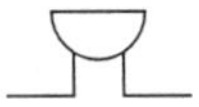
connecting wire	
	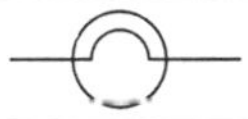
	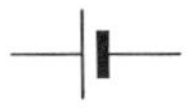

10f Energy changes

16 We use electrical energy in many different ways. A light bulb is an appliance that changes electrical energy into light energy.

Name an appliance that changes electrical energy into:

a sound energy
b heat energy
c movement energy.

17 Here are some sentences from a pupil's homework. They are all wrong! Change the words in bold to put them right.

a The electric current flowing out of a light bulb is **less than** the current flowing into it.
b A light bulb changes **electric current** into light energy.
c You can measure the electrical energy carried by an electric current using **an ammeter**.

18 In this circuit, a battery is making a motor spin round. Copy the diagram, and add:

a labelled arrows to show the flow of electric current
b labelled arrows to show where electrical energy is being carried by the current
c energy transfer diagrams to show the energy changes in the battery and in the motor.

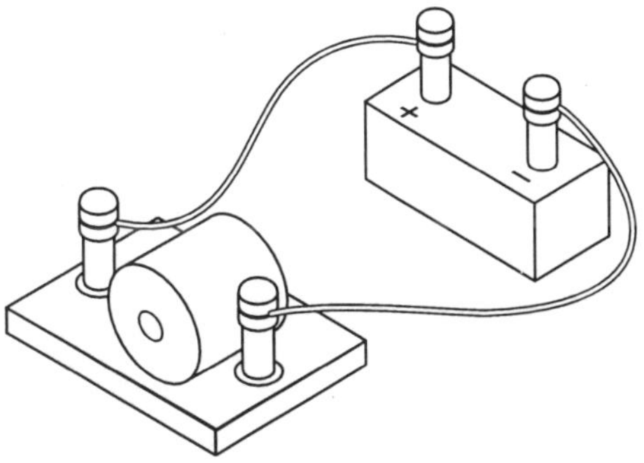

11a In the beginning . . .

1 **a** Copy this picture of a slice through a volcano.

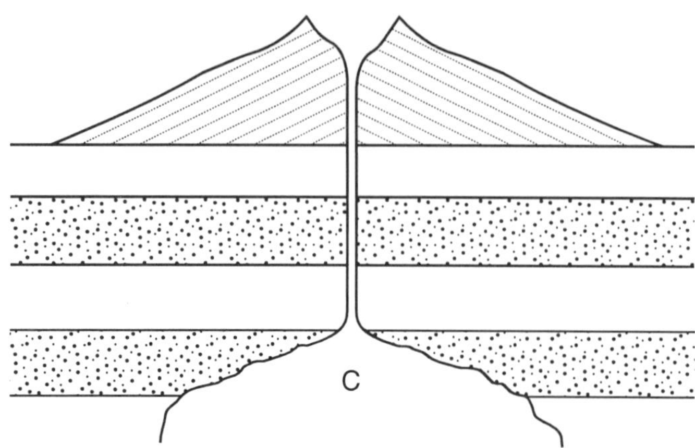

b Label the volcano and crater.
c What is the chamber C full of when the volcano is active?
d What is the most likely temperature in chamber C when the volcano is active:
15 °C, 100 °C, 500 °C, 1000 °C or 10 000 °C?
e What is the most likely temperature at the surface between eruptions:
15° C, 100 °C, 500 °C, 1000 °C or 10 000 °C?
f When the volcano erupts, material from C pours out onto the surface.
  i What is this material called?
  ii What happens to it?
g If the volcano becomes extinct, the liquid in C will harden to form rock.
  i How will this rock differ from the rock that forms on the surface?
  ii Why is it so different?
h What are rocks called that form from liquid rock like this?

2 Edinburgh is built on an outcrop of dark rock. Under the microscope, you can see that this rock is full of tiny crystals. What must have happened there a long time ago to make this rock?

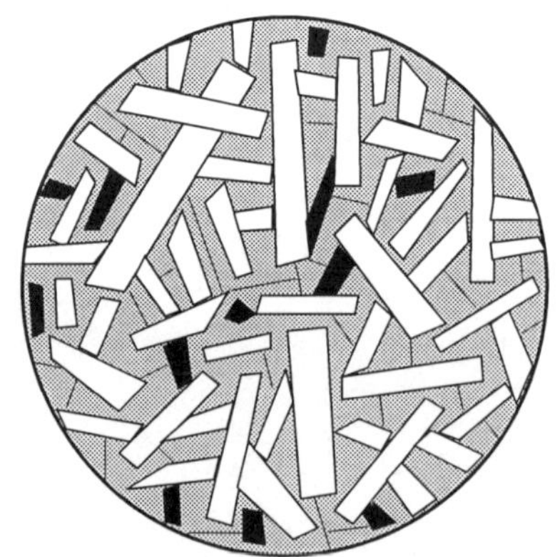

3 Dartmoor is dotted with outcrops of granite. This rock shows large crystals.
a How did this rock form?
b What do you think has happened to expose this rock on the surface today?

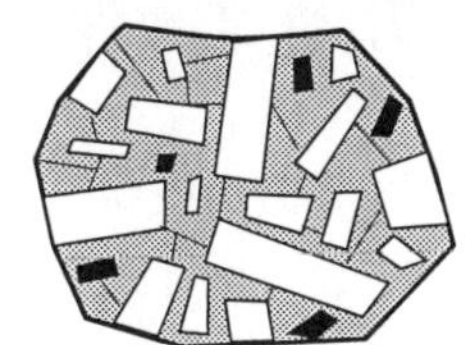

11b Breaking the rocks

4 Many people keep plants growing in pots in their gardens. These pots sometimes break if they are left out over the winter. This often happens if there is a hard frost after heavy rain. Why do the pots break?

5 When rocks break, the pieces are large and have sharp edges. If they get rolled along by a river, the sharp edges get knocked off and so the pieces get smaller and rounder. Finally, on a beach, waves wear them down more, leaving small, round pebbles.

Look at these pictures of rock pieces. Which was probably found:
a at the foot of a cliff in the mountains
b half-way along a river
c on a beach?

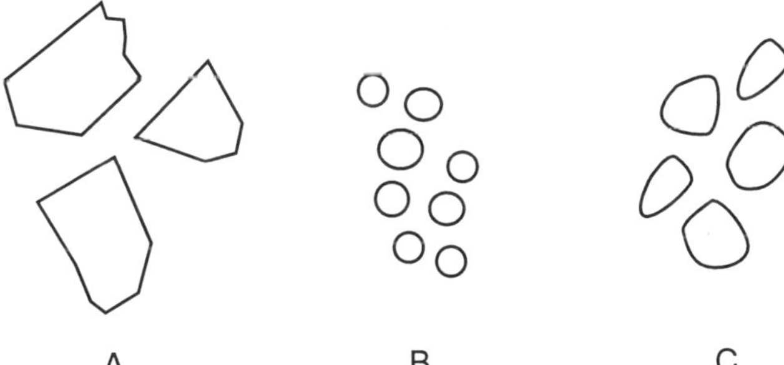

6 It takes rocks hundreds of years to weather down to soil. Around some mines and quarries, you may find heaps of rock waste. You can often tell how old they are by how many plants are growing on them. Fresh tips are bare of plants, while ancient tips are covered in them.

Why is there this difference in plant cover?

7 Old quarries and mines can spoil the environment with their barren waste tips. Today new quarries have to strip off the soil and save it. When they have finished digging, the soil is pushed back over the waste rock.

Why do you think the quarries have to do this?

11c Move it!

8 The beds of mountain streams are often covered with pebbles that are not moving. How did they get there?

9 Sometimes houses are flooded if a river breaks its banks after heavy rain.
The water will eventually go down, but a thick layer of mud and fine sand is often left in the houses.
a Where does all this sediment come from?
b Why is it left behind?

10 A river flood plain is the flat area around a river that gets covered in water in times of flood. In the past, flood plains were always very fertile and crops grew well there. Today, raised banks have been built around many rivers to stop them flooding, but the crops now need extra fertiliser to make them grow. Why is this?

11 At the seaside, you often see pebbles and sand on the beach where the waves crash in, but the sea bed in deeper water is covered with clay. Why is this?

11d New rocks from old

12 Limestone forms from the remains of shells.
a What are these shells found in ancient rocks called?
b The limestone found high up in the Pennines contains shells of ancient sea creatures.
i Where must this limestone have formed?
ii How do you think it got up into the mountains?

13 In general, the older a rock is, the more cemented it is. Sandstone is found all over Britain, but the sandstone in the south-east is much younger than that further north and west. Why is sandstone not usually used for building in the south-east?

14 If you dig a deep hole down into the rocks, it will probably fill up with water. (You have dug a well!) This water comes out of the rocks.
a How can a solid rock such as sandstone contain water? Draw a close-up diagram to explain your answer.
b Soft sandstones can often carry more water in them than hard sandstones. Why is this?

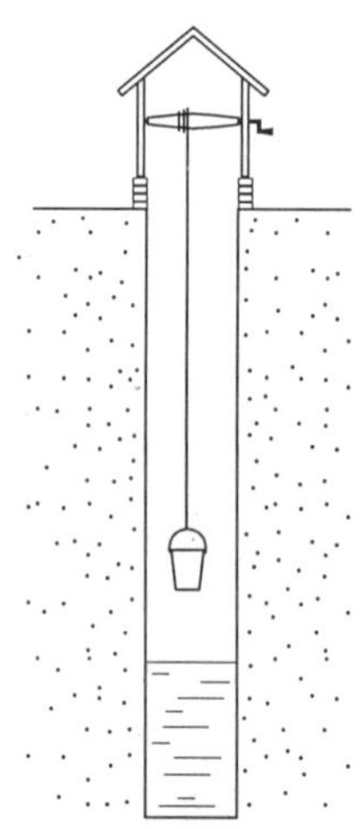

15 **a** Fossil ammonites found in shale are often squashed flat. Why is this?
b Sometimes patches of the shale become cemented into hard lumps called nodules. The ammonites in these nodules are not squashed.
Did the nodules form before or after the mud turned to shale? Explain your answer.

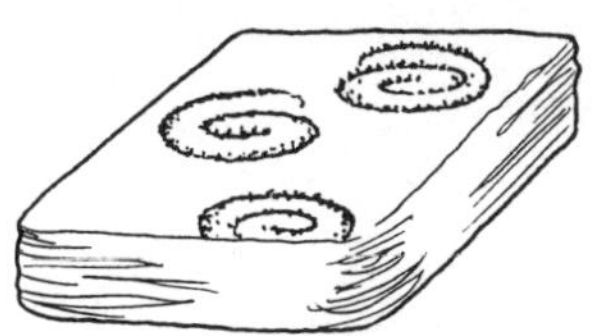
ammonites in shale

ammonite in a nodule

11e All change

16 Shale is a soft sedimentary rock.
a If shale is covered by molten lava, it becomes very hard. What has caused this change?
b What kind of rock has it turned into, sedimentary, igneous or metamorphic?
c If shale is squeezed and heated deep in the Earth, it can turn to schist, which has wavy layers.
Why does the rock formed in **a** not show these layers?
d What do you think happens to any fossils in shale that turns to schist?
e If the rock was heated so much that it melted, and then cooled and set solid again, would it become sedimentary, igneous or metamorphic?

17 **a** Draw a diagram to show the rock cycle.
b You can use pottery to model the rock cycle.

- You start with clay.
- This is fired in a kiln to make it hard.
- It is then covered with glaze powder and fired again. The powder melts and then sets again as it cools.
- Old pottery left lying around for a long time starts to crumble away.

Match each of these stages to one of the stages in your rock cycle.

18 Most metamorphic rocks are formed deep in the Earth's crust. Metamorphic rocks are found at the surface in the Highlands of Scotland.
a What has happened in Scotland to expose these rocks?
b Scientists think that some of the Scottish rocks formed 40 km down in the Earth's crust, about 400 million years ago. On average, how much of Scotland must have been worn away every year since then?

Getting started

Book coverage

Key ideas	PoS
You need an unbalanced force (a push or a pull) to start something moving. Forces are measured in newtons (N), using a newtonmeter.	AT4.2c

Notes: This spread concentrates on horizontal forces; weight is dealt with in the next spread. Measuring weight is also reserved for the next spread.

Answers

Activities in text

a There are other examples included in the spread. Obvious day-to-day examples include pulling open doors and drawers, pushing prams, etc. It is best to focus on examples of humans pushing and pulling at first.

b

c1 D, B, A, C.
c2 A the lorry, B the door, C the lorry, D the drawing pin.
c3 A pulling along the road, B pulling towards the woman, C pushing along the road, D pushing into the board.
d1 Spring-balance type newtonmeter (A).
d2 Bathroom-scales type newtonmeter (C). (A dynamometer can measure either.)

What do you know?

1

two types of force	pull, push
units forces are measured in	newtons (N)
used to measure forces	newtonmeter, dynamometer

2 a

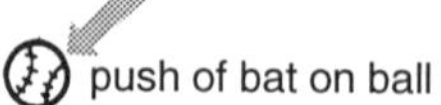

b

3 a

b Your pull is bigger.

c The forces are unbalanced.

Extras: The trolley Derby

This requires pupils to use the ideas of mass and force together. The question avoids the distinction between speed and acceleration by imagining the trolleys racing over a fixed length of course.

1 Pupils should grasp intuitively that, for two trolleys with the same mass, the one with the greater force will win the race; for two equal forces, the smaller mass will win the race.

2 In the final, pupils need to juggle both of these factors. The greatest value of force/mass wins.

Assignments

1 a

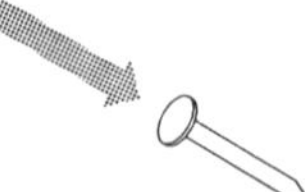

b

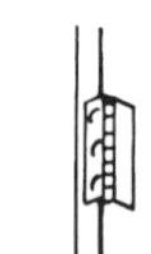

2 a

b The forces are unbalanced.

c The scrum will start to move in the direction of the Stripes' push.

3 The rubber band is attached to the top of the ruler. (It may be best to cut the rubber band.) There is a hook at the bottom (paper clip). You could ask pupils to make a newtonmeter at home and hang different objects on it, marking the end of the rubber band each time. Then discuss and carry out the calibration in the lesson.

Measuring forces

Running the activity

Show the pupils different types of newtonmeter, how to use them and how to read them. Discuss whether each newtonmeter measures pushes or pulls, and their different ranges. If a dynamometer is available, pupils could find out which is greater, their biggest push or their biggest pull. Avoid embarrassing the 'weaklings' by public exhibitions of strength. Avoid competition, in case some pupils are tempted to over-exert themselves.

Differentiation

Sheets available are:

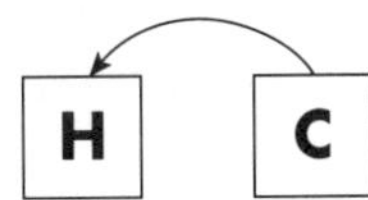

AT1 programme of study

Core: Invites pupils to predict greatest and smallest forces (**AT1.1c**). Pupils design their own table to record results (**AT1.2e**). They check their prediction (**AT1.3f**). This sheet also introduces the idea of meters with different ranges (**AT1.1h**).

Help: For pupils who need help to read newtonmeter scales, some practice examples are given. A table for results is also provided (**AT1.2e**).

Risk assessment

Ensure that benches used for the pushing a friend activity are clear and dry. If pupils mistreat newtonmeters, there is a danger of flying hooks. If you consider that this is a potential problem, pupils should be instructed to wear eye protection.

Equipment needed

For the class

- newtonmeters: a number, having different ranges (0–10 N, 0–20 N, 0–50 N), and bathroom scales type
- dynamometer (optional)
- loops of string for attaching to objects being pulled

The pull of gravity

Book coverage

Key ideas	PoS
The weight of something is the pull of the Earth's gravity on it. The weight of something is a force measured in newtons. The mass of something tells us how much matter is it made of. The mass of something is measured in kilograms (kg).	The idea of weight as a force arising from gravity is implicit in AT4.

Notes: The distinction between mass and weight is not easy to get hold of. The spread relates mass to the amount of matter ('stuff') of which something is made. Later, when a particulate model of matter is established, it will be more obvious that mass is constant because the number of particles of which a body is made remains constant when it is taken to (for example) the Moon.

Answers

Activities in text

a1 **a2**

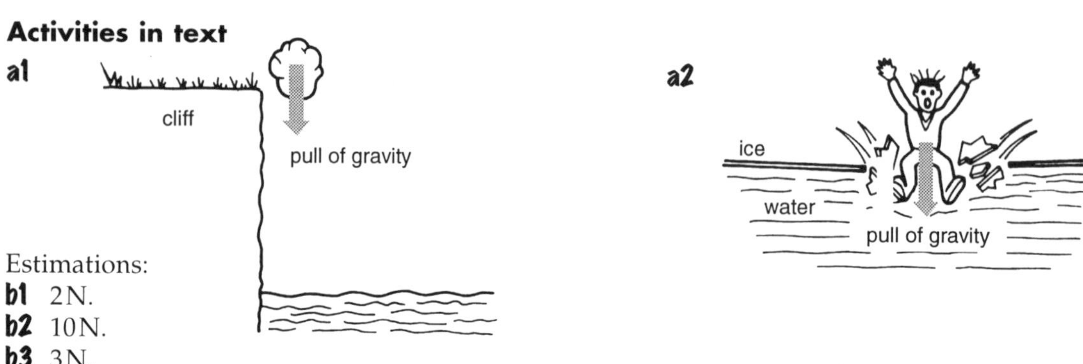

Estimations:

b1 2N.
b2 10N.
b3 3N.
b4 About 500N.
c The trolley with the smaller mass wins the race. The same stretch gives the same force to make it a fair race.

What do you know?

1 The **weight** of an object is a **force**, caused by the **pull** of the Earth's **gravity** on it.

2

measured in kilograms	**mass**
a force	**weight**
less on the Moon	**weight**
measured in newtons	**weight**
stays the same in empty space	**mass**

3 **a** 20 N. **b** 500 N.

Extras: On the Moon

1

Place	Explorer's mass	Explorer's weight
Earth	60kg	600N
Moon	60kg	96N
Mars	60kg	198N

Assignments

4 Person B: 100 N pushing 20 kg.
5 "Hi! I'm Pat the space hero. My **mass** is 70 kilograms. That means that the **pull** of the Earth's gravity on me is 700 **newtons**. When I went to the Moon, my **weight** was much less than on the Earth."
6 **a** N for newton, kg for kilogram. **b** Its weight. **c** 200 N.

Light and heavy

Running the activity

Weight is a force, and we measure it using a newtonmeter. However, we can judge weight by feeling the weight of something by hand; this is useful for making comparisons. (Our sensitivity allows us to distinguish weights which are within about 10% of one another.) It is better to make accurate measurements. You may need to explain the scoring system: 1 = lightest, 2 = next lightest, and so on.

Differentiation

Sheets available are: 

Pupils are given approximately five objects to look at, then compare by hand, then weigh. The objects should be chosen to include two which look very similar but which have different weights, and two which look different but have similar weights that can only be satisfactorily distinguished by weighing.

There are several ways to make this more or less demanding:
- more or fewer objects
- more similar or more different weights: introduce the idea of an improved (more sensitive) newtonmeter
- the need to choose appropriate newtonmeters.

AT1 programme of study

This activity can lead to a discussion of the need for measuring instruments, and for making measurements to an appropriate degree of precision (**AT1.2b**).

Equipment needed

For each group
- set of five objects (see below) labelled A–E, on a tray
- newtonmeter(s) for weighing the objects

Note
The objects might be: a wooden block; a stone; a lump of Plasticine; a book; a metal rod. The Plasticine might conceal a heavy metal object, or be hollow. All objects need string or hooks to allow weighing using a newtonmeter.

Moving along

Book coverage

Key ideas	PoS
Friction is a force which happens when one surface rubs against another. Friction can make it difficult to start something moving. Friction can make it difficult for something to move fast. Friction can also be useful, because it gives you grip.	AT4.2e

Notes: Friction between solid surfaces is treated here as a phenomenon. There is a hint of an explanation in the illustration of smooth and rough surfaces.

Friction can be a problem, but it can also be useful.

Answers

Activities in text

a This is an open-ended question at this point. Could pull it downhill, or use a machine, or reduce friction.

b Ball bearings have a smooth surface for less friction.
Oil is a lubricant – it gives a smoother surface for less friction.
Air is also a lubricant of sorts – it reduces contact between rough surfaces.

What do you know?

1 Friction is a **force**. It is found when one **surface** rubs against another.
Friction can make it difficult to **start** something moving. Friction is **less** when surfaces are smooth.

2 Friction as a problem – pulling a heavy object.
Friction being useful – giving grip.

3 **a** Water acts as lubricant, reducing friction.
b Grit increases friction to give grip.
c Wet roads have low friction, giving poor grip, so drivers should drive slowly to avoid sliding.

Extras: Braking

1 a

c

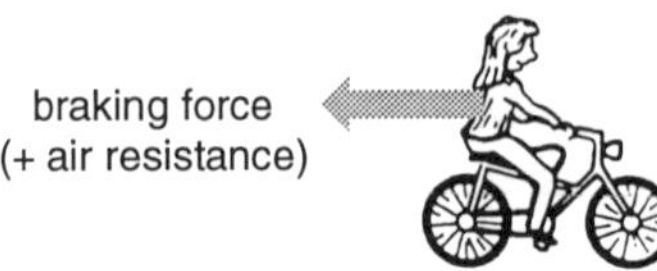

b

d

Assignments

7 **a** Using the brakes on a bicycle.
b Putting oil into a car engine.
c Hammering a nail.
d A skater sliding easily over the ice.
(Pupils could be asked to think of another example of each.)

8 **a** There is low friction on a smooth polished floor.
b Carpet is rough, so high friction slows the car down.
c A tennis court has a rough surface with high friction for good grip; if you fall, friction causes a graze.

9 **a** The forces are unbalanced, because the push is greater than the friction.
b She will succeed.
c She could reduce friction by oiling the floor, using rollers or wheels, etc.

Easy going

Running the activity

Since friction is a force, we can measure it. This activity explores the effect of changing the nature of two sliding surfaces. Note that only horizontal forces are involved; experiments with sloping planes are an added complication at this stage.

Differentiation

Sheets available are:

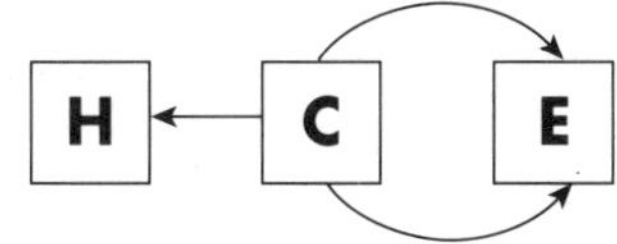

AT1 programme of study

Core: Pupils investigate the effect of changing surfaces on the force of friction. They may need assistance in thinking about the factor that they are changing. The experiment is an automatic fair test, in that they are unlikely to change the area of the block, or its mass. This should be discussed (**AT1.1e**).

Help: This provides support for pupils who need help with systematic organisation of their measurements, and with recording their results and drawing conclusions (**AT1.2e, 1.3e**).

Extension: The nature of the surface is a qualitative factor. This sheet encourages pupils to think about the other variable factors involved in friction, namely the area of contact and the mass of the block (**AT1.1e**). It is surprising that the area has little effect on the force.

Risk assessment

If polystyrene beads get on the floor, they can make it dangerously slippery. The beads must be confined to a tray of some sort and their use carefully supervised. Whilst oil is not so dangerous, it can still make the floor slippery, so care is needed.

Equipment needed

For each group

- newtonmeters
- wooden blocks (approximately 5 × 8 × 10 cm), each with a hook fitted
- wooden tray
- polythene sheet and sandpaper, with some means of fixing to block
- access to oil, water, small polystyrene beads
- 1 kg, 2 kg, 5 kg masses

Notes

1. Alternative lubricants include washing-up liquid and baby powder (talc). To limit the mess, it is possible to set up plastic trays with and without lubricants.
2. To limit the demands on materials, it may be desirable to set up one set of equipment for each test around the room, to allow pupils to circulate. However, this limits pupils' opportunities for making their own decisions.
3. Those following the Extension sheet will need their own block, board, masses and newtonmeter.

Moving in water and air

Book coverage

Key ideas	PoS
There is friction when something moves through air or water. Friction can be reduced by using a streamlined shape.	AT4.2e

Notes: Frictional resistance to movement is more obvious in water than in air; pupils will have experienced this in trying to walk through water in the sea or at the swimming pool.

It is simple to demonstrate the effect of changing shape on air resistance, using a flat sheet of paper and then crumpling it into a ball.

Answers

Activities in text

a Running into the sea – as the water gets deeper, the resistance gets greater, so it becomes harder to walk. In the swimming pool – the resistance of water makes walking difficult.
b Otter, dolphin, sea lion, etc.
c Concorde is shaped for minimum air resistance. (Concorde flies higher than other aircraft, where the atmosphere is very thin, to reduce the effects of friction.)
d The sugar glider uses air resistance – flaps between its limbs act like a parachute to keep it airborne as it glides between trees. It cannot glide upwards.

What do you know?

1 a Air resistance is the force of **friction** when something moves through air.
b There is **less** friction in air than in water.
c Friction makes it **harder** for cars and planes to travel fast.
2 a A is designed to go faster.
b The car body is streamlined – it has no bumps, is low and sleek-looking.
3

Extras: Speedy cyclists

1 These points give pupils an opportunity to show their knowledge of the effects of friction in air and between solids. They should make it clear whether friction is desirable or undesirable in each situation, and use correctly the terms **grip**, **air resistance** and **lubrication**.

Assignments

10 a It is streamlined, so there is little fiction and it can move fast.
b There is less friction in air so they can move faster.
c The friction in water is bigger.
d There is no air in space, so no friction (air resistance).
11 a The gannet's shape is streamlined to give low friction, so that it is moving fast when it hits the water. This means it can travel fast and deep in water.
b The puffin's wings are outstretched to increase air resistance. This provides a braking force so that it stops at its nest.
12 a Air resistance means that they fall slowly.
b Thistledown will travel furthest. It is the fluffiest and so has most air resistance, staying in the air longest.

Down the tube

Running the activity

To appreciate friction in water, it helps to think about how changing shape can reduce or increase this force. Pupils might be encouraged to think about their own experience in water. Divers make themselves into a streamlined shape to reduce the resistance of the water. A belly-flop is very painful! Creatures that swim fast through water are streamlined; creatures that live on the sea bed are not.

Differentiation

Sheets available are:

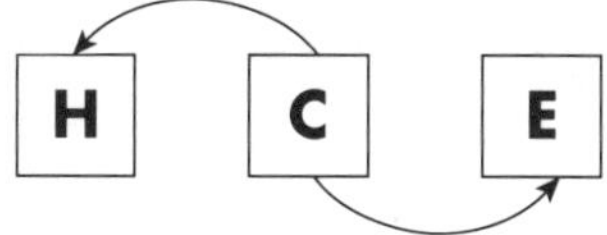

Core: Pupils are given approximately five equal sized pieces of Plasticine to shape. They time these as they fall through the water. It is best to mark a starting line 10 cm from the top of the tube, because of the difficulty of starting the stopclock at the moment of release. A finishing line 10 cm from the bottom also helps.

The idea of speed is implicit in this activity, but it is not made explicit on the Core sheet.

Once pupils have found the fastest shape, they are asked which is the slowest, and why. They should be encouraged to use the word 'friction' in their account.

Help: To avoid difficulties in timing, reading stopclocks, and recording results, pupils may stage a race between two shapes of Plasticine. The sheet also suggests some possible shapes to test.

Extension: This sheet provides two different extensions. One involves calculating speed (which is dealt with later in the Key Stage 3 course). This is no longer in the Programme of Study for Science at Key Stage 2, but pupils may have come across it in Maths.

The second extension involves comparing spheres of Plasticine of different sizes. It may be intuitively obvious that the smaller spheres will fall slowest, but why? The explanation involves considering both weight and friction; friction is more significant for an object of smaller weight. (This is because friction depends on area, whereas weight depends on volume. The ratio friction:weight is greater for small weights.)

AT1 programme of study

Although this experiment is rather messy, it is a suitable activity in which to emphasise the desirability of repeat observations and measurements (**AT1.2d**). Releasing the Plasticine shapes in a controlled way requires care, if repeatable results are to be obtained.

Pupils could also be asked to consider how the experiment could be improved, for example, by using taller measuring cylinders (**AT1.4c**).

In the Extension sheet, a table is suggested as a neat way of performing repeated calculations (**AT1.3a**).

In the second Extension activity, a Year 7 pupil might be expected to say: "For a small object, there is more friction for its weight than for a larger object." The idea of a ratio is implicit in this statement (**AT1.3g**).

Alternative approach

Another way to tackle the idea of streamlining is to start with a non-streamlined object, such as a large metal nut. How can Plasticine be used to change its shape and make it fall faster or slower?

Equipment needed

For each group

- 5 identical pieces of Plasticine, approximately 2g
- tall measuring cylinder or similar
- tray
- digital stopclock or watch
- beaker for emptying the measuring cylinder to recover the Plasticine
- access to water, marker pen, metre rule

Note

In order to increase the time taken for the objects to fall through the water, you may wish to make up a dilute solution of wallpaper paste, the viscosity of which will slow them down significantly.

With a little help

Book coverage

Key ideas	PoS
A lever is useful for lifting heavy objects. The further a force acts from the pivot of the lever, the more it can lift.	AT4.2f

Notes: The emphasis of this spread is on the idea of lifting, and the use of levers as an aid to lifting. The principle of moments is dealt with in Book 3.

Answers

Activities in text

a Various ways might be suggested: levers (obviously), ropes and pulleys, car jacks, etc.
b Anywhere at equal distances from the pivot.
c Downwards on the man's side.
d Towards the pivot.

e1, 2

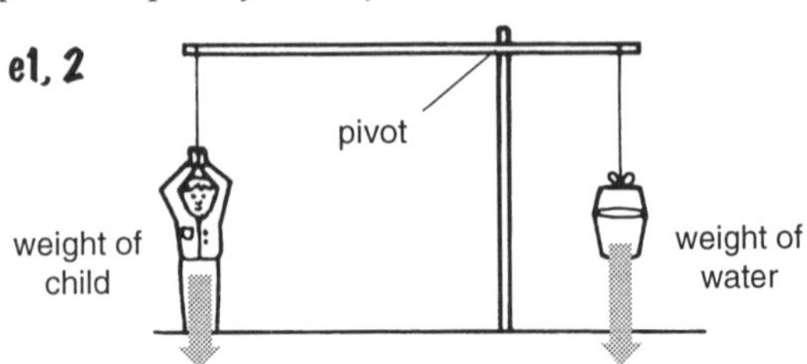

e3 Extend the pole at the child's end. Perhaps add a counterweight at the child's end.

What do you know?

1 A lever can be useful if you are not **strong** enough to lift a **heavy** load.
You can lift more if you push down near the **end** of the lever.

2 **a** 300N.
b The biggest lifting force that she can manage is 300N.
c

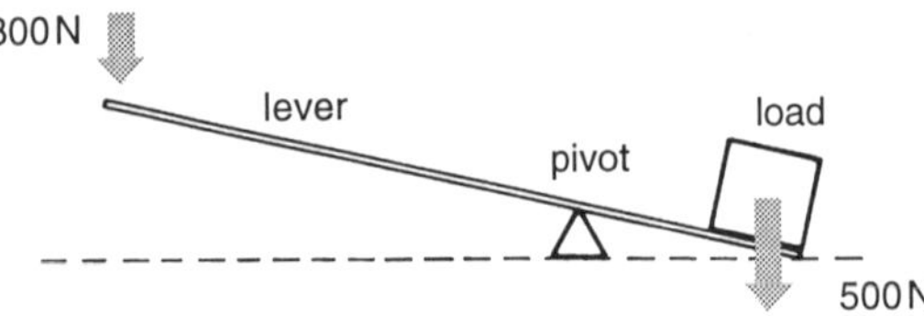

3 **a** Weigh one against another at the ends of the balance. If one goes down, it is the fake. If they balance, the third is the fake.
b Divide the coins into two threes plus one. Weigh three against three, and proceed as in **a**.

Assignments

13

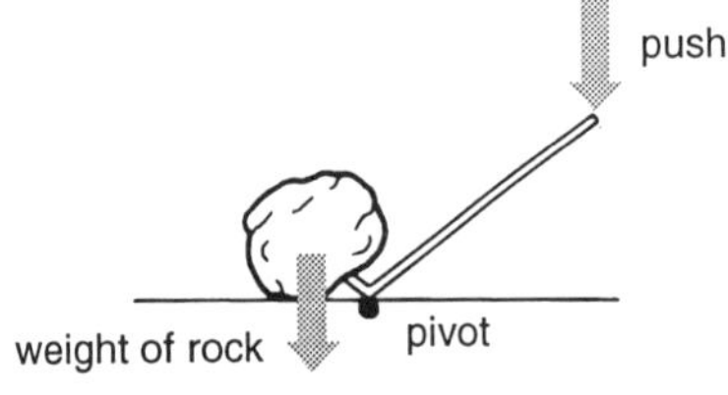

14 The see-saw is not balanced. The adult's end will go down.
15 The soil or sand being lifted is the load. The pivot is the point where the blade of the spade rests on the edge of the soil or sand. Your effort is pressing on the end of the handle.

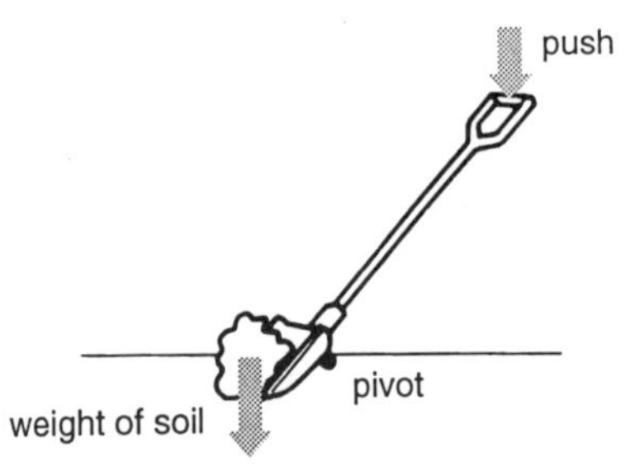

Teeter totter

Running the activity

This activity allows pupils to explore the forces needed to balance a see-saw. They will already have an intuitive grasp of how a see-saw balances; in this activity, they will deal with forces quantitatively.

Differentiation

Sheets available are:

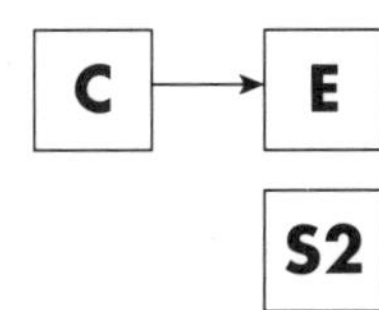

AT1 programme of study

This activity is useful for discussing the idea of changing a variable systematically, in even steps. The distance from the load to the pivot should be decreased in steps of, say, 10 cm, from 50 cm to 10 cm, (**AT1.1f**). The gradual decrease in balancing force as the distance from the pivot is decreased should then be obvious.

Pupils might be left to make their own measurements before this point is discussed.

Core: This provides basic instructions for setting up the model system, together with broad questions about the factors involved.

Extension: For pupils who are already capable of making measurements in a consistent way, this sheet encourages them to look for the pattern in their results (**AT1.3a–e**). It could be given out to all pupils after a discussion of how to be systematic in making measurements.

Equipment needed

For each group
- metre rule
- prism-shaped block of wood for use as pivot
- weight, e.g. 1 kg, to use as load, labelled with weight in newtons
- 1 N weights

Note
Pupils may find a long ruler difficult to manage; a half-metre ruler may be preferred.

Alive or . . . ?

Book coverage

Key ideas	PoS
Living organisms all carry out the common features of life. Non-living material may have some of the features of life, but not all.	AT2.1a

Answers

Activities in text

a This is a stimulus for thought – the answer is given in the text that follows.

b

	Roundabout horse	Real horse
Does it move some or all of itself?	**yes**	**yes**
Does it need food for energy (fuel)?	**yes/no***	**yes**
Does it need oxygen?	**yes/no***	**yes**
Does it produce and get rid of waste?	**yes/no***	**yes**
Can it grow?	**no**	**yes**
Can it have offspring (babies)?	**no**	**yes**
Does it feel things and respond to them?	**no**	**yes**

c

Apple tree	Tiger	Toy robot	Fossil
yes	**yes**	**yes**	**no**
yes	**yes**	**yes/no**	**no**
yes	**yes**	**no**	**no**
yes	**yes**	**no**	**no**
yes	**yes**	**no**	**no**
yes	**yes**	**no**	**no**
yes	**yes**	**no**	**no**

* These answers depend on whether pupils consider the roundabout as a whole or the horse alone. Both are quite acceptable.

What do you know?

1 A waxwork is not a living thing because it does not need to **breathe** air, eat **food** or **excrete** waste. If you stick a pin in it, it does not **respond**, and it cannot **move** even its eyes. It does not **grow/reproduce** and it cannot **reproduce/grow**.
2 A car needs fuel (food), needs oxygen, produces and gets rid of waste and moves. It is not alive because it doesn't grow or reproduce and isn't sensitive.
3 Suitable examples chosen – there may be more than one correct choice.

Extras: Living, non-living or dead?

1 Non-living: metal bicycle frame, plastic bowl (it could be argued that this is made from oil-based chemicals and was therefore once alive – if this is raised, discuss!), waterfall. Part of organisms: wooden spoon, woolly jumper, rubber wellingtons, coal fire.

2

Material	Came from
wood	trees
wool	**sheep**
rubber	**sap of rubber tree**
coal	**prehistoric plants**

3 Suitable examples.

Assignments

1 Living things: holly tree, hamster, worm, oyster, potato, snake.
Non-living things: watch, bicycle, computer, popcorn, aeroplane.
2 **Movement**, Respiration, Sensitivity, **Growth**, **Reproduction**, Excretion, **Nutrition**.
3 The living fish shows all the features you would expect from a living thing, the non-living fish has never carried out all the features of a living thing, the dead fish has been living but has stopped respiring, moving, etc.

Is it living? Does it respire?

Running the activity

Core: Most of the characteristics of living things are fairly obvious to pupils, but respiration, as opposed to breathing, can be more difficult for them to grasp. This experiment helps them get to grips with the idea that respiration occurs in all living things, even those that do not 'breathe'. The amount of carbon dioxide in normal air is insufficient to cause limewater to turn cloudy, but the levels of carbon dioxide produced by the respiration of living organisms will cause a change. Thus limewater turning cloudy can be used to indicate that respiration has taken place.

Extension: As pupils complete the Core practical, they can begin planning this experiment. Draw from pupils their ideas for using limewater to demonstrate that maggots or woodlice respire. Using the best ideas, set up the experiment with a control (the same set-up without animals). Also introduce a similar experiment using soaked seeds (peas or cress) with the idea "What about plants? Do they respire too?". Depending on preference and timing, this experiment could be done in advance by the technician, by the teacher as a demonstration, or the class could be divided into four groups to set up the four flasks.

Differentiation

Sheets available are:

AT1 programme of study

Core: Careful observation can be developed here (**AT1.2b, c**) and familiarisation with laboratory equipment as the basis for future choice.

Extension: This activity provides an opportunity to develop planning of experiments and how that relates to the observations made (**AT1.1e, f, 1.2c, d**).

Risk assessment

In the Core activity, pupils should blow gently to prevent wet benches, books, etc. and take care not to suck on the straws. Dispose of straws immediately after use.

Equipment needed

For each pupil or pair (Core)
- 2 test tubes or boiling tubes
- approximately 3 cm³ of limewater
- drinking straw

For the class (Extension)
- 4 flasks
- 4 pieces of muslin
- 4 pieces of cotton thread
- rubber bungs to fit flasks
- 100 cm³ of limewater
- live maggots or woodlice
- soaked seeds starting to germinate
- dry seeds

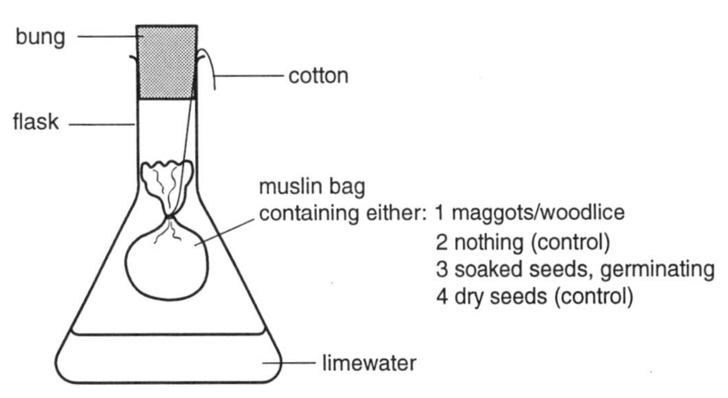

Notes

Four flasks are set up, each containing about 25 cm³ of limewater. Suspend the maggots or woodlice above the limewater in a muslin bag in one flask. Put an empty muslin bag in a second flask as a control. To show respiration in plants, suspend germinating seeds in a muslin bag above limewater in one flask, with dry, dormant seeds in a second flask as a control. Leave until the next session. After a day or two the limewater in the flasks containing the living organisms should turn cloudy (give it an occasional swirl). If there is no change before the next lesson, 'cheating' by blowing into the limewater in the two experimental flasks might avoid confusion amongst the pupils at this early stage in their scientific careers!

Moving matters

Book coverage

Key ideas	PoS
Living organisms can be divided into two large main groups. These are the plants and the animals.	AT2.4d

Answers

Activities in text

a Suitable examples.

b1,2

Animals	Plants
B hedgehog	A cactus
C butterfly	D orchid
F snake	E vine

b3 Suitable examples.

What do you know?

1 Animals move **quickly** and they often move their whole **bodies**. They need to **eat** to get energy. **Plants** move only **part** of their bodies and usually move very **slowly**. They can make their own **food**.

2 This question is looking for ideas about putting plants on a windowsill, or with light shining on them from one side, and showing that they bend towards the light or turn their leaves towards the light. If measurements are suggested, for example the angle of the stem or counting the leaves facing in a particular direction, this should be given extra credit. A different suggestion, for example observing flowers as it gets dark to see if they close, is equally valid, even if it ignores the clue in the question!

3 If it was a flower, it would not move, and would quickly wilt and die disconnected from the plant. An animal would move and probably attempt to get away – it would flutter, wriggle, walk, etc. If kept captive an animal would eat and drink if suitable food was offered.

Assignments

4 a Suitable examples, e.g. cows, sheep, chickens, etc. and cabbages, lettuces, tomatoes, potatoes, etc.
b Suitable examples, e.g. cats, dogs, goldfish, spiders, worms, etc. and house plants and garden plants.

5 a We divide the living world into two groups to make it easier to understand and to talk about.
b Animals are living organisms that move their whole bodies around and move fast. They feed on other animals or plants. Plants are living organisms that move only parts of their bodies very slowly. They can make their own food.

6 a Moving is a way of responding to the world around an organism.
b This should include movements of both plants and animals, including insects, birds and mammals. Movements range from flowers opening and leaves moving to face the sun, to respiratory movements of animals, animals finding food/escaping being eaten/flying/singing, etc.

Do plants move?

Running the activity

This is a fairly simple activity which many pupils will be able to work out for themselves. There is a Help sheet for those who have difficulty. The pupils should get clear results, with the cress grown in one-sided light bending over in the direction of the light, whilst the other cress grows straight up.

Differentiation

Sheets available are:

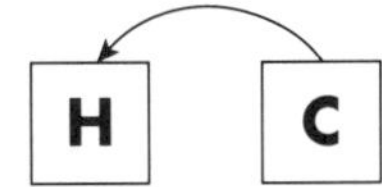

AT1 programme of study

Core: Pupils are making choices about control aspects of the activity and the selection of apparatus (**AT1.1e, f, h**). They also are invited to evaluate their experiment (**AT1.4c**).

Help: Pupils are helped to learn about what apparatus to select and led to drawing conclusions (**AT1.1h, 1.3e, g**).

Equipment needed

For each group
- 2 beakers or Petri dishes
- cotton wool
- cress seeds
- black paper
- scissors
- labels
- Sellotape
- access to water

Note
The most important point to ensure success with this activity is to make sure that the cotton wool is very wet before the seeds are added. If the seeds dry up, nothing will happen. It is worth asking the technician to check that the experiments are watered at least once during the week if necessary.

The Lego of life

Book coverage

Key ideas	PoS
A cell is the small single building block of any living thing. All animal cells have a membrane, nucleus and cytoplasm. All plant cells have a membrane, nucleus, cytoplasm and cell wall. Cells in the green part of the plant have chloroplasts.	AT2.1b,c,d

Answers

Activities in text

a A plant cell (cell wall, membrane, cytoplasm, nucleus, chloroplasts), B animal cell (membrane, cytoplasm, nucleus).

What do you know?

1 Drawings should be copied carefully from the book with simple labels.

2

Part of cell	Job done
cell membrane	The skin of the cell. Substances move in and out through the membrane.
cytoplasm	the jelly where all the important jobs of the cell take place
nucleus	the 'control room' of the cell
cell wall (plant cells only)	a tough wall which helps the plant stay upright
chloroplasts (plant cells in the green parts of plants, the leaves and stems)	special packets of green colour which help to make food for the plant

3 Because chloroplasts help plants to make food from light, air and water. Animals can't do this.

4 Animals move their whole bodies fast; plants move parts of their bodies slowly.
Plants can make their own food; animals have to eat food.
Plant cells have a cell wall and often chloroplasts; animal cells don't.

Extras: Planimals?

1

Things found in all cells	Things found in animal cells	Things found in plant cells
nucleus cytoplasm membrane	flagella	chloroplasts

They could be animals because they move fast. They could be plants because they have chloroplasts. Being able to make their own food could be argued as making them definitely plants. But there is a strong case for giving them their own kingdom, along with any other similar organisms. This has actually been done – they are called Protoctista.

Assignments

7 a Animal cells have a membrane, cytoplasm and nucleus.
b Plant cells have all the above, with a tough cell wall and chloroplasts as well.

8 a Cell wall and chloroplasts.
b Plant cells need extra support, provided by cell walls. They make their own food, and chloroplasts are needed to make this possible.

9 Posters should be clear and colourful, with a thoughtful layout which includes drawings of the different types of cells and the similarities and differences between them.

Looking at cells

Running the activities

The two Core activities for this book spread are so closely related that it would be useful to do both in the same lesson.

Before the pupils begin looking at cells, they should work through the Skills sheet on using a microscope. This will begin to familiarise them with the parts of the microscope, and focusing the lenses. It might be convenient to follow the introduction to the microscope with this work on cells, as the pupils can immediately reinforce their skills in a practical situation. There is an Extension task for more able pupils who have completed the Core activities. This requires a section of a green leaf to be set up at the side of the laboratory under a binocular microscope or using a projector, so that the pupils simply go and look at it.

Differentiation

Sheets available are:

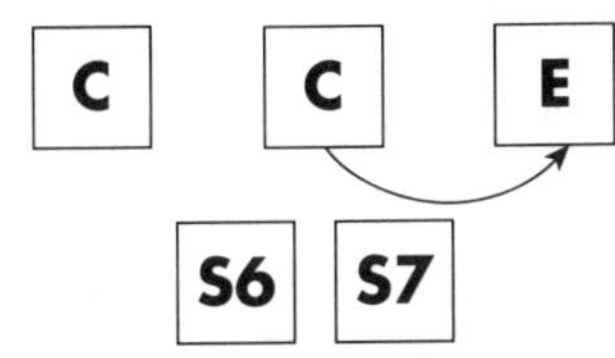

Differentiation is by outcome – the more able pupils will manipulate the microscopes and materials more readily, draw more accurately, etc. In the Extension activity, pupils need to transfer knowledge from their practical experience and their theory work into a new situation.

AT1 programme of study

This activity requires careful observation of a specimen and recording, skilful use of a microscope and manipulation using instruments (**AT1.2a, b, e**).

Equipment needed

For each group (Core)

- microscope
- glass slide
- coverslip
- pipette
- mounted needle
- tissue paper
- slice of onion or rhubarb (handle with sharp forceps)
- prepared slide of human cheek cells

Note

It can be quite tricky pulling off a layer of onion or rhubarb tissue one cell thick, so it is worth practising if you haven't done it before!

For the class (Extension)

- prepared slides of plant leaf set up under binocular microscope or projector so that palisade mesophyll cells are clearly visible

What is an animal?

Book coverage

Key ideas	PoS
We classify animals and plants by putting them into groups. The animal world is divided into two main groups, vertebrates and invertebrates. Vertebrates are animals with backbones. Invertebrates are animals without backbones.	AT2.4d

Answers

Activities in text

a Pupils could come up with any number of different groupings, which are acceptable as they will subsequently see whether they are in common use or not.

What do you know?

1 Vertebrates are animals which all have **backbones**. They have **skeletons** inside their bodies. Examples of vertebrates are **horses**, **rabbits** and **sparrows**. Invertebrates are animals **without** backbones which often have very **soft** bodies. Examples of invertebrates are **snails** and **beetles**.

2 These lists will obviously vary, although probably not a great deal! It can be an interesting exercise to see how the numbers of invertebrates to vertebrates works out in each list. If the information is collected, e.g.
- number of pupils with (only 1/up to 2) invertebrates in the list
- number of pupils with (2/3–4) invertebrates in the list, etc.

then the pupils can display the data for homework (see Assignment 11). The exact form in which you group the pupils will depend on how they make up their lists in the first place.

Assignments

10 The vertebrates have an internal bony skeleton with a backbone; the invertebrates have no backbone.

11 The bar chart will depend on the data collected. The usual tendency is for pupils to choose many more vertebrates than invertebrates. There are lots of possible reasons for this: vertebrates tend to be bigger and so more noticeable; we are vertebrates and so notice our own kind; most of the animals we keep as pets, farm animals or see in zoos are vertebrates, etc.

12 Invertebrates do not have a rigid structure inside the body, and so they can take up a great variety of forms. When the body shape is governed by a bony structure with a rigid spine, then the possibilities are much more limited.

All creatures great and small

Running the activity

The class is divided into groups, and each group is provided with several specimens of a particular animal group. These specimens could include preserved specimens, living specimens and parts of specimens, such as skeletons, shells, feathers, etc., depending on what is available within the department. The course textbook should be available to pupils. Ideally, other reference books including animal encyclopaedias, a computer program on animal groups or (even better) a CD-ROM animal encyclopaedia would also be available.

Give the pupils a period of time (start with 15–20 minutes, which can be shortened or lengthened as necessary) in which to prepare a two-minute talk about their animal group. Each group of pupils should then have the opportunity to present their talk to the class, using their specimens and any visual aids they have produced to illustrate it.

Differentiation

Sheets available are:

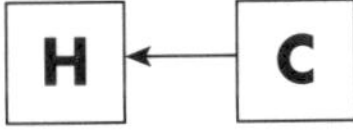

More able pupils will be able to make much fuller use of secondary sources. The type of talk produced will vary in how well it addresses the target audience.

The Help sheet is available for pupils who will have difficulty in planning a talk. It contains some structured questions to help them build up a description of their animal group.

AT1 programme of study

Careful observation specimens is required, as is recording of evidence as pupils carry out the work (**AT1.2b**, **e**). There is an opportunity for using secondary sources of information, and for presenting information for a specific audience.

IT potential

The use of a CD-ROM animal encyclopaedia would enhance this activity.

Equipment needed

For each group
- tray
- selection of specimens of an animal group

Note
Groups should be relatively specific, e.g. insects, crustaceans, birds, reptiles or spiders. Specimens should include preserved and part animals as well as (in some cases) living organisms. The specimens chosen will depend on the selection available within the department but, for example, insects could comprise a butterfly or moth, a wasp or bee, an ant or termite, a large beetle, ladybirds.

For the class
- reference books on animals suitable for Key Stage 3 pupils
- computer program on animal groups/classification e.g. CD-ROM animal encyclopaedia

Creepy crawlies

Book coverage

Key ideas	PoS
Invertebrates can be divided into several large groups: jellyfish, starfish, flatworms, roundworms, segmented worms, molluscs and arthropods. Each large group can be split into smaller ones. For example, arthropods are divided into spiders, crustaceans, insects and centipedes and millipedes.	AT2.4d

Answers

Activities in text

b

Crustaceans	Spiders	Insects	Centipedes and millipedes
C, G	A, E	B, F, H	D

What do you know?

1 Jellyfish – jelly-like bags for bodies with stinging tentacles.
Roundworms – thin rounded bodies.
Arthropods – jointed limbs and hard outer skeletons.
Segmented worms – long bodies divided into segments.
Molluscs – muscular body with a shell inside or outside the body.
Starfish – bodies with five 'arms'.
Flatworms – simple flattened bodies.

2 Arthropods: dung beetle, tarantula, moth.

Extras: Molluscs and more molluscs

Ways of classifying will vary. The most likely will be squid and octopus (eyes, tentacles, large, live in water); snails and sea slug (soft, muscular body, often shell on the back, some can live on land); shellfish (soft body protected by double shell, live in water). Pupils' names for groups will obviously be their own, but the real names are cephalopod molluscs, gastropod molluscs and bivalves.

Assignments

13 A segmented worm, B starfish, C flatworm, D jellyfish, E roundworm.

14 Again, this could be done in a number of ways. Judge the designs on the ease of use and the thought given to ways of making it weather-proof and user friendly.

15 This will depend on the group or individual animal chosen, but it is important that pupils stick to the word limit to prevent wholesale copying out of other books!

Where do we belong?

Book coverage

Key ideas	PoS
Vertebrates are divided into five main groups: fish, amphibians, reptiles, birds and mammals. Each group contains many smaller groups.	AT2.4d

Answers

Activities in text

a

Whales and dolphins	Herbivores	Carnivores	Primates
C, D	E, F	A, G	B, H

What do you know?

1 **a** Reptile.
b Mammal.
c Bird.
d Amphibian.
e Fish.

2 Mammals – we are warm blooded, have hair, sweat, produce live young, produce milk to feed our young.
Primates – we have forward facing eyes, and hands that can use tools.

Assignments

16 **a** Fish, amphibians, reptiles, birds, mammals.
b Fish – live in water, have gills, fins and scales.
Amphibians – have smooth moist skin, return to water to lay eggs.
Reptiles – have dry scaly skin, lay eggs with leathery shells.
Birds – have feathers, wings, are warm blooded, lay eggs with hard shells and often care for their young.
Mammals – have hairy skin, are warm blooded, sweat, give birth to live young, make milk to feed their young.

17 **a** They are warm-blooded vertebrates which give birth. Most important point – they feed their young on milk produced by the mother.
b They have fur. They feed their young on milk produced by the mother.
c The platypus lays eggs, a feature typical of a reptile or bird rather than a mammal.

18 Suitable account.

Pick a plant, any plant

Book coverage

Key ideas	PoS
The plant world is divided into four main groups: mosses, ferns, conifers and flowering plants. The biggest group is the flowering plants. They range from big flowering trees to small grasses.	AT2.4d

Answers

Activities in text

a Suitable examples/groupings. Pupils might choose flowers, fruits or edible plants as groupings.
b Pupils should show that they understand that mosses and ferns need damp, shady conditions, and that children playing need tough plants to withstand knocks. Common sense and the application of information from the spread are required.

What do you know?

1 **Mosses** are small plants that produce spores and need a damp place to live. **Ferns** have a water transport system, but also produce spores. Conifers are often **evergreen**. They produce seeds which are carried in **cones**. The **flowering plants** produce **seeds** which grow in fruits.
2 Conifers, flowering plants.
3 Mosses have thin leaves and can't transport water through the plant, so they need to live in damp places. Marshes suit them very well, but on drier hillsides they find it more difficult to survive.

Assignments

19 A moss, B conifer, C fern, D flowering plant.
20 **a** Apples, strawberries, raspberries, tomatoes, etc.
b Pine tree, fir tree, Christmas tree, etc.
c Grass, some trees.
d Tree, preferably a specific example such as oak, ash, giant redwood, etc.
21 Look for a clear design with simple information about plant groups. Mosses and ferns need to be kept moist and not in too bright a light. Flowering plants also need to be watered but can stand being drier, and need plenty of light.

Organise your organisms

Book coverage

Key ideas	PoS
A key identifies organisms using simple questions about the differences between them.	AT2.4c,d

Answers

Activities in text

a A panther, B lion, C tiger, D snow leopard, E lynx.
b A gulper eel, B manta ray, C dragon fish, D plaice, E sargassum fish, F turbot.

What do you know?

1 **a** A panther is **black**.
b The fur of a tiger has **stripes**.
c A lynx has **ear tufts**.
d A snow leopard has **patterned/spotted** fur.

2

1 Has it got patterned fur?	Yes: go to question **2**	No: go to question **3**
2 Has it got stripes?	Yes: tiger	No: snow leopard
3 Is it black?	Yes: panther	No: go to question 4
4 Has it got ear tufts?	Yes: lynx	No: lion

3 The keys produced here will vary, depending on the features selected by the pupils.

Extras: Extinct is forever

1 Molluscs; flowering plants; birds; mammals; reptiles (dinosaurs).
2 Judge keys on merit.
3 Destroying the places where organisms live, hunting and eating organisms and pollution are the three most likely answers. Any others which seem valid should be accepted.
4 Look for thought put into the preparation of the talk, and the ability to organise arguments and present ideas.

Assignments

22 A spider plant, B holly, C variegated ivy, D cheese plant, E oak.
23 This will vary depending on the organisms chosen by the pupils.

The key to the class

Running the activity

The pupils devise a key to identify each individual member of the class. Pupils work individually or in small groups first, to develop ideas about how to break the class down into smaller groups. The whole class can then discuss those ideas with the teacher, who can help to direct the class to the most coherent way of identifying themselves. If time and resources allow, it can be very successful to make a large wall-chart version of the key and then get other adults, such as staff, technicians or visitors, to use the key to identify particular children.

It is important that the teacher states clearly at the beginning of the exercise that rude, personally abusive or unkind differences will not be tolerated.

Once the class has been at least partially subdivided (by sex, eye colour, etc.), each child could be allowed to decide what would make it possible to identify him or her from the others in the sub-group. This means that no one ends up with a description imposed on him or her by others. The child who doesn't mind his or her glasses might comment on them, while children who are self-conscious about glasses can choose freckles or short ear lobes or any other distinguishing feature they are happy with.

The point about relatively permanent differences needs to be drawn out, so that features such as sex, hair colour and type, eye colour and ear lobes are used rather than wearing a watch or having a ponytail. If skin colour is a clear distinguishing feature, then 'pale' and 'dark' are less emotive terms than 'black' and 'white'. If impermanent features have to be used in the final identification chosen by the children, this should be used to point out how difficult it is to separate people from one another. Even hair colour and texture is not really accurate in the long term. Not only do people dye their hair, but everyone eventually turns grey and many of the boys will end up at least partially bald!

Differentiation

Sheets available are:

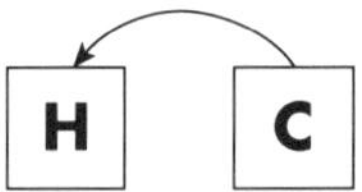

The Help sheet provides a much simpler key, broken down and explained. This could be used as an alternative to the class key.

AT1 programme of study

This activity develops concepts and skills involved in the use and construction of keys, which are described in **AT2..4c, d**. Careful observation and recording of evidence are required (**AT1.2b, e**).

Solids, liquids and gases

Book coverage

Key ideas	PoS
Solids have a fixed volume and a fixed shape. Liquids have a fixed volume but no fixed shape. Gases have no fixed shape and no fixed volume. These are called properties of solids, liquids and gases	AT3.1a

Answers

Activities in text

a Answers such as: the 'solid' watch seems to flow; the liquid water has a sharp edge.
b Observation and description of a range of common solid objects to determine the key characteristics of solids. Answers appropriate to objects chosen. The two key characteristics are fixed shape and fixed volume.

What do you know?

1 Your bones are **solid**. They are hard and have a **fixed** shape which supports your body. Your blood is a **liquid**. It changes its **shape** as it flows through your veins. The air you breathe is a **gas**.
2 Answers to include: a solid glass container keeps its shape; liquid lemonade pours out; bubbles are full of gas, which escapes into the air.

Extras: Liquid brakes

1 **a** Liquids flow and change shape around the pipes, but they cannot be squashed.
b Bubbles of gas will be squashed when the brake pedal is pressed, so the full force will not be passed to the brakes.
c The water might freeze. Solid ice could not move through the pipes, so the system would jam. (Breaking of the pipes because of ice expansion is not expected here.)

Assignments

1 **a** Cut the cake into equal sized pieces. She can do this because cake is solid and so keeps its shape.
b Pour the cola into equal sized glasses. The cola pours because it is a liquid, but it keeps its volume.
c Gases cannot be poured out like liquids because they spread out to fill whatever space they are put into.
2 Water is a liquid, so it can flow and change shape. Bread is a solid, so it cannot flow or change shape.
3 Water is a liquid, so it can flow and change shape. Beans and carrots are solid, so they cannot flow and change shape. They get stuck in the plughole.
4 Blood is a liquid, and so cannot be compressed. When the heart squeezes, the blood is forced out. Air is a gas and becomes squashed when the heart squeezes it.

Pouring liquids

Running the activity

Core: One set of containers is needed per bench or large group for the ordering part of the exercise. The actual measurement of volume could then be done in pairs.

Note that cubic centimetres have been used to maintain continuity of units in the text. Many measuring cylinders are now graduated in cubic centimetres, but others still use millilitres, as do many measuring jugs. This confusion is mentioned on the Skills sheet but pupils may need to be told that 1 ml = 1 cm^3.

Extension: This sheet may be used purely as a paper exercise to develop planning skills. Alternatively, if time permits, the investigation could be carried out by groups of 2–4 pupils. In this case, the plans would need to be discussed and agreed in advance.

Differentiation

Sheets available are:

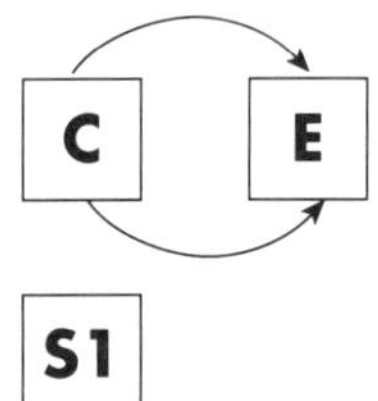

Core: This is a simple experiment which may well be appropriate for all. If used in this way, the Extension sheet is a paper exercise only, which gives scope for simple fair test planning.

Extension: Instead of carrying out the Core experiment, more able pupils may jump straight to the Extension sheet as a complete AT1 activity, giving limited scope for planning, observation and measurement, and simple analysis and display.

AT1 programme of study

Core: This is a simple activity to reinforce the idea of conservation of volume and to practise measuring liquid volumes (**AT1.2a, 1.3a, f**).

Extension: This sheet further develops an understanding of the properties of liquids (**AT1.1d–f, 1.2a-e** if carried out).

Risk assessment

The containers used for the Core activity must be clean. Meths is highly flammable – ensure there are no naked flames around. If the liquid paraffin used is medicinal paraffin, there is no hazard, but kerosene is flammable.

Equipment needed

For each group (Core)
- Assorted containers (uncalibrated, or with volumes on labels masked) will need to be collected, and washed if necessary, in advance. One set of containers may be shared by three or four groups if they work in rotation.

For each pair
- measuring cylinder(s) of suitable size(s) to match the containers used

For each group (Extension, if experiment carried out)
- stopclock or watch
- assorted liquids to give a range of viscosities, such as water, various oils, glycerine, meths, liquid paraffin, or made up sugar solutions of different strengths
- filter funnel of suitably small bore
- clamp stand
- beakers
- measuring cylinder

Notes
1 The filter funnel will need to be flushed through with water between runs, so a running order for liquids could be suggested, ending with oil.
2 A trial run might be useful to check the time taken. The volume of liquid used each time should be matched to the funnel bore, to give a reasonable time period for water. If water runs through too fast, it will be difficult to time.

Solid properties

Book coverage

Key ideas	PoS
Solids have many different properties. They may be heavy or light, weak or strong, hard or soft. To choose the right material, you have to match these properties against those needed for the job.	AT3.1a

Note: A fuller treatment of density comes in spread 3d, so keep to the simple idea of 'light' and 'heavy' materials here.

Answers

Activities in text

a Answers such as: glass is transparent; iron is heavy/strong; concrete is hard and lasts a long time; waterproof cotton is attractive and flexible; leather 'breathes'; rubber is hardwearing and flexible.

b This gives a real-life experience of the range of hardness of materials in a puzzle-solving context. It might be worth demonstrating a simple case of scratching first to show the idea. Drag a quartz crystal lightly across the surface of an unimportant piece of glass. (This could lead to the context of glasscutting.) The mark cannot be rubbed off. Repeat with a crystal of calcite. A mark may be made, but this will rub off, leaving the glass undamaged. Some pupils may confuse scratching with shattering, caused by pressing extremely hard.
There is a possible cross-link with geography, where pupils might wish to identify these minerals in the field.
Quartz is not scratched by anything.
Fluorspar is scratched by a knife only.
Calcite is scratched by a coin and a knife, but not by a fingernail.
Gypsum is scratched by everything, including a fingernail.

What do you know?

1 Bricks and concrete are used for building because they are **strong** solids. Aluminium can be used to build aeroplanes as it is very **light** for its size. Pencils use a solid called graphite which is so **soft** that it rubs off on paper.

2 **a** Aluminium (strong and 'light').
b Brick (strong and not too 'heavy').
c Diamond (very hard).
d Steel (strong and 'heavy').

Assignments

5 The £50 note, as it is much lighter. The coins would be very heavy.
6 Superglue is too strong. You would never get the brace off without damaging your teeth,.
7 He could cut the glass with his diamond ring. Diamond is the hardest material of all.
8 **a** Rubber is harder/tougher/more hardwearing.
b Iron is too hard. It would damage the floor, and be too heavy to walk in.

Which bag is best?

Running the activity

This can be tackled in small groups or pairs as the simple apparatus should be readily available. Although the problem is posed in an open-ended way, the pupils are clearly directed towards the strips and masses model. Alternative methods should give way under subtle questioning (such as: "Where will we get the 50 tins of baked beans you need?").

Differentiation

Sheets available are:

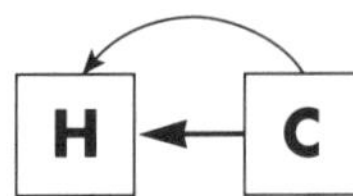

AT1 programme of study

Core: This sheet gives an opportunity to follow a simple AT1 activity through to a full investigation. There are sufficient prompts to target the investigation at level 4. Question **5** links in with economic awareness.

Help: This sheet takes the investigation down to levels 2/3. The Help sheet could be used by appropriate pupils from the start, or all pupils could be allowed to plan their own experiments, with the Help sheet held in reserve for those who could not develop a viable plan on their own.

Risk assessment

There is no specific risk in the suggested model. If any groups were to attempt to fill whole bags until the handle breaks, then there would be a danger to toes from falling heavy masses. A waste bin or similar container packed with crumpled paper placed under the bag will prevent this.

Equipment needed

For each group

- plastic bags of different strengths
- clamp stand
- 100 g masses and holder
- split corks
- Sellotape
- splints
- paperclips

Note

Plastic bags need to be gathered in advance. It may be advantageous to have equal-sized strips ready prepared.

Useful metals

Book coverage

Key ideas	PoS
Metals are useful because they are hard and strong and can be shaped in many ways. Metals are good conductors of both heat and electricity. Non-metals are insulators of both heat and electricity. Metals are shiny. Some metals such as iron and steel are magnetic.	AT3.1j(k,l)

Answers

Activities in text

a Any appropriate answers.
b Steel cans may be picked up with a magnet. Aluminium cans will not be affected.

What do you know?

1 In an electric plug, the pins are made of brass because this metal is a good **conductor** of electricity. The case is made of **plastic** as this is an **insulator** and so stops you getting an electric shock.
2 Solids, good conductors of heat and electricity, shiny, easily bent or shaped, often 'heavy for their size', often hard and strong.
3 Graphite is a non-metal, but is unusual as it conducts electricity.

Extras: More solid properties

1 **a** Polythene – cheap and flexible.
b Nylon – very strong and flexible.
c Melamine – strong and heat resistant.
d Polystyrene – reasonably strong/reasonably cheap.
e Polystyrene – acid resistant.
(Other answers may be acceptable if well argued, e.g. nylon for carrier bags as it is strong and flexible.)
2 **a** Is the material transparent?
b Is the material resistant to bleach?

Assignments

9 Answers appropriate to objects chosen, e.g. knife – hard and strong, aluminium can – easy to shape, etc.
10 **a** A is plastic (non-metal) – light, 'weak', not bendable. B is copper (metal) – heavier, stronger, bendable.
b Copper conducts electricity, but plastic does not, so electrical wiring may not be earthed if connected to a plastic pipe.
11 Steel is strong enough, but is a poor conductor of electricity. Aluminium is a good conductor, but is too weak on its own.

Metals and non-metals

Running the activity

(Note that, strictly speaking, 'non-metal' is a term used to describe only elements. However, at this level, the use of 'non-metal' for compounds should not cause problems.)

Start with a demonstration: get pupils to feel the difference in temperature between a metal can and an expanded polystyrene cup each containing hot (not boiling) water. Use this demonstration as an exemplar for the displays. The property is heat conduction: the metal is a good conductor; the non-metal is a poor conductor (insulator).

The main part of the activity is designed as a circus, with each station containing a selection of metals and non-metals that show a particular difference in properties. Each station should be made as simple and clear as possible, with the metals in one (labelled) box or tray and the non-metals in another. Check regularly to make sure that they have not been mixed up. At each station, the metal property is kept the same, but the non-metals are more variable. Pupils should be warned that non-metals can be very variable. The stations are as follows:

- station 1: shiny metal/dull (variable) non-metal
- station 2: electrically conducting metal/insulating non-metal
- station 3: bendy metal/rigid or brittle non-metal
- station 4: solid metal/variable non-metal
- station 5: strong metal/weaker non-metal
- station 6: 'heavy' metal/'lighter' non-metal.

The test circuit for electrical conductivity should be set up ready. A full understanding of the circuit is not required here. If time is pressing, pupils need not visit every station. The full set of results can be pooled towards the end of the lesson so that the exercise can be completed by all.

Differentiation

Sheets available are:

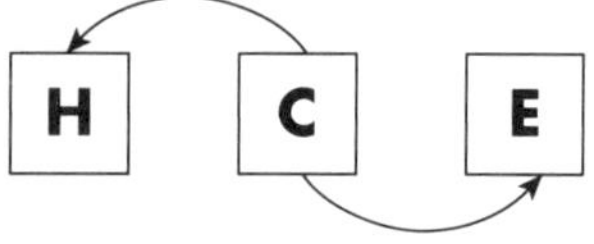

Core: Differentiation is by outcome in terms of quality of observations and the ability to generalise from the examples given.

Help: This sheet gives the table drawn ready to fill in.

Extension: This is a data analysis exercise which can be completed by those who finish the Core work early.

AT1 programme of study

There are limited opportunities for **AT1** in terms of use of apparatus and pattern recognition.

Risk assessment

Wash hands carefully after handling lead.

Equipment needed

For the class
- metal can
- expanded polystyrene beaker
- source of hot (but not boiling) water

Station 1
- pieces of bright, shiny metal such as aluminium foil, fresh copper or brass, stainless steel, freshly cut lead
- dull or glassy non-metals, such as graphite, rock sulphur, glass, pebbles

Station 2
- battery in holder
- bulb in holder
- 3 connecting wires
- 2 crocodile clips

▶ continued

▶ continued

- strips or pieces of different metals
- pieces of non-metals such as wood, glass, plastic

Station 3
- sheets or wires of metals such as copper, aluminium, iron or brass that are thin enough to be easily bent
- non-metals such as matchsticks, charcoal or chalk that will snap rather than bend

Station 4
- solid pieces of any obvious (shiny) metal
- examples of solid, liquid or gaseous non-metals

Station 5
- pair of pliers
- small blocks of metals such as iron, bronze or brass that will fit in the jaws of the pliers but show no effect when squeezed
- non-metals such as marble chips, small pieces of wood or rock sulphur that can be crushed in the pliers

Station 6
- range of larger pieces of metals of various densities from aluminium to lead
- various solid non-metals, from wax to glass

Note
Each station should have three or four different metal samples in one (labelled) box or tray, and three or four non-metal samples in another. The examples given have been chosen to highlight the difference in properties clearly. They may be substituted by others that show the same property.

'Heaviness'

Book coverage

Key ideas	PoS
Density is used to compare the 'heaviness' of materials. The density of a material is the mass of 1 cubic centimetre. The density of water is 1 gram per cubic centimetre.	AT3.1a

Answers

What do you know?

1 To find out if one material is heavier than another, you have to weigh pieces of the same **volume**. The mass of a 1 cm cube is the material's **density**. **Liquids** have lower densities than solids and **gases** have the lowest densities of all.

2a,b Lowest to highest density: water 1 g/cm³, aluminium 2.7 g/cm³, lead 11.6 g/cm³, gold 19 g/cm³.

c

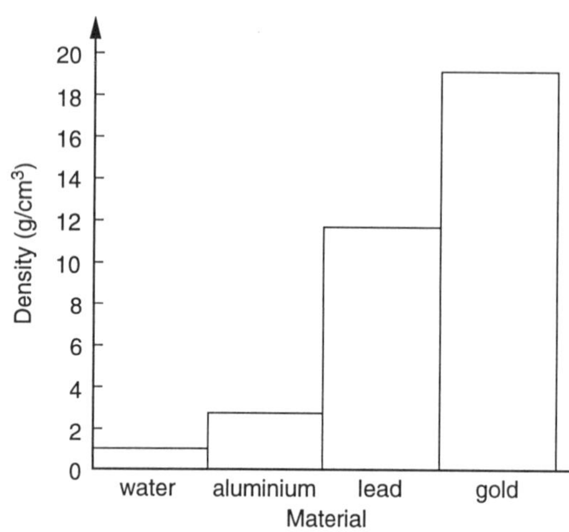

3 Gold bars are almost twice as heavy as lead bars (of the same size).

Extras: Changing the weight

1 **a** Filled with a denser material, the base will be heavier and so more stable.
b Unlike water, the concrete could not be removed if the sunshade base needed to be carried.

2 25 litres.

Assignments

12 Wet sand is more dense than dry sand. The increased load could have caused the tyres to burst.

13 **a** Aluminium has a lower density. Lighter planes need less fuel.
b Dried beans and dried milk. They have a lower density so weigh less, leading to lower fuel costs.

14 **a** 20 g. **b** 10 cm³.

15 5.5 g/cm³ is an approximate average between the surface rocks (2.7 g/cm³) and the denser iron core (7.8 g/cm³).

'Heaviness' and volume

Running the activity

This activity looks at the relationship between mass and volume, but it is designed so that it can be used to introduce or reinforce the concept of fair testing. Pupils should work in groups of three or four so that they can share their ideas.

The first part sets out the problem to lead deliberately to an unfair test that produces muddled results. 'A fair test' then shows how fair testing leads to a successful conclusion, with a clear relationship. This direct relationship should then be described in terms of: "volume and mass go up together".

As the calculation of volumes can be problematic for many pupils, a crib sheet giving the volumes of the numbered blocks should be available. If prepared in advance, these volumes could be pre-printed on the Help sheet.

The sheet uses the term 'factors' to describe the variables, in keeping with the National Curriculum document.

Differentiation

Sheets available are:

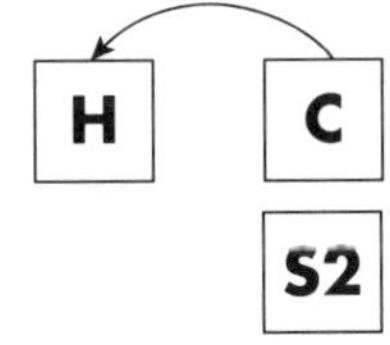

Core: Most pupils would follow 'A first try' and then move on to 'A fair test'. Those pupils who are able to should pursue this by plotting the points on a graph and drawing the best fit line. They might then consider the accuracy of their readings and make mass or volume predictions from the graph.

Exceptional candidates might choose to bypass 'A first try' completely, as they immediately spot that it is not a fair test and so will not give meaningful results. If they can clearly justify this then they should be allowed to plan and carry out their own test. They will have the time (and presumably the skill) to calculate the block volumes for themselves.

Help: This sheet follows the path of the Core sheet, but in a more structured way.

AT1 programme of study

The activity develops an understanding of the need for fair testing (**AT1.1c–e, 1.3a, b, d**).

Risk assessment

Wash hands carefully after handling lead. Take care if glass blocks are chipped.

Equipment needed

For each group
- access to 10 numbered rectangular blocks of different sizes and materials
- balance

Note
One set of blocks could be shared by three or four groups, as they can be weighed in rotation. The calculated volumes of the blocks should be available on a crib sheet, or pre-printed on the Help sheet.

Each set of blocks should contain:
- 5 blocks of different volumes made of the same material (if this is wood, make sure it is the same wood with the same density)
- 5 blocks of different sizes and materials, such as lead, brass, iron, glass, Perspex.

If possible, make sure that the block that obviously has the largest volume is not the heaviest, and that the block that is obviously the smallest is not the lightest.

A model for materials

Book coverage

Key ideas	PoS
Scientists use a simple particle model to explain how materials behave. In solids, the particles are closely packed, and are stuck together. In liquids, the particles are closely packed but are free to move. In gases, the particles are far apart.	AT3.1b

Answers

What do you know?

1 Particle diagrams drawn as in the pupils' book.
2 a Gas particles are far apart, so they can be pushed closer together. The particles in liquids are already closely packed.
b The particles in liquids are free to move against one another, but the particles in solids are stuck together.
c The particles are stuck together in solids.
d The particles in hard solids are stuck together more strongly.
e The particles are broken apart from one another.
3 a The model would fall apart if you moved it.
b You could stick the balls together, for example.

Extras: A brief history of comfort

1 You can't squash metal tyres, so every jolt is passed through to the saddle. You can squash air, so the tyres absorb the bumps.
2 You can't squash liquids either.

Extras: Moving particles

3 The 'smell' particles collide with the moving air particles and become mixed up with them.

Assignments

16 Posters should be clear and colourful, illustrating the small size of the particles of matter in an imaginative way.
17 a There are fewer water particles in a litre of steam, as they have spread out.
b There is the same number of particles in each.
18 The particles would become closely packed again, and the gas would become a liquid.

Making sandcastles

Running the activity

Designed primarily to give a practical example to support the abstract particle model, this exercise also gives opportunities for AT1 investigations at the lower levels. The context is given by using cement, but in practice use plaster of Paris which is easier and safer.

The preparation of the sandcastles will need to take place a lesson in advance of the testing, to give the plaster time to set. The activity can be messy if not carefully controlled – newspaper will help to contain plaster dust and liquid spills. Pupils will need to consider the size of the 'spoonfuls' used, to ensure fair testing.

The testing procedures are left open ended in the Core sheet, though dropping a ball bearing is given in the second Help sheet. Pupils trying alternative methods will need to talk over their ideas and have them agreed before they proceed.

Differentiation

Sheets available are:

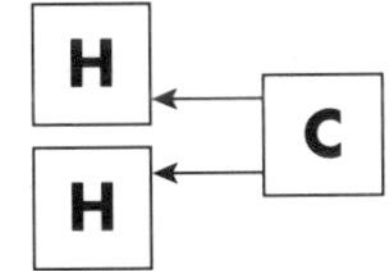

The Core sheet is open ended, while the Help sheets direct pupils in a step-by-step approach. All may start on the Core sheet, keeping the Help sheets in reserve for those that cannot come up with a viable plan. Alternatively, some pupils may be targeted directly with the Help sheets, having read the introduction from the Core sheet.

AT1 programme of study

The Core sheet gives an opportunity for a full AT1 investigation.

Risk assessment

Falling ball bearings could damage benches or hurt toes. Thrown ball bearings could cause more damage. Give appropriate warnings. Use mats or thick newspaper to protect benches, or perform the experiment on the floor.

Equipment needed

For each group (making sandcastles)

- 6 clean yoghurt pots or disposable paper cups
- sand
- plaster of Paris
- spoon or spatula of suitable size (a level teaspoonful is suitable for small yoghurt pots, a rounded teaspoonful for paper cups
- bowl large enough to mix one sandcastle
- access to water
- plenty of newspaper

For each group (testing)

- large steel ball bearing
- metre rule
- clamp stand
- mats or newspaper

Notes

1. Plaster, no matter how watery, should never be washed into sinks as it will set and block the drains. Waste material should be allowed to stand and set in a bucket or bowl. It can then be disposed of safely as solid waste.
2. The drop height needed to break the sandcastle will vary with the sizes of the ball bearings and the sandcastles, as well as with the mix. A technician may wish to trial the richest mix to determine the most suitable ball bearing size.
3. Other methods of testing could involve stacking masses onto the sandcastle, though a large number might be needed, which could be a problem if they fell.

Getting moving

Book coverage

Key ideas	PoS
Everything that is moving has energy. We call this type of energy movement energy or kinetic energy. To make something move, you have to transfer energy to it.	AT4.5f

Notes: This unit develops the idea of different forms of energy, and changes between them. It starts by assuming an intuitive understanding of the word energy; this spread deals with kinetic energy, one of the most obvious forms.

Answers

Activities in text

a Answers involve various terms for transferring kinetic energy: A pedalling, B pushing, C blowing, D throwing, E kicking.
b1 The motorbike (and rider).
b2 It is moving.
b3 The engine.

What do you know?

1 The hamster is **transferring** energy to the wheel by **turning** it. The wheel has **movement** energy. The scientific name for this kind of energy is **kinetic** energy.
2 They get their kinetic energy from the moving water. Redirecting the jets would make them turn the other way.
3 The bicycle and rider have the most kinetic energy (big and fast). The windmill toy probably has the least kinetic energy (small and slow).

Assignments

1

Transferring movement energy by ...	To ...
kicking	a ball
pedalling	a bicycle
throwing	a ball
pushing	a pram
pulling	a trailer

2 In order from most to least kinetic energy:
a a high-speed train travelling at top speed
b a 40-tonne truck travelling at 100 kilometres per hour
c an eagle soaring at 50 kilometres per hour
d a butterfly fluttering by at 2 kilometres per hour
e a pupil sitting still at her desk (has no kinetic energy).
3 Kinema or kinematograph means moving pictures.

Capturing kinetic energy

Running the activity

This activity reinforces the idea of kinetic energy. Anything that is moving has kinetic energy, and anything that is being made to move is being given kinetic energy. The activity may be run as a competition to see whose sculpture collects the most energy from hot air rising above a radiator, hotplate or radiant heater. One design is suggested on the activity sheet, but pupils may be able to suggest others. The important thing is to have tilted surfaces which will be pushed by the rising air.

An alternative approach would be to use hair driers as standard sources of moving air. Pupils might then make horizontal or vertical axis sculptures.

Pupils could make and test their models at home.

Differentiation

Sheets available are: C

AT1 programme of study

Although this activity is intended to reinforce the concept of kinetic energy, it can also be used to develop the idea of fair testing. In the competition, pupils might be required to make sculptures from standard circles of card, and their sculpture should use all of the card. They should carry out trial runs of their sculpture (**AT1.1b**). They should also think about how to measure the rate at which their sculpture spins (**AT1.1h**). The faster it spins, the more kinetic energy it has.

Risk assessment

Pupils should be warned about the dangers of touching hotplates or of allowing their sculptures to touch them. Pupils should not be allowed free access to hotplates; the teacher should demonstrate the pupils' models if these are used. Pupils can safely use radiators and modern, low voltage, radiant heaters.

Hair driers should *not* be brought from home. They should be subject to the normal regular electrical safety check. Beware of the risk of overheating.

Equipment needed

For each pupil

- circle of card
- scissors
- mounted pin or thread, e.g. in cork
- clamp
- stopclock or watch
- access to hot air rising from radiator, hotplate or radiant heater (but see risk assessment)

Getting warm

Book coverage

Key ideas	PoS
Something hot has heat energy. A fuel is a store of chemical energy. When it is burnt, heat energy is released. We use energy transfer diagrams to show how energy is transferred.	AT4.5a,f

Answers

Activities in text

a Various heaters, e.g. electric, gas, paraffin; also the sun, etc.
b Heat energy (of the water) → heat energy (of the person).
Heat energy (of the burning gas) → heat energy (of the beans and pan).
c In pictures: camping gas, petrol, coal, wood. Also: natural gas, paraffin, diesel, fuel oil, etc.

What do you know?

1 Fuels and uses depend on those used in pupils' homes.
2 **Chemical** energy (stored in C**harcoal)** → **heat** energy (in the **food**).
3 A fuel is a useful **store** of **chemical** energy. The energy is **released** when the fuel is **burnt**.
4 Camel dung: used by camel owners in Arabia, N Africa.
Candle wax: used for lighting when electricity fails, and where electric light is not available.
Seal oil: used by Arctic people for cooking and lighting.

Extras: Barbecuing yourself

1 It heats the air and other surroundings.
2 Heat energy in food → heat energy in the surroundings.
3 Only a tiny fraction of the heat is used for cooking; most is wasted, so it is very inefficient.
4 Cook lots of food; try to confine the heat using a cover, brick surround, etc.

Extras: Fuel to go

5 Chemical energy (stored in the fuel) → heat energy (in the engine) → kinetic energy (of the car).
6 Kinetic energy (of the bicycle) → heat energy (of the rubber pads).

Assignments

4

Type of fuel	Use of fuel
petrol	car
charcoal	barbecue
wax	candle
wood	bonfire
coal	steam train

5 **a** Something that is moving has **kinetic** energy.
b Something that is hot has **heat** energy.
c A fuel is a store of **chemical** energy.
6 **a** Chemical energy (stored is petrol) → kinetic energy (of the car).
b Chemical energy (stored in wood) → light energy
→ heat energy.

Fuels: chemical energy stores

Running the activity

This activity asks pupils to consider the question: "Which fuel stores the most energy?". They have not reached the point where they can measure the energy content of a fuel, but they can at least think about some possible approaches to the question. In following the activity, you might proceed as follows:

1 Allow pupils to examine samples of the suggested fuels. They can guess which they think is the best store of energy, giving reasons to support their answer.

2 Now, focus the question: "Which fuel stores the *most* energy?". The sheet suggests several approaches*. The most suitable involves using standard amounts of fuel to heat water. You could discuss why this is the best method. (You might draw an analogy with the question: "Which pupil is the most hard-working? The one who makes the most noise, or the one who gets hottest, or the one who works longest, or the one who produces the most work?".)

3 The sheet then directs pupils to consider the detail of such an investigation.

4 Finally, you might compare two fuels by setting up a fair test as a demonstration. Place equal masses of two different fuels in suitable metal containers. Ignite them so that they burn beneath identical beakers each containing 100 cm^3 of water. Place a thermometer in each beaker, and observe the temperature rise. Which heats the water more? The results are never perfect, as heat is inevitably lost to the surroundings. Pupils should be expected to comment on this.

Differentiation

Sheets available are: 

This activity is designed as a discussion–demonstration session, in which you can draw on responses from the class. More able pupils could be asked to produce written answers to the questions, prior to discussion.

AT1 programme of study

Pupils can make suggestions about the idea to be investigated (**AT1.1a**), make predictions (**AT1.1c**) and devise a fair test (**AT1.1d, e**).

Risk assessment

Care should be taken in igniting and burning the fuels. Eye protection should be worn by both teacher and pupils, and screens used between the demonstration and the pupils. Pupils should not crowd round the demonstration. If seated, they should be at least 2–3 metres away.

Equipment needed

For the class

- samples of the following fuels: coal, charcoal, candle wax, cooking oil, wooden splints
- 2 containers for safe burning of fuels
- matches
- 2 250 cm^3 beakers
- 2 thermometers, 0–110 °C
- water
- 250 cm^3 measuring cylinder
- balance
- means of cutting fuels
- tongs
- 2 tripods and gauzes

*none of which are satisfactory

Body fuel

Book coverage

Key ideas	PoS
Food is the fuel that our bodies need for movement and to keep warm. Foods have different energy values. Energy values are measured in kilojoules.	AT4.5a,f

Answers

Activities in text

a Our bodies use food as fuel.
b1 Milk has only 200 kJ for each 100 g.
b2 1120 kJ.

What do you know?

1 a Fuel, food.
b Chemical energy.
c Heat energy, light energy.

2 Chemical energy (stored in food) → kinetic energy, heat energy.

3 a Pupils may need help in drawing this bar chart.

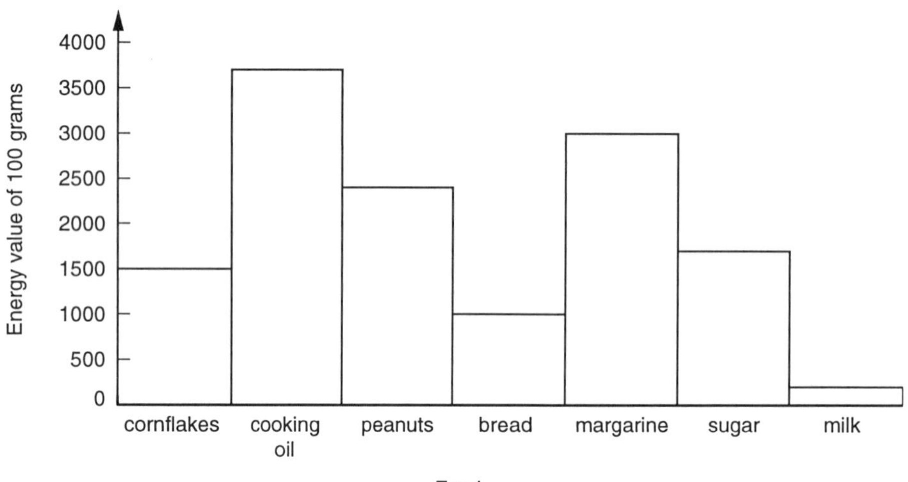

b To give a fair comparison.
4 a You need less energy to keep you warm in a hot country.
b They get less exercise, and live in warmer houses.
c She has to supply energy for her baby, too.

Assignments

7 a Chemical energy.
b Heat energy, kinetic energy.
8 a Cocoa.
b Cola.
c Soda water.
9 Cocoa, because most energy is stored for each 100 cm^3.

Camping supplies

Running the activity

This is a written activity, to emphasise the idea of foods as energy stores. You might start by discussing the requirements of a camping trip: what activities would lead to pupils feeling hungrier than usual? Just keeping warm out of doors can require more energy. The figure of 15000 kJ has been chosen to be greater than the standard daily requirement for people of this age.

Differentiation

Sheets available are:

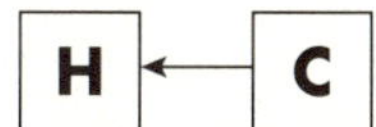

Core: Pupils are given the energy values of a range of foods. They have to select suitable foods to give 15000 kJ. It may help to suggest about 4000 kJ per meal, plus 3000 kJ for snacks.

Help: Pupils can cut out food items whose area represents the energy value. They then stick them on to a 10 cm × 15 cm rectangle, to achieve the required total. They may need to make copies, or simply draw on to the squared paper.

Extension: No one wants to carry a heavy load that does not supply much energy. Pupils could be asked to think about how concentrated the different foods are, in terms of energy per unit mass. They could consider the examples in the table in the book, or look at some food packets. Does Kendal Mint Cake really provide a lot of energy for little weight? Would they change their selection to make it lighter?

AT1 programme of study

This is not an investigative activity. However, it does relate to aspects of the Introduction to the Programme of Study: relating scientific knowledge to things that are used every day (**2a**), and presenting their ideas through the use of diagrams, tables, etc. (**4c**).

IT potential

The information about the energy values of foods could form the basis of a database or a spreadsheet. The latter could allow the calculation of energy values of various selections of foods, and the calculation of energy:weight values.

Equipment needed

For each pupil
- scissors
- glue
- squared paper (1 cm squares)

Electrical mysteries

Book coverage

Key ideas	PoS
We can change electrical energy into several other kinds of energy. Most of our electricity is generated in power stations that get their energy from burning fuels.	AT4.5a,c

Answers

Activities in text

a Desk lamp, electric fire, radio.
b A sound energy, B kinetic energy, C light energy, D kinetic energy.

What do you know?

1 a Electrical energy.
b Light (and heat) energy; heat (and light) energy; sound energy.
2 a Kinetic energy.
b Kinetic energy.
c Electrical energy.
d Kinetic energy (of the wind) —kinetic energy (of the turbine)→ electrical energy.

3 a You don't have to pedal so hard.
b You don't have to replace and pay for batteries.

Extras: The great escape

1 Boiler: heat and kinetic energy; dynamo: sound energy; bulb: heat and light energy.
2 No, paraffin is a fuel, a store of energy.
3 No, but it has heat energy and kinetic energy.
4 Chemical energy (stored in the fuel) → (heat energy and kinetic energy) → (sound energy) → electrical energy.
5 There is more energy in the fuel at the start than electrical energy supplied by the dynamo.

Assignments

10 a Electric fire/heater.
b Lamp, television set.
c Fan heater, motor, etc.

11

stores of chemical energy	batteries, oil, coal
fuels used at power stations	oil, coal
they spin round to generate electricity	dynamo, generator
where electricity is generated	power station

12 Kinetic energy (of the wheel) —electrical energy (in the wires)→ light energy.

Electrickery

Running the activity

This activity is an opportunity to summarise pupils' understanding of energy transfers, as developed in the first four spreads of this unit. A series of demonstrations should be positioned around the room, or you may prefer to demonstrate each one, with class discussion.

Differentiation

Sheets available are:

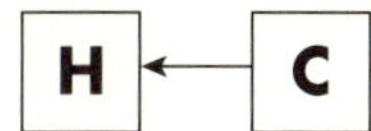

Core: This sheet sets up the activity, and asks pupils to draw energy transfer diagrams for each demonstration. If they can do this, then they will have shown that they can identify different forms of energy, and their transfers. The sheet includes a list of relevant forms of energy.

Help: This sheet provides energy transfer diagrams, which pupils have to match to the demonstrations. Each demonstration could be accompanied by a card giving a brief description, and pupils could then copy this into their tables.

Extension: Pupils might be asked to show appropriate demonstrations to the whole class, giving their explanation of the energy transfers involved.

Pupils might also be asked to identify situations in real life where each of these energy transfers is important.

AT1 programme of study

This is not an investigative activity. However, it does give pupils an opportunity to present their ideas through diagrams, using appropriate scientific terminology (Introduction to Programme of Study, **4c**).

Risk assessment

Only low voltage radiant heaters should be used. Pupils may be involved in these demonstrations; however, steam engines should be used only by the teacher, and only solid fuels should be used.

Equipment needed

One set of each demonstration

1 battery, bulb, connecting wires
2 battery, buzzer, switch, connecting wires
3 power supply, motor, wheel or fan turning, connecting wires
4 steam engine, dynamo, lamp, connecting wires
5 power supply, radiant heater, connecting wires
6 solar cell, mini-motor, wheel, connecting wires

Hidden energy

Book coverage

Key ideas	PoS
Gravitational energy is stored by anything that is lifted off the ground. Elastic energy is stored in anything that is stretched or squashed. Chemical energy is a third type of stored energy.	AT4.5f

Answers

Activities in text

a Ignite the fireworks, burn the paper, drink the milk, put the petrol in a car.

What do you know?

1

Chemical energy	Elastic energy	Gravitational energy
a battery	a stretched spring	a raised hammer
food	a squashed rubber ball	a skier at the top of a slope
fuel	a stretched rubber band	water behind a dam

2 Gravitational energy (stored by water) —kinetic energy (of moving water)→ electrical energy.

3 a A springboard stores energy, which helps you jump high. A pogo stick also helps you go high. In bungee jumping, you have gravitational energy at the start, and you rely on the rope to stretch (elastic energy) and so bring you to a halt.

b More and more elastic energy is stored each jump.

Extras: Energy for clocks

1 a Chemical energy.

b Elastic energy stored in a spring.

c Gravitational energy stored in a raised weight.

Assignments

13 a A battery stores **chemical** energy.

b A stretched rubber band stores **elastic** energy.

c Water behind a dam stores **gravitational** energy.

14 Elastic energy (stored in spring) → kinetic energy (of car).

15 a Gravitational energy.

b Chemical energy stored in the battery.

c Elastic energy stored in the stretched bow for archery.

Jumpers

Running the activity

A jumper is a bit like a pogo stick. It stores elastic energy when the rubber band is stretched, and jumps up in the air when you let go. There are several factors that affect the height to which it jumps: the length and elasticity of the band; how much it is stretched; the mass of the jumper. The Core sheet focuses on the degree of stretch.

Differentiation

Sheets available are:

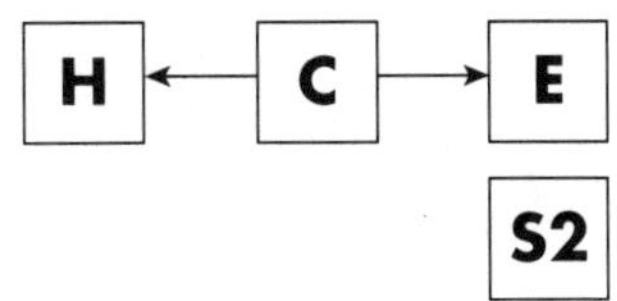

Core: This sheet shows how a jumper is constructed. Pupils then investigate how the height of the jump depends on the degree of stretching. They could be asked to draw a line graph of their results. The outcome is not itself surprising, but the relationship between amount of stretch and height is unlikely to be one of direct proportionality. (In principle, the height should be proportional to the square of the extension for a particular band.)

Help: This sheet shows how to mark a scale on the stick, and how to measure the height of the jump. It also includes a table for results.

Extension: This sheet asks pupils to measure the force needed to stretch the band. They should stretch each band by the same amount. They should find that the height jumped is roughly proportional to the force (assuming the different bands are stretched by the same amount).

AT1 programme of study

This investigation allows pupils to develop their skills in many aspects of **AT1**. They might be encouraged to make repeat measurements, as the height to which a jumper rises is not reliable (**AT1.2d**). In drawing conclusions, they should be encouraged to use the ideas of elastic energy and gravitational energy to explain their findings (**AT1.3g**).

Risk assessment

Pupils should be instructed not to lean over the jumper when it is about to be released, and to make sure that nobody else is leaning over it. Wear eye protection.

Equipment needed

For each group
- cotton reel
- 25 cm length of dowel (to pass through reel)
- Sellotape
- metre rule
- half-metre rule
- rubber bands
- newtonmeter (for Extension)

Sun for supper

Book coverage

Key ideas	PoS
Plants make food by photosynthesis. In photosynthesis, plants use light energy to turn water and carbon dioxide into carbohydrates and oxygen. Photosynthesis takes place in the leaves of a plant.	AT2.3a,b,c

Answers

Activities in text

a Light, air, water, warmth. Pupils may give oxygen/carbon dioxide. No credit to be given for 'food'.
b To capture as much sunlight as possible, for making food.
c The leaves turn green so the whole leaf can capture sunlight and make food. The plant may die if it can't make enough food to survive.

What do you know?

1 Plants make their own food in their **leaves** using **water**, **carbon dioxide** and energy from **light**. The way plants make food is called **photosynthesis**. They produce **carbohydrates** and **oxygen**.
2 It would die once it had used up all its stores of food, as without light it could not photosynthesise.
3 Plants need carbon dioxide to make food. When you talk you breathe out, so by talking to plants you give them extra carbon dioxide. This might let them photosynthesise more and so make them grow better.

Extras: Who grows tallest?

1

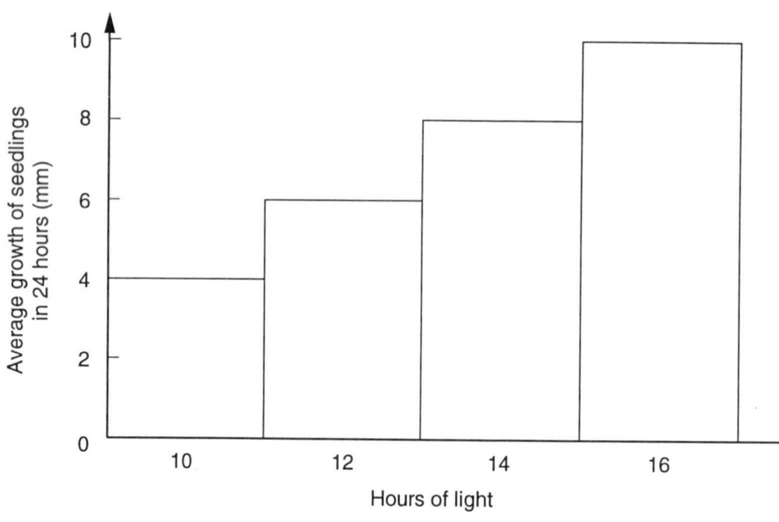

2 The plants exposed to light for the longest grew tallest.
3 The more light the seedlings had, the more they could photosynthesise. This means they could grow more, as they made more carbohydrate.

Assignments

1 **a** The sun or electric light.
b The soil.
c The air/the plant's own respiration.

2 **a** The caterpillars will eat most of the leaves, so the plant will die.
b Because the plant makes food in its leaves. More specifically:
- Carbon dioxide gets in through holes in the leaf surface.
- The special green colour in the leaves catches light energy.
- Carbohydrates are made in the leaves.

3 Paragraphs should demonstrate understanding of the process of photosynthesis.

Testing for starch

Running the activity

This is a relatively simple activity. Pupils could work individually or in pairs. The yellow-brown (sometimes described as red-brown) iodine solution turns a deep blue-black colour when exposed to starch.

Pupils will come up with the idea of using iodine to show that a plant has made starch in its leaves, but when they try this there will be no change in the colour of the iodine. The waterproof cuticle of the leaf prevents the iodine from getting into the leaf, and within the leaf the starch is contained inside the cells. It is also difficult to see the colour of the iodine against the green of the leaves. Therefore the fact that the iodine does not change colour does not show that the leaves have not produced starch.

Differentiation

Sheets available are: C

The ability of pupils to hypothesise as to why iodine does not show the presence of starch in their leaves may act as a differentiating mechanism.

AT1 programme of study

This activity requires pupils to follow a simple procedure with some degree of accuracy and to think about how their testing technique for the leaf can be improved (**AT1.2a**).

Risk assessment

Wear eye protection. Remind pupils about the dangers of eating in a laboratory. They should not nibble the bread provided for the experiment.

Equipment needed

For each pupil or pair

- test tube containing about 3 cm^3 of standard starch solution
- test tube containing about 3 cm^3 of water
- white tile
- small piece of bread
- bottle of iodine solution (less than 1 mol dm^{-3}) with dropper
- plants that have been exposed to light

Has your plant made some food?

Running the activity

Pupils plan an experiment in which they compare starch production in plant leaves that have been kept in the light and in the dark. They may decide to follow the procedure for testing leaves with:

a one leaf from the plant in the light followed by one leaf from the plant in the dark
b both leaves together (quicker, but provides less skill practice, and also the risk of confusing the leaves – they would need to choose distinctly shaped or sized leaves)
c a leaf with areas that have been in the dark and areas that have been in the light.

The experiment is probably best done with pupils working in threes, testing one leaf each. Where pupils find practical procedures difficult, it is best done as a teacher demonstration.

Differentiation

Sheets available are:

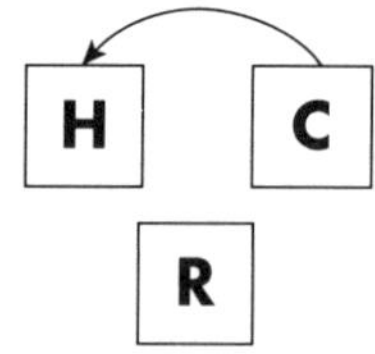

The Help sheet does not have the planning element, the practical instructions have been slightly simplified and there are questions to prompt the conclusions.

AT1 programme of study

The procedure described shows pupils what factors are considered in this experiment (**AT1.1d, e**). Pupils could be encouraged to consider the apparatus used and the scientific knowledge underpinning the procedures.

Risk assessment

Wear eye protection. It is very important that Bunsen burners are switched off before ethanol is brought into the laboratory, as it catches fire easily. The test tubes of ethanol should be heated in a large water bath at the front of the laboratory. At the end of the practical, waste ethanol should be collected in a large beaker for disposal down an outside sink, not tipped down sinks in the laboratory.

Equipment needed

For the class
- 2 plants that have been exposed to light for at least 2 days, labelled A
- 2 plants that have been kept in the dark for at least 2 days, labelled B
- 2 plants with parts of their leaves covered and kept in the light for at least 2 days, labelled C
- ethanol
- water bath at 90 °C

For each group
- tripod and gauze
- Bunsen burner
- heat-proof tile
- 100 cm^3 beaker
- 3 test tubes or boiling tubes
- tongs
- iodine solution with dropper
- white tile

Notes

1 Before the practical, it is worth getting a technician to check the experiment on a leaf that has been partly covered, because the results are not always clear.
2 After boiling in ethanol, the leaves need only a very brief wash in the hot water, otherwise they tend to fall to bits. Fragments can be collected and tested with iodine if the worst happens!

The root of it all

Book coverage

Key ideas	PoS
The roots of a plant hold it firm in the soil, and take up water and other substances that the plant needs to make its food. People add fertiliser to soil to help plants grow well. Nitrogen, phosphorus and potassium are important ingredients in fertiliser.	AT2.1e, 2.3d,e

Answers

Activities in text

a To hold plants in the soil, to get water from the soil, to get minerals from the soil (discourage 'food'), to hold the soil together.
b1 Nitrogen is needed for strong stems and good leaf growth (for protein formation).
b2 Phosphorus is needed for good root growth.
b3 Potassium is needed for root growth and green leaves.

What do you know?

1 Anchoring the plant, taking up water from the soil, taking up minerals from the soil.
2 Posters should be clear and colourful, with a thoughtful layout and clear explanations.
3 The investigation should include:
- a plant (preferably several plants) given house plant food in their water compared with a similar number of plants not given the food
- the plants observed over a period of time
- the plants to be as similar to each other as possible (ideally, cuttings of similar sizes from a single plant)
- other conditions (amount of light, water, etc.) to be kept the same for both sets of plants
- the average growth (increase in height) of all the plants in each group compared
- other measures of growth, e.g. number and size of leaves, might be suggested.

Assignments

4 a The roots hold the plant in the soil, so without roots it can't support the shoots. The roots take up the water and minerals that the plant needs to photosynthesise.
b Water enters the plant from soil through the root hair cells, and then moves across the root into the plant's transport system.
c Roots anchor the trees and hold the soil together. When trees are blown over, the soil can be blown and washed away.
d Large numbers of trees died, leaving the soil unprotected.
5 a Plants need minerals to be healthy and to grow, particularly to make proteins.
b Nitrogen, phosphorus and potassium.
c They replace the minerals used up by the plants by putting fertiliser on the fields, using either manure or chemical fertilisers.

Why roots?

Running the activity

This is a simple activity which the pupils should be able to set up very easily for themselves. It involves leaving the growing bean plants with their roots in water containing a red or blue dye for a few days. The plants can then be cut open and the dye seen in the tissue throughout. The main organisational point is that the beans need to be set up to grow for up to two weeks before the activity, as seedlings with roots and shoots are necessary.

Differentiation

Sheets available are:

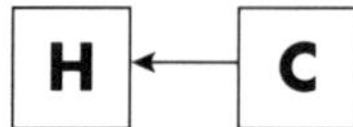

Differentiation is by outcome. The Help sheet gives an experimental set-up for pupils who do not know how to start.

AT1 programme of study

This is an opportunity for pupils to plan and carry out an investigation given a selection of apparatus. They will use apparatus safely and have to consider how to display their data clearly (**AT1.1a, 1.2a, 1.3a**).

Risk assessment

Pupils must not eat the beans.

Equipment needed

For each group
- well developed bean seedling with roots and shoots
- beaker
- blotting paper
- label
- water coloured with red or blue dye

Animals need food too

Book coverage

Key ideas	PoS
Animals cannot make their own food. They eat plants or other animals. The important parts of a person's diet are carbohydrates, fat, protein, vitamins and minerals. These must be eaten in the right amounts for a healthy, balanced diet.	AT2.2a,b,c

Answers

Activities in text

a Appropriate animals and diets.
b1 Thomas: fresh fruit and vegetables. Claire: fat.
b2 Suggestions for Thomas should include reducing his fat intake and increasing the amount of fruit and vegetables. Suggestions for Claire should include increasing her fat intake and possibly her overall food intake.

What do you know?

1 Human beings and other animals must eat **food** to get energy. If we eat too much we get **fat** as our bodies **store** the energy in the food. To be healthy, we need to eat a **balanced** diet with the right amounts of fat, **proteins**, carbohydrates, **vitamins** and **minerals**.
2 a Children are growing quickly and making new body tissue, which is mainly protein.
b Fatty foods contain a lot of energy in a small amount of food, so eating fewer fats is the easiest way to reduce the energy intake.
3 Because plants make their own food, in the quantities that are needed. They do not overproduce and they don't store extra food as fat.

Extras: Slimmer of the year

1 People get fat by eating more food than they need for energy. The excess is stored by the body as fat.
2 a Cut down the amount of fat and carbohydrate in her diet, as these are the high energy foods.
b If she did more exercise she would use up more energy, and this would help her to lose weight.
3 The points raised here will obviously vary quite a bit. They may well cover: wanting to look nice and fit into clothes, worrying about what other people think, wanting to look trendy or fashionable, trying to look like the models and actresses on television and in magazines.

Assignments

6 The table should be filled in with the sort of examples talked about in class.
7 The dietary record will be made with varying degrees of accuracy and diligence! Credit should be given to those who complete the task well, and a few minutes' discussion could be introduced as a follow-up. This could include ideas about making choices about what to eat, how those choices are largely managed by parents at home and peers at school, etc.
8 Carbohydrates are high in energy that the body can use very easily.
Protein is needed for growth and repair.
Fat is full of stored energy – it is much higher in energy than carbohydrates. We need some fat in our diets to keep us healthy, but too much can make us overweight and give us heart problems.
Vitamins and minerals are needed in minute amounts to keep us healthy.

Looking at labels

Running the activity

The success of this activity depends on having plenty of food packages (either full or empty). Pupils can bring in a couple of examples each, but have your own adequate supply handy in case everybody forgets! In the Core activity, you may need to explain to some pupils that labels may give more information than they need. The questions are designed to make the pupils think about what they are doing, and link it to the theory of food and diet, so that it is not merely a mechanical exercise of tabulation. In question **6**, the answer looked for is water.

Differentiation

Sheets available are:

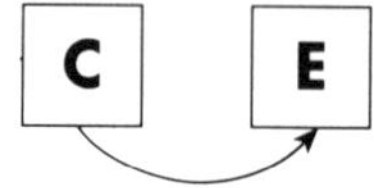

For pupils who work rapidly through the labels, and the more able pupils generally, the Extension sheet looks at other information that may be on labels, and why. Pupils produce a bar chart comparing the levels of information given on labels, and consider why such differences exist.

AT1 programme of study

This activity gives pupils the opportunity to handle data from the commercial world. They look for relevant information and for patterns in data and express it qualitatively or quantitatively (**AT1.3d**). At Extension level they use graphs to display data (**AT1.3b**).

Equipment needed

For the class

- as wide a range as possible of food containers (tins, packets, etc.) with labels
- plain paper
- graph paper

Healthy eating

Book coverage

Key ideas	PoS
Tiny amounts of minerals and vitamins are needed in the diet for a healthy body. Fibre is needed in the diet to keep food moving through the gut properly.	AT2.2a,b

Answers

Activities in text

a

Iron	Calcium	Vitamin C	Vitamin D
beef	milk	oranges	cod liver oil
lamb	cheese	lemons	cream
apricots	cream	cabbage	oily fish
spinach	yoghurt	cranberries	

We need iron for the blood to carry oxygen.
We need calcium for strong bones.
We need vitamin C for healthy gums and skin.
We need vitamin D for strong bones and teeth.

What do you know?

1 **Vitamins** and **minerals** are substances which our bodies need in **tiny** amounts to be able to work properly. Milk is a good source of **calcium** for healthy bones and teeth, but it does not contain the iron the **blood** needs to carry oxygen properly. **Fibre** is food material which we cannot use. It keeps food moving through the **body** regularly.

2 Leaflets and posters should be clear and colourful, demonstrating an understanding of the rolè in the body of the food substance chosen.

Extras: In at one end, out at the other!

1 Food made of plant material contains lots of fibre. This can't be digested. It is bulky and absorbs water, so there is a lot of waste material. It also keeps the muscles of the gut working.

2 The plant-based diet, with its high fibre levels, is more likely to make it possible for waste to pass quickly through the body.

3 Elephant – large amounts, soft, as it eats plants.
Lion – small amounts, hard, as it eats animals.
Rabbit – relatively large amounts, soft, as it eats plants.
Zebra – large amounts, soft, as it eats plants.
Cow – large amounts, soft, as it eats plants.
Owl – small amounts, hard, as it eats animals.
Fox – small amounts, hard, as it eats animals.

Assignments

9 **a** Calcium is needed to make our bones hard and strong.
b Iron is needed for the blood to carry oxygen properly.

10 **a** Because the bones are soft and cannot take the person's weight without bending.
b Vitamin D.

11 The message can be illustrated. It needs to get across the point that food moving regularly through the gut is a good thing for health. This movement of the gut is encouraged by plenty of fibre (indigestible material) in the diet. Thus changing to Branno, a high-fibre cereal, will help to prevent constipation and be better for all-round health.

Food for thought

Book coverage

Key ideas	PoS
Plants are producers. They produce new food from carbon dioxide and water using energy from the sun. Consumers are living organisms that need to eat other organisms to get their energy to grow. Food chains are the links between different animals that feed on each other and on plants.	AT2.2b, 2.3g, 2.5d

Answers

Activities in text

a Appropriate foods in plant and animal boxes.

What do you know?

1 **a** Plants are called producers because they produce new food from carbon dioxide and water using energy from light.
b Consumers need to eat other living organisms to get their energy.
2 Posters should be colourful and clear, demonstrating understanding of food chains.
3 Food chains always start with a plant, and plants are dependent on energy from the sun to make their own food.
4 Appropriate food chains.

Assignments

12 **a** Pine cones → squirrels → pine martens.
b Plants are producers because they make food by photosynthesis using energy from the sun. Animals are consumers because they either eat plants or eat animals that eat plants.
13 The presentation should be brief, accurate and punchy. The examples chosen should illustrate feeding relationships well, and be feasible for handling in the television studio.

What the owl had for dinner

Running the activity

If you can get hold of real owl pellets for the pupils to see, that would be ideal. The main difficulty with real pellets is that it takes pupils a long time to extract the bits and pieces! For this reason, four different possible 'owl pellets' are provided so the pupils can cut out and identify the remains on their sheet.

Differentiation

Sheets available are:

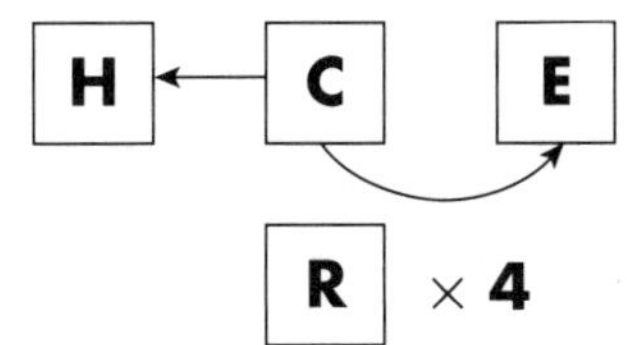

Differentiation will come about by the pupils' competence at identifying and then constructing chains and webs. The Help sheet gives support for the construction of the food web. The Extension sheet considers the scientific validity of looking at just one pellet, and encourages the pupils to think about how they could improve the reliability of the exercise.

AT1 programme of study

Pupils handle information and look for patterns in the data. They draw conclusions and make predictions building upon their knowledge (**AT1.1c**). They go on in the Extension to suggest how the situation could be investigated, the sufficiency of the data for firm conclusions and how the experiment could be improved (**AT1.1a, 1.4a, c**).

IT potential

There are lots of programs based on the ecological theme of food chains and food webs. Some of these could be used at their simpler levels at this stage in the course, leaving more complex ideas of limiting factors and human intervention until later.

Risk assessment

If real owl pellets are used, disinfect the pellets, wear gloves and use forceps.

Equipment needed

For each pupil

- Petri dishes (for the piles of cut-out paper remains)
- scissors

The web of life

Book coverage

Key ideas	PoS
The habitat of an animal or plant is the place where it lives. A food web is a model of a habitat showing how the animals and plants in it are connected.	AT2.5d

Answers

Activities in text

a,b There are a variety of ways of linking the chains to make webs.

What do you know?

1 Producers: land plants (grasses, reeds); water plants (tiny water plants, pond weed).
Consumers: all animals.

2 Food chains are too simple. Animals don't eat just one type of animal or plant, so lots of different organisms are connected together. Food webs show this more clearly.

Extras: Woodland webs

1 It appears that the rabbits will die because there is no grass. Then the foxes will die because the numbers of rabbits will fall.

2 If a disease kills off the rabbits, the foxes will die as they will be short of food. The grass will grow very long as there is nothing to eat it.

3 There should be as many links as possible in the web.

4 If the amount of grass in a web increased because rabbits had died out, the grass would soon be eaten by the many other animals that also feed on grass. If the number of rabbits changed, then there would be a slight effect on the grass and the foxes, but not much because so many other organisms are involved. If there was an increase in rabbits, more grass would be eaten and so the numbers of other small grass feeders might fall. The number of foxes might go up as there would be plenty of rabbits to feed the cubs. If the number of rabbits fell, fox numbers might drop a little, but not for long as other small grass feeders would increase with their greater share of the grass.

Assignments

14 This food chain is actually part of a much bigger food web. Foxes eat lots of other animals as well as rabbits, so when rabbits disappear they just eat more of the other animals. Rabbits aren't the only animals to eat grass, so when the rabbits disappeared, other animals just ate more grass and did better – ready to feed the foxes!

15 Appropriate webs. They should contain as many links as possible.

Separating mixtures

Book coverage

Key ideas	PoS
Solids of different sizes can often be separated through a sieve. If a mixture of liquid and solid settles out, the liquid can be decanted off. Solids can be filtered out of liquids.	AT3.1i

Answers

Activities in text

a Each example here would involve picking out by hand. The simple examples given could be tried out if required.
b1 Separate in water – snooker balls sink, ping-pong balls float.
b2 Separate with a magnet – iron filings are picked up, copper filings are not.
b3 Separate with a carpet tile – the aniseed balls roll off, the Polo mints do not.

What do you know?

1 You would need to pick them out individually, which would be a very slow process.
2 **a** To trap any dust or dirt.
b The dust or dirt gets trapped by the fibres, as in filter paper.

Extras: Separating oil and water

1 Any method that involves pouring off from below, for example, using a separating funnel or partitioned jug.

Assignments

1 **a** The sand and some of the mud settle out. The clearer water can then be decanted off.
b To filter out the suspended mud.
c The filter bed gets clogged up with mud.
2 **a** Cafetiere – the plunger is a sieve. The grouts are trapped beneath it while the liquid coffee passes through.
b Filter method – the grouts stay in the paper as the residue, and the coffee is the filtrate.
3 The seeds are 'heavier' than the chaff. The seeds fall back to the ground, but the chaff is blown away.

Calamine alert!

Running the activity

Working in pairs or individually, most pupils should be able to tackle the Core sheet given the information in the pupils' book. Sufficient sets of the standard apparatus should be visibly available. Help sheets could be given out to targeted pupils, or left openly available in case they get stuck.

The Extension sheet continues from the Core sheet. Pupils should be able to view different filter papers under a microscope. They can then plan their experiment and answer the associated questions.

Differentiation

Sheets available are:

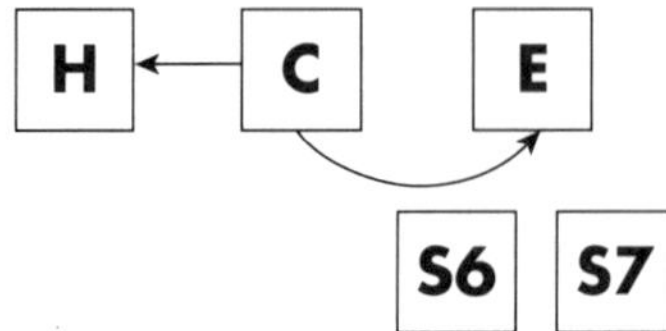

Help: This sheet gives a step-by-step breakdown of the techniques needed.

Core: Most pupils will probably need to practise standard filtration techniques. The Extension activity will then follow on.

Extension: Any pupils who do not need practice in simple filtration could move on directly to this sheet. They could then perform their experiment and write it up before moving on to the questions.

AT1 programme of study

Core: This covers practice of a standard technique (filtration) in a new context. It is presented as a problem-solving exercise (**AT1.1h, 1.2a, 1.3e**).

Extension: This opens out the idea of filtration, giving pupils the opportunity to consider different types of filter paper in terms of the effectiveness of filtration versus speed and/or cost. Pupils need to isolate factors to consider and ensure fair testing at a low level (**AT1.1a, d**).

Risk assessment

General risks concerning the use of glassware and chemicals apply. Wear eye protection.

Equipment needed

For each pair or each pupil (Core)
- calamine lotion (about 200 cm^3)
- filter paper and funnel
- clamp stand
- beakers

For each pupil (Extension)
- open weave 'food use' standard grade and finest grade filter papers
- access to a microscope

Note
Water down the calamine lotion to make filtration simpler, and give the instruction to shake it before use. Commercial calamine lotion often has chemicals added to keep it in suspension, so a specially prepared slurry of zinc carbonate powder (or even chalk) may be easier to use.

Solids in liquids

Book coverage

Key ideas	PoS
Some solids dissolve in water to form a solution. The water is the solvent, and the solid is the solute. If no more solid will dissolve, the solution is saturated. You can speed up dissolving by crushing the solid or heating or stirring the liquid. Solids that will not dissolve are insoluble.	AT3.2a,b

Note: The idea of speeding up dissolving is introduced as a precursor to the more advanced work regarding the speed of chemical reactions, in a more accessible context.

Answers

Activities in text

a Washing powder, salt, sugar, instant coffee are soluble.

What do you know?

1 The solid that dissolves – solute.
The liquid that does the dissolving – solvent.
It's full up with solid – saturated.
It will dissolve – soluble.
It will not dissolve at all – insoluble.
A liquid with solid dissolved in it – solution.
2 Heat, stir, crush.
3 The sugar frosting has smaller grains than granulated sugar and so dissolves faster, giving a rapid 'flavour burst'.

Extras: More about solvents

1 **a** Gloss paint is insoluble in water.
b White spirit (or turpentine).
2 Test it with different solvents such as meths, white spirit, etc.

Assignments

4 **a** The salt dissolves.
b You can still taste it.
5 **a** Heat or stir it.
b Caster sugar has finer grains – like crushed granulated sugar.
c It is more expensive.
6 The residue from the cola is water soluble, but the varnish is not.

Dissolving

Running the activity

Core: Use this sheet as an introduction for all pupils. The thinking points could be discussed as a class or in groups of three or four. Ideas could be pulled together before the practical starts, or through teacher discussion with the groups.

Help: Any groups having problems with their own plan could be directed to this sheet, which gives a step-by-step approach.

Extension: This sheet is designed as a planning exercise for groups to move on to as they finish the Core sheet. It could, however, be expanded into an additional practical exercise for all. In this case, the lead-in exercise could be class or group discussion with feedback. The findings of the different groups could also be summarised at the end.

Differentiation

Sheets available are:

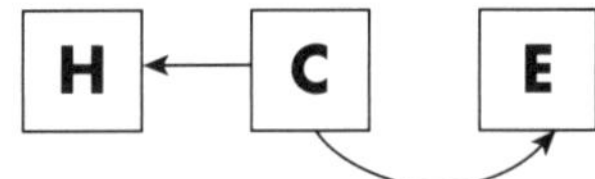

AT1 programme of study

Core: This is a simple practical exercise to look at variation in solubility, with the emphasis on recording and displaying results (**AT1.1d–f, 1.2e, 1.3a**). The Core sheet gives scope for individual planning, while the Help sheet provides the step-by-step support needed by the less able.

Extension: This is an opportunity to plan an investigation, concentrating on the recognition and isolation of controlling factors (**AT1.1c–f, 1.3e, f**).

Risk assessment

Copper sulphate is harmful if eaten. Potassium nitrate is an oxidising agent. Do not taste the sugar. Check any specific problems with any alternative chemicals used. Wear eye protection.

Equipment needed

For each group (Core)
- water-soluble solids such as copper sulphate, salt, potassium nitrate and sugar, to give a range of solubilities
- small spatulas
- test tubes and rack
- corks or bungs
- access to water

Note
A trial run by a technician to find the number of spatula measures needed for the given combination of test tube size and spatula size might be useful.

For each group (Extension, if developed to a full practical)
- large crystals of copper sulphate
- small crystals of copper sulphate
- pestle and mortar
- balance (one or two decimal places)
- Bunsen burner, mat, tripod and gauze
- beakers
- measuring cylinders
- glass rods

Separating solutions

Book coverage

Key ideas	PoS
Water will slowly evaporate into the air. Evaporation happens faster if the water is heated. If a solution is allowed to evaporate, the solute is left behind.	AT3.1i

Answers

Activities in text

a B: Sand and mud settle out.
C: The clear liquid is decanted off.
D: Any remaining solid is filtered out.
E: The water is evaporated off to leave the salt.
Given the information in the text, pupils could perform their own purification of rock salt:

Equipment needed (per group)
- rock salt
- filter paper
- filter funnel and stand
- flask
- evaporating basin
- Bunsen burner
- tripod and gauze

Notes
If the brine is boiled rather than left to evaporate, care should be taken to avoid the salt spitting at the end. Turn off the Bunsen burner when most of the water has boiled away. Wear eye protection.

What do you know?

1 Salt is **soluble** in water, but mud and sand are not. Rock salt is mixed with water to **dissolve** the salt. Most of the mud and sand **sink** to the bottom, so the salt **solution** can be run off. It is then **filtered** to get rid of the rest of the mud. The clear salt solution is **evaporated** to give crystals of salt.
2 It evaporates into the air.
3 The water evaporates, keeping the air around the fern moist.
4 Let the water evaporate away.

Extras: Where does sugar come from?

1 Suitable flow diagram showing the stages which must include dissolving the sugar, filtering the liquid and evaporating the filtrate.
2 It disposes of the waste and saves energy (and therefore money), making the sugar cheaper.

Assignments

7 The water had evaporated.
8 The plant takes in water vapour from the air.
9 **a** The salt is brought in by the rivers. The water evaporates away, but the salt is left behind.
b **i** The water has completely evaporated away, leaving behind the salt.
ii The Great Salt Lake is getting smaller as more water evaporates and more salt is left behind.

Dissolving in reverse!

Running the activity

Start with a discussion of what happens to the solute as a solution evaporates.

In 'Slowly does it', groups of two to four pupils dissolve some copper sulphate, pour it into a small dish, label it and leave it until the next lesson.

'As fast as you can' can be a teacher or technician demonstration (or series of demonstrations) running concurrently with 'Slowly does it'. Alternatively, if pupils are already familiar with the use of a microscope (and enough microscopes are available), they could perform this in groups.

Differentiation

Sheets available are: **C**

S6 **S7**

This is a simple experience of crystal growth at high or low speed for all.

AT1 programme of study

The activity involves the use of standard equipment plus a microscope (**AT1.2a**).

IT potential

A CD-ROM search could be carried out for examples of naturally occurring crystal shapes to compare with the pupils' crystals.

Risk assessment

General risks concerning the use of glassware apply. Copper sulphate is harmful if eaten. Wear eye protection. The warm slides should not be allowed to get so hot that they could burn pupils.

Equipment needed

For each group ('Slowly does it')
- copper sulphate
- water (distilled water is preferable)
- small dishes (Petri dishes are ideal)
- labels or paper

Notes
1 Only relatively small amounts of copper sulphate are needed per group – 2 or 3 small spatulas in a test tube half-full of water.
2 The activity could extend to other crystals such as alum, chrome alum, nickel sulphate or copper chloride.
3 A safe area is needed to leave the solutions between lessons.

For each class or group ('As fast as you can')
- the solution(s) above
- simple, low-powered microscopes
- warm (but not too hot) slides
- glass rods for transferring a small amount of liquid onto the slide

Evaporation

Running the activity

Use the thoughts on the Core sheet as an introduction. Possible factors can be discussed as a class or in groups of three or four pupils. Factors can be selected either by discussion with groups or by class allocation.

Plans will need to be agreed, checking that other factors are controlled. Groups having difficulty could be given one of the two structured experiments on the Help sheet.

Groups should summarise their findings and report back to the class at the end.

Differentiation

Sheets available are:

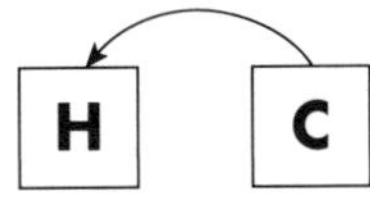

Start with the Core sheet at first, giving verbal support where required or providing the structured Help sheet for those who need it.

AT1 programme of study

Core: This concentrates on evaporation, in the context of a full AT1 investigation, based on the recognition and isolation of controlling factors. The familiar context (washing drying on the line) allows some early development of the higher level ideas of speed of reaction (**AT1.1c–f, 1.2a, b, e, 1.3e, f**).

Help: This sheet splits the Core activity into two structured activities, either of which may be undertaken by less able pupils.

Risk assessment

If beakers of very hot water are used, appropriate warnings should be given. Propanone is highly flammable. Wear eye protection.

Equipment needed

For each group (Core)
The possibilities here are very wide. Some pupils may wish to dry out small pieces of cloth on a miniature line, for example. Equipment needed might include:

- apparatus as in 'Help' below
- measuring cylinders
- balance (2 decimal places)

For each group (Help)

- propanone
- 3 watch-glasses
- 3 beakers
- access to hot water
- access to hair drier with a cold jet, or fan (electric or hand-powered!)

Getting the water back

Book coverage

Key ideas	PoS
Water vapour in the air can condense back to form liquid water if it is cooled. Pure water can be collected from a solution by distillation.	AT3.1i

Answers

Activities in text

a The 'cloud in a jar' can be demonstrated. Be careful pouring hot water into the jar if it is not Pyrex.
b To cool and condense the vapour.
c The 'surviving in the desert' apparatus can also be demonstrated, using a bowl of damp soil instead of a hole in the ground.

What do you know?

1 Water can slowly turn into a gas at low temperatures by **evaporation**. If you keep heating, the water starts to **boil**. A lot of **water vapour** escapes into the air. If it is cooled it **condenses** and turns back to **liquid** water.
2 Water condenses on the windows and runs down onto the wood.
3 Distil the ink to give water.

Extras: The water cycle

1 Moist air rises and cools over the mountains. As it cools, some of the water vapour condenses to form clouds. The droplets collect and fall as rain.

Assignments

10 Warm, moist air from the house was chilled on the cellar walls. Some of the water vapour condensed to form water droplets.
11 Water condenses from the air as it is cooled.
12 **a** Water vapour.
b Water droplets.
c He is right. The clouds are masses of liquid water drops. Steam is the invisible gas that forms when water boils.
13 **a,b** The sea water is heated. Water evaporates from it. The water vapour condenses on the pipes full of cold water. The drops of distilled water are collected.
c It is very expensive. Our water is distilled naturally as rain (solar powered).

Pure water from ink

Running the activity

This activity follows on from the pupils' book. Groups of two or three pupils should set up the apparatus and use it to obtain clear distilled water. It may be worth talking through the key steps on the sheet and stressing the safety aspects.

Discuss progress with the groups, giving hints and pointers where required.

Bring the class together to discuss problems with the apparatus. It is difficult to get all the steam to condense. Then demonstrate distillation with a Liebig condenser. Try to get the class to explain why this is better and how it works.

Diagrams are available on the Resource sheet for labelling, and these could be used to introduce the idea of annotation.

Differentiation

Sheets available are:

C

S4 **R**

Differentiation is by outcome – clarity of the distillate.

AT1 programme of study

The simple distillation exercise described in the book (water from ink) is carried out to practise basic heating and distillation skills (**AT1.2a**).

Risk assessment

Issue warnings about Bunsen burners and steam. Take care that the apparatus is sound and that no tubes have become blocked. Watch out for sucking back if the delivery tube dips into the distillate. Wear eye protection.

Equipment needed

For each group
- standard distillation apparatus
- ink

Note
Try to match the diagram set-up as closely as possible to avoid confusion. Ensure that some form of anti-bumping agent is used.

For the class
- standard Liebig condenser set-up

What's in that mixture?

Book coverage

Key ideas	PoS
You can separate out mixtures of dyes using chromatography. Chromatography can be used to identify the dyes in a mixture.	AT3.1i

Answers

Activities in text

The chromotography used to check the dyes used in sweets shows that A and C contain yellow dye and should be avoided. The ink-on-filter paper chromatography demonstrates the basic process in a simple but enjoyable way, which could be performed by the class. All that is needed is a supply of filter papers, some felt-tipped pens, dropping pipettes and water.

a Inspector Chroma sets up an AT1.1 plan for an investigation, which could be carried through in a practical session. (In practice, this exercise is quite delicate and fiddly to do well, so a simpler practical version involving the standard technique is outlined in Activity 6e.)

What do you know?

1 When water soaks up into paper, it carries any soluble **dyes** with it. The different dyes are carried along at different **speeds**. This **separates** out the different dyes. The process is called **chromatography**.

2 a Dissolve the dye in water, put spots of the dye on chromatography paper (as in text), put the paper in water. Mixtures will separate out into different colours.

b Compare the named dye chromatographs to those from the Smarties.

Assignments

14 a To release the substances from the leaves and start to dissolve them.

b To (safely) speed up the dissolving of the substances.

c To remove any solid material, giving a clear filtrate.

d To concentrate the substances.

e Chromatograpy separates out any different substances dissolved in the alcohol.

Shadow challenge

Running the activity

This activity gives pupils an opportunity to observe shadows and how they change. It requires them to think about how light is intercepted by objects to give shadows. It can also serve as a reminder of how the Sun's apparent position changes during the day; you may wish to remind pupils of how this relates to the Earth's rotation.

The sheet suggests that they might demonstrate the changing shadow length to other members of the class. A group of pupils could be asked to prepare this in advance of the lesson.

It may help pupils to show them the circle–square shadow. The Plasticine should be moulded into a cylinder whose length equals its diameter. Pupils might demonstrate their Plasticine models to the whole class: you will need to set up a greaseproof paper or cloth screen, with a projector or overhead projector behind it.

Differentiation

Sheets available are:

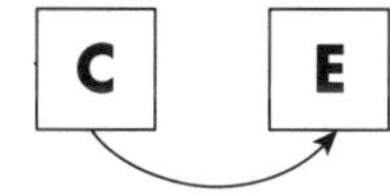

AT1 programme of study

Core: Pupils should record their observations (**AT1.2b, e**); they should note how the length, width and angle of the shadow changes as the 'Sun' moves across the sky (**AT1.3a**).

Extension: This sheet requires pupils to draw single rays from the Sun to show where the shadows will be formed. They then need to be able to carry out simple scaling calculations to deduce the length of the shadows.

Risk assessment

Take care with flexes if electric lamps are moved around the laboratory.

Equipment needed

For each group
- cardboard
- scissors
- movable lamp, e.g. torch, anglepoise lamp (properly earthed), microscope lamp
- lump of Plasticine, approximately 5 cm cube
- greaseproof paper screen

For the class (optional)
- projector or overhead projector
- large cloth or greaseproof paper screen

Long time coming

Book coverage

Key ideas	PoS
Light travels very fast, much faster than sound.	AT4.3a,c

Answers

Activities in text

a You see the bat hit the ball because light rays come to your eyes. You hear the bat hit the ball because sound vibrations come to your ears. Light rays travel faster than sounds.

What do you know?

1 Fastest to slowest: a ray of light; a sound; a racing car; an Olympic sprinter; a tortoise.
2 **a** The Moon is closer.
b The Sun.
c A very distant star.
d Further at night – surprisingly!
3 **a** The light travels faster than the sound.
b 2 kilometres.

Assignments

10 Slowest to fastest: a toy train; a cyclist; a sound; a ray of light.
11 **a** Light doesn't travel instantly from the Sun to Earth. It's a long way, and so the light takes 8 minutes (approximately).
b The Moon is closer to us than the Sun is.
12 You light the blue touchpaper and stand well back.
You see the rocket rush up into the sky.
You see the rocket explode into many stars.
You hear the rocket explode.
You hear the dead rocket land on the greenhouse next door.

A message to Planet Mirth

Running the activity

This activity is intended to reinforce the idea that light travels very fast, but that it can take a significant amount of time for light to travel over astronomical distances. Each group of pupils should be divided in two; some stay behind on Earth, the others go to Planet Mirth. Mirth is an imaginary planet, closer to Earth than Mars but further away than the Moon. (This is because messages to Mars take 5 minutes, which is too long, and the Moon is too close.) Mirth should be located at some place within the school or its grounds which is visible from the laboratory. Pupils can send messages back and forth without being able to speak directly to one another. The light messenger must take 1 minute to get from Earth to Mirth.

As an introductory demonstration, you might like to show the 'cricket-match' effect: have a pupil bang two blocks of wood together at the other side of the playing field (if available). Is there a noticeable time difference between seeing the action and hearing the bang?

Differentiation

Sheets available are:

This sheet includes instructions for the activity, together with questions which can form the basis for a discussion to sum up the activity. Pupils should conclude that the appreciable time taken by messages between Earth and Planet Mirth could be a problem when quick reactions are needed. For example, the spacecraft could not be controlled directly from Earth.

AT1 programme of study

This is a simulation, rather than an investigative activity.

Equipment needed

For each group
- paper and pencils for messages

For the class (optional)
- wooden blocks

Bouncing light

Book coverage

Key ideas	PoS
Light reflects well from smooth surfaces. It is scattered by rough surfaces. A source of light gives out its own light. We can see other things because light scatters off them into our eyes.	AT4.3d

Answers

Activities in text

a Users of mirrors include actors, drivers, security guards, dentists, etc.
b1 B is the rougher side.
b2 A is the smooth side.
b3 A will reflect the light better.

What do you know?

1

mirror	a good reflector of light
firework	a bright source of light
pencil	not a source of light
scattering	happens when light rays fall on something rough

2

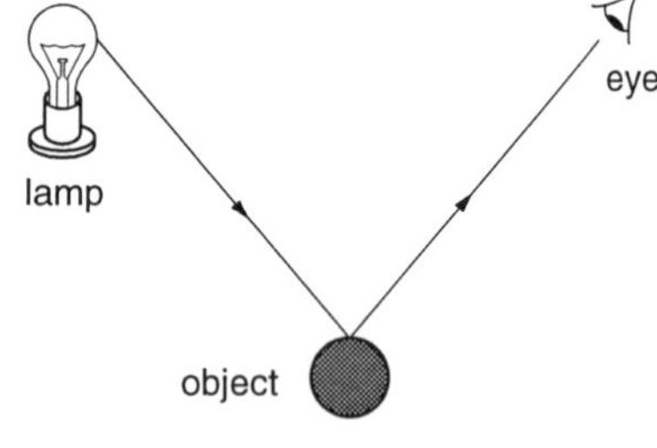

3

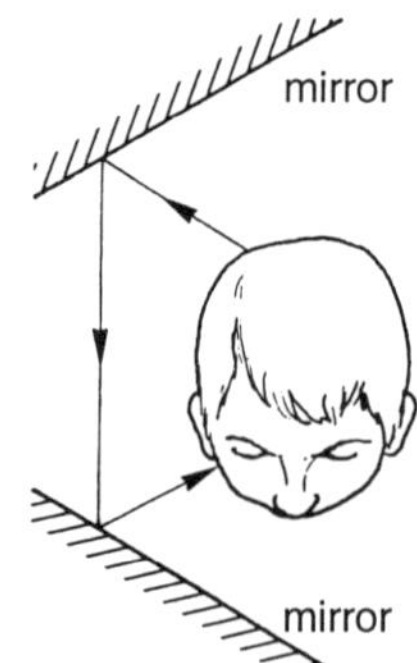

Extras: Source or scatterer?

1 Find a pitch-black room. Take the things in individually. If you can see them, they are sources of light.

Assignments

13 a A smooth surface is a **good** reflector of light.
b A **flame** is a source of light.
c Light is scattered by a **rough** surface.
d The Moon is a **reflector** of light.

14 a Seeing inside, upstairs, behind the bus.
b Putting on costumes and make-up.
c Security mirror.
d Reflecting the Sun's rays to attract the attention of passing ships.

15

Mirror, mirror

Running the activity

This activity gives pupils a chance to try out some interesting effects using mirrors. They are asked to record the position of the mirror and to measure some angles, skills which will be useful when they tackle the law of reflection later.

Differentiation

Sheets available are: 

All pupils will benefit from being shown how to draw a pencil line marking the position of the mirror, and how to measure angles with a protractor.

What they should observe:

1 The angle required is 90°.
2 With the mirror across the diagonal of the figure, they will see a square. Reversing the mirror shows an equilateral triangle.
3 The word DIOXIDE is symmetrical along a horizontal axis, as are the letters B, C, H and K, if written appropriately.
4 Normally we appear to be left–right reversed in a mirror. With two mirrors, we are reversed again; the effect is disconcerting.
5 For a six-fold kaleidoscope, the angle between the mirrors must be 60°. Pupils could be encouraged to work out how to produce a circle, a hexagon and a star.

AT1 programme of study

While pupils are developing skills in observation and measurement (**AT1.2b, c, d**), they should be encouraged to make accurate observations and measurements.

Equipment needed

For each group

- 2 plane mirrors
- wooden block, bulldog clip or Plasticine for stand
- protractor
- plain paper

More about water

Book coverage

Key ideas	PoS
Water can be a solid (ice), a liquid or a gas (steam), depending on its temperature. Ice melts at 0 °C (its melting point) and water boils at 100 °C (its boiling point). These changes are reversible.	AT3.2c,d

Answers

Activities in text

a Water is ice when cold, liquid water at ordinary temperatures and gas when hot.
b1 A: the ice is melting. B: the water is boiling.
b2 Ice warms up until its melting point, and then melts. The water gets hotter until its boiling point, and then boils.
b3 Boiling water in the container.

What do you know?

1 0°C and 100 °C should be marked.
2 **a** Water turns to ice at 0 °C.
b Water is liquid at 37°C, body temperature.
3 Gas.
4 Water boils at 100 °C, not 80 °C.

Extras: Rain, snow and hail

1 Ice crystals fall, warm up and start to melt, picking up a new layer of water. They are then blown back up into the cloud and the new layer freezes, and so on.
2 **a** Successive layers of ice.
b It has been back up through the cloud five times.
3 Hailstones build up from layers of frozen rain. On a freezing cold day, there will be no liquid water.

Assignments

1 **a** No.
b She can heat up the snow and turn it to water.
2 **a** No.
b Boiling water stays at 100 °C, no matter how long it is boiled.
3 It turns to ice, forming frost.
4 You can boil water in a paper cup as the paper will be kept cool by the water. The water will never get above 100 °C, so it will not reach the 200 °C needed to set the paper alight.
(Note: if the cup is not full, the paper above the water level may burn. Warn pupils not to try this for themselves.)

From ice to steam

Running the activity

This activity leads on from the pupils' book, and introduces the practicalities of using a thermometer to monitor temperature changes. Groups of two or three pupils can select apparatus and perform the experiment. Bring the groups together at the end to compare graphs and draw conclusions.

Differentiation

Sheets available are:

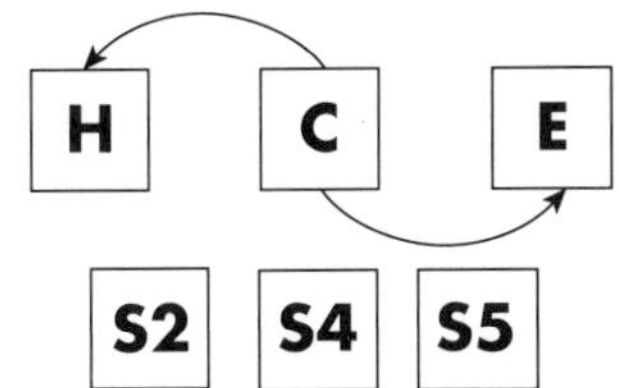

Core/Help: All pupils can try this experiment, either from the Core sheet or the more structured Help sheet. Differentiation will be by outcome, according to accuracy of readings, quality of graph produced and level of understanding of the significance of the results.

Extension: This sheet could be used as an 'extra', or the opening information could be used to prepare one or two groups for the Core experiment.

AT1 programme of study

The activity covers data collection, presentation and analysis (**AT1.2a–c, e, 1.3a, b, d, e, g**).

IT potential

Data-logging equipment could be used for the Core experiment by one or more groups, either for those pupils skilled in the use of standard apparatus, or in rotation so that all pupils have the opportunity to use the data-logging apparatus over a series of experiments.

The suggestions for alternative experiments are designed to show the advantages of IT usage in less straightforward situations.

Risk assessment

Safety warnings should be given concerning the use of boiling water. Wear eye protection. Beware of unstable apparatus (a beaker on a tripod with a tall thermometer in it). For the Extension activity, make sure probes and wires are suitable for use in an oven if that option is taken up.

Equipment needed

For each group (Core/Help)
- Beaker half-full of freshly crushed ice from the freezer
- thermometer (-10 – 110°C)
- tripod and gauze
- Bunsen burner
- stopclock
- paper towels

For each group (Extension)
- temperature sensor(s)
- interface box
- computer and appropriate software

Note
A printer is useful. 'Obsolete' computers are often perfectly adequate for data logging. Ideally, the software should display the incoming data in graphical form during the experiment. Free-standing data-collection units such as the Vela or Emu are not ideal for this, as they do not give the immediate visual feedback.

Changing state

Book coverage

Key ideas	PoS
Different substances have different melting and boiling points. Whether something is a solid, liquid or gas at room temperature depends on its melting and boiling points. Changes of state such as melting and boiling are reversible. Only pure substances have definite melting and boiling points.	AT3.2c,d

Answers

Activities in text

a Examples such as oil or paraffin (not substances such as vinegar or bleach which contain water).
b Wax, as it melts cleanly at one temperature.
Melting chocolate and wax over a beaker of boiling water is a good practical to show pupils the difference between pure substances and mixtures. If carried out, pupils should describe what they see happening as carefully as they can. Use stearic acid for the 'wax' and test the chocolate in advance – some turn to a soft mousse-like texture rather than forming a more liquid part. Both substances can be melted on the same watch glass, but put the wax in the centre and the chocolate higher, near the edge, otherwise the wax will melt and run into the chocolate.

What do you know?

1 **a** Solid.
b Liquid.
c Gas.
d Gas.
e Liquid.
f Solid.
2 The metal is heated until it melts. It is poured into a mould, where it cools and takes the shape of the mould, making a cast.
3 Below 40 °C, the solid grease sticks to the plates and is hard to remove. Above 40 °C, it melts and this loosens it from the plates.
4 No. Pure alcohol would boil at a constant temperature (approximately 80 °C).

Assignments

5 **ai** Tungsten. **ii** Iron would melt.
bi Mercury. **ii** It is liquid at room temperature.
c It would melt.
di It melts. **ii** It does not melt.
e You can melt solder with the soldering iron. The solder then cools and sets, joining the wires together.
f The zinc boils before the copper melts.
6 The sugary sap does not freeze at 0 °C, so the plants are not damaged by a slight frost.

Working with wax

Running the activity

'Sealing a letter' and 'A shell cast' are simple experiments to give experience of the practical uses of change of state (compare the casting of metals). These experiments could be skipped by stronger candidates, though they might object to missing the fun.

'The melting point of wax' gives the opportunity to produce a straightforward plan. This could be done in groups of two or three pupils. The Help sheet could be held in reserve unless obviously required.

Differentiation

Sheets available are:

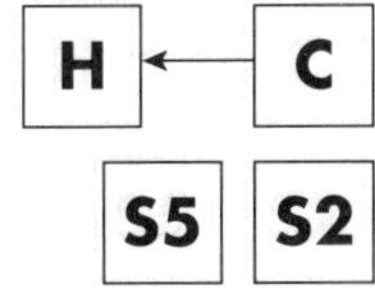

Differentiation is by degree of support needed. Pupils should be allowed to develop their own plan if possible. The Help sheet provides structured support for those that need it.

AT1 programme of study

The activity covers data collection, presentation and analysis (**AT1.2a–c, e, 1.3a, b, d, e , g**).

IT potential

This activity gives another opportunity for one or more groups to use a temperature sensor and data-logging equipment in place of the standard thermometer.

Risk assessment

Safety warnings should be given concerning the use of boiling water, and guidance on how to deal with burns. Wear eye protection.

Equipment needed

For each group ('Sealing a letter')
- paper
- candle
- 'sealing ring', e.g. a coin

For each group ('A shell cast')
- shell or raised ornament to make the mould
- Plasticine

For each group ('The melting point of wax')
- test tube half-full of wax
- thermometer
- beaker of suitable size
- tripod and gauze
- Bunsen burner

Note
Use fresh stearic acid if possible to ensure purity and so a constant temperature over the change of state.

Fudge

Running the activity

Use the introduction about boiling point and purity from the pupils' book. Plans need to be carefully thought out and approved before this activity is attempted. Stress the safety aspects of boiling sugar solutions at well above 100 °C.

Groups need to spend some time planning. This could be homework, but discussion with the teacher is useful. You may wish to encourage convergence to the 'boil and weigh' model. Once plans have been approved, groups can begin their experiment.

Differentiation

Sheets available are:

Differentiation is by outcome – detailed planning, more accurate readings, more care in practical work overall, more detailed conclusions, etc.

AT1 programme of study

The activity covers data collection, presentation and analysis (**AT1.2a–c, e, 1.3a, b, d, e, g**).

IT potential

This activity gives another opportunity for one or more groups to use a temperature sensor and data-logging equipment in place of the standard thermometer.

Risk assessment

There are specific dangers given the high temperatures reached by the boiling solution. Great care should be taken as spilt boiling sugar solution could burn very badly. Specific consideration should be given to how the hot beaker could be reweighed at intervals – remove the heat source and lift by gripping the rim with an oven glove or something similar. Wear eye protection. Pupils must not taste sugar, milk or fudge in the laboratory, or that has been in laboratory containers. If the fudge is to be consumed, use the HE room and equipment.

Equipment needed

For each group

- 100 cm^3 beaker
- sugar (about 20 g)
- 0–200 °C or 0–360 °C thermometer
- tripod and gauze
- Bunsen burner
- oven glove
- access to balance (2 decimal places)

Notes

1 A suitable starting solution is 20 g of sugar in 40 cm^3 of water, heated over a low–medium Bunsen burner flame.

2 Warn pupils to watch the temperature carefully when it gets above 120 °C, as it starts to rise rapidly. They should stop heating before it reaches 150 °C, otherwise the sugar will char and ruin the beaker.

Expansion and contraction

Book coverage

Key ideas	PoS
Most materials expand when they are heated and contract when they are cooled. The amount of expansion is small, but it can cause problems for large objects such as road slabs. Liquids expand and contract in the same way.	AT3.2e

Answers

Activities in text

a They would buckle and crack.
b1 Steel and concrete expand and contract by the same amount. Brass expands too much.
b2 The less expansion, the less the chance of breakage.

(The water thermometer idea is developed in Activity 8c.)

What do you know?

1 When railway lines get hot, they **expand**. There is a **gap** between the lengths of rail to give room for this. If the rails were laid end to end, they would **buckle** when they got hot. When the rails **cool** down again, they contract.
2 The iron hoops were made to be slightly small. The hoops were heated so they expanded and put into place, where they cooled and contracted.
3 **a** Mercury (alcohol would boil and break the thermometer).
b Alcohol (mercury would freeze solid).

Extras: Tricks with expansion

1 Set the contact on the other side, so that the circuit is broken as the bimetallic strip bends when it is heated.

Assignments

7 The rails contract, so the gaps are larger.
8 **a** They are expanding and scraping against the woodwork.
b To allow for expansion.
9 To allow for expansion – one end moves over the rollers.
10 Platinum alloy – it expands by the same amount as glass.

Looking at expansion

Running the activity

The organisation of this work will depend on how many sets of equipment are available. 'Ball and ring', 'Flask of air' and 'Are all solids the same?' could be part of a circus with two or three stations of each. The ball in 'Ball and ring' needs to be plunged into water to cool it before the next group arrives. 'Flask of air' could be run as a class competition: who can get the most bubbles by hand heat alone? The flask needs to be emptied of water between pupils.

'What happens to liquids?' is best demonstrated unless multiple sets are available, as it would take too long to cool down between groups. This would probably be best at the end of the circus, as it develops the idea of relative expansion further, and gives a lead-in for the Extension sheet.

Differentiation

Sheets available are:

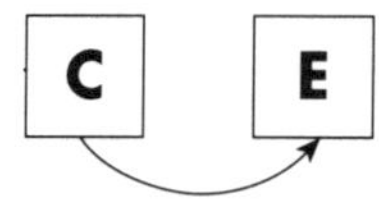

Core: Differentiation is by outcome from common experience.

Extension: This sheet extends the ideas.

AT1 programme of study

The primary purpose of the Core sheet is to give practical experience of expansion and contraction. The Extension sheet develops data analysis and inference (**AT1.3e, g**).

Risk assessment

Metals heated in the Bunsen burner flame will be hot enough to burn for some time after they have stopped glowing. Use heat-proof mats to stand hot object on, perhaps with warning cards nearby. Avoid highly flammable liquids in 'What happens to liquids?', as Bunsen burners are being used.

Equipment needed

For each group/class

- standard ball and ring plus Bunsen burner and tongs, with bowl of water for cooling
- round-bottomed flask with bung and tubing inverted into trough of water, with damp cloth
- series of bottles with bung and tubing, each filled with different liquids such as water, oil, paraffin, glycerine, etc. and labelled
- hot water
- bimetallic strip and Bunsen burner

Make a water thermometer

Running the activity

Groups of two or three pupils should make the water thermometer and calibrate it during the lesson. It could be used to measure the room temperature in the laboratory, and perhaps one other temperature, before moving on to questions **7–10**.

Differentiation

Sheets available are: C

Differentiation is by the quality of the water thermometer produced, and the understanding of its limitations.

AT1 programme of study

This is a simple experiment to give insight into the workings of liquid-filled thermometers (**AT1.3e, g**).

Equipment needed

For each group
- glass bottle
- Plasticine
- narrow-bore tubing
- water
- tape for 'flags'

For each class
- bowl of hot water at a known temperature (adjusted to 50 °C)
- bowl of crushed melting ice

Physical or chemical?

Book coverage

Key ideas	PoS
Changes of state are physical changes. Physical changes are reversible. Chemical changes are not reversible. Burning is an example of a chemical change. Chemical changes often give out heat, light or electricity.	AT3.2l

Answers

Activities in text

Possible demonstration: a beaker of cold water held over a candle flame will become covered in soot. The positioning of the beaker over the flame and the coldness of the water are crucial if condensation is to be seen. (If the beaker contains iced water, you may well get condensation occurring whether it is near the flame or not!)

a A physical, B physical, C chemical, D chemical, E chemical, F chemical, G chemical, H physical.

What do you know?

1 When wax melts, it is a **physical** change. The wax turns back to a **solid** when it cools. Burning wax is a **chemical** change. New substances are formed and you cannot get the **wax** back.

2 a Physical – the water vapour could be condensed back to liquid water.
b Physical – dissolving of the dyes is reversible.
c Chemical – the paper burns and makes new chemicals. This is not reversible.

3 a Carbon, as with a candle.
b Heat is given out by the burning gas.

Assignments

11 Chemical changes are not reversible, they make something completely new, they may give out (or take in) energy.

12 a Physical – dissolving (you could get the grease back).
b Chemical – a new substance forms (the gas).
c Physical – it melts again if you heat it.
d Chemical – it has changed into a new substance (not reversible).
e Chemical – it has changed into a new substance (not reversible).
f Physical – it melts again if heated.

13 Chemical change – heat is given off.

14 a Copper sulphate.
b Chemical – a new substance has been formed.

Physical or chemical?

Running the activity

Start by recapping the differences between physical and chemical changes. Stress the importance of careful observation. Also stress the safety aspects of working with laboratory chemicals (see risk assessment below). Explain how to heat a test tube in a Bunsen burner if this has not been done before.

Pool the results at the end and get pupils to give their reasons for deciding between physical and chemical changes.

Differentiation

Sheets available are:

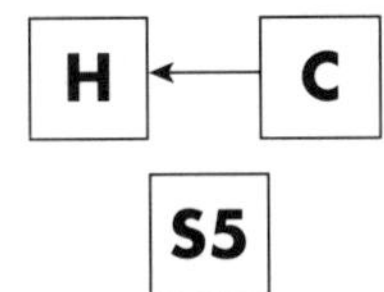

Differentiation is by outcome from a common task. A Help sheet is available with a table to fill in if required.

AT1 programme of study

The activity covers the use of simple apparatus, observing and drawing conclusions (**AT1.2a, b, 1.3e, g**).

Risk assessment

Ensure the correct method is used for heating a test tube. When toxic substances are used, there should be no eating, wash hands afterwards. Take care in pouring acids. Acids will damage clothes, so wear a lab coat. Wash acids off hands immediately if spilt. Wear eye protection. Copper carbonate and the oxide formed are harmful. Lead nitrate solution is toxic above 0.01 mol dm^{-3}; harmful above 0.001 mol dm^{-3}. Hydrochloric acid is irritant above 2 mol dm^{-3}.

Equipment needed

For each pupil

- test tubes
- spatulas
- bungs
- test tube holder and rack
- Bunsen burner
- dilute hydrochloric acid (less than 2 mol dm^{-3}) in small bottle
- small labelled jars of zinc oxide, copper carbonate, powdered wax (stearic acid), sugar, sodium carbonate crystals (small)
- small labelled bottles of salt solution, lead nitrate solution, oil

More chemical changes

Book coverage

Key ideas	PoS
Some chemical reactions are very useful. Fuels burn in oxygen and give out energy. The oxygen usually comes from the air. One-fifth of the air is oxygen. Some chemical reactions are not useful. Iron turns to rust in the presence of air and water.	AT3.2n,o

Answers

Activities in text

a Both air and water are needed for iron to rust.
The rusting of iron experiment can be performed, but it will need to be left for a week to show the effect. Pupils could set up their experiments either the week before this spread is studied, or at the end of the spread for discussion the following session. Air-free water is best obtained by boiling a kettle for several minutes and then letting it cool to a safe temperature before use.
As an extension, pupils could suggest ways to stop rusting (by stopping air and/or water getting to the iron). They could plan and/or perform these experiments.

What do you know?

1 When a **fuel** burns, it reacts with the **oxygen** in the air. Burning is a **chemical** reaction which gives out **energy** as heat and light.
2 **a** Water and carbon dioxide.
b Fuel + oxygen → water + carbon dioxide + energy.
3 In the rainy tropics. The desert has air, but no water.

Extras: Chemical change and food

1 Any suitable plan.
2 The starch in bread is partly broken down to sugars when it is toasted.
3 They eat their food twice, by re-eating their 'first run' faecal pellets.

Assignments

15 **a** Energy.
bi Oxygen. **ii** The air. **iii** One-fifth.
16 Carbon dioxide and water.
17 **a** From the air.
b There is no air in space.
c To burn the hydrogen fuel.
18 **a** Air and water can't get to it.
b There is no air or water.
c There is no water.
d There is no liquid water.

Burning fuels in air

Running the activity

Core: First give a reminder of the importance of oxygen in the air for burning. The activities can then be run in sequence by groups of three or four pupils. There is a Help sheet for those needing assistance with 'How much air does a candle need?'.

Extension: If jars wide enough to cover a floating candle are not available, the wording can be amended appropriately. Tall candles fixed to a base and standing in the water will work, but the proportional loss is not so easy to see. Fixed tapers may be preferable.

Differentiation

Sheets available are:

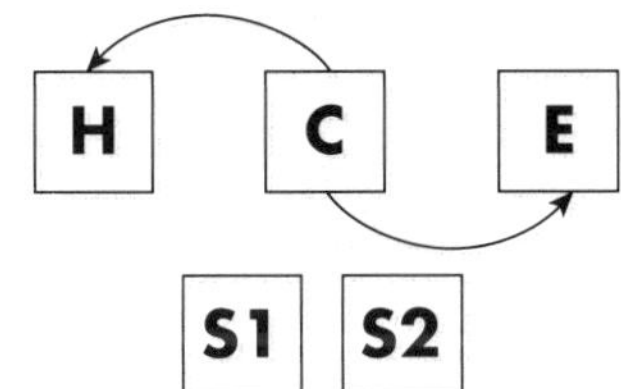

Differentiation is by outcome from a common task.

Core: For 'How much air does a candle burn?', higher level pupils should be edged towards considering the best factor to measure for the size of the jar – height or volume? (This is given in the Help sheet.)

Extension: This sheet develops the idea that burning is a chemical change with a new substance (carbon dioxide) being formed as the oxygen is used up.

AT1 programme of study

The activity gives basic experience of two standard practicals, followed by the opportunity to plan a simple investigation (**AT1.1a, c, h, 1.2e, 1.3b, c**).

Risk assessment

Ensure candles are used sensibly.

Equipment needed

For each group (Core)
- candle
- jars of different sizes
- stopclock

For each group (Extension)
- gas jar or other tall jar
- floating candle (or tall candle or taper fixed to a dense base)
- trough
- limewater

More of the same

Book coverage

Key ideas	PoS
All living organisms need to reproduce. Sexual reproduction involves special female and male sex cells. The female sex cell in a plant is called an ovule. In an animal it is called the ovum. The male sex cells in plants are called pollen. In animals they are called sperm. Fertilisation is the joining of a male and female sex cell.	AT2.1a,e, 2.2j, 2.3f

Answers

What do you know?

1 All living things need to **reproduce**. Many use **sexual** reproduction. This means two special cells called **sex** cells join together to form a new animal or **plant**.

2 The male sex cells in animals – sperm.
The female sex cell in animals – ovum.
The male sex cells in plants – pollen.
The female sex cell in plants – ovule.

3 Plants: insects or the wind carry the pollen.
Animals: mating – sperm are placed inside the female; or sperm and/or eggs are released into water.

Assignments

1 **a** Male – pollen, female – ovule.
b Male – sperm, female – ovum.
c The sex cells join together to make a new cell.

2 **a** Deer, peacock, stickleback.
b Grass, sea anemone.

3 **a** The eggs could be washed away, they could fail to meet up with the male cells, they could be eaten by other animals.
b Plaice – in the sea, the eggs can be washed huge distances away and there are lots of fish that could eat them.
Pike – the eggs could be washed away or eaten, but this is less likely as rivers are smaller than the sea, and the pike themselves reduce the numbers of other fish about, so fewer eggs are produced.
Stickleback – these lay eggs in a nest and defend them, so the eggs are much more likely to be fertilised and much less likely to be washed away. Fewer eggs are needed.
Bluetit – the eggs are cared for by the parents, so not many eggs are needed, but lots of bluetit eggs and young are eaten.
Crow – again the eggs are laid in a nest and the parents look after the eggs and young. Crows are big birds which nest high up, so few eggs or young are taken, therefore fewer eggs are needed.
Elephant – the egg is kept within the body of the mother until the baby is fully formed, so there is little risk of damage and so only one egg is produced at a time.

Flowers

Book coverage

Key ideas	PoS
Flowers contain the sex organs of plants. Pollen made in the stamen of one plant lands on the stigma of another. This is called pollination. The pollen fertilises the ovules in the ovary. The fertilised ovules develop into seeds containing a new embryo plant.	AT2.3f

Answers

Activities in text

a

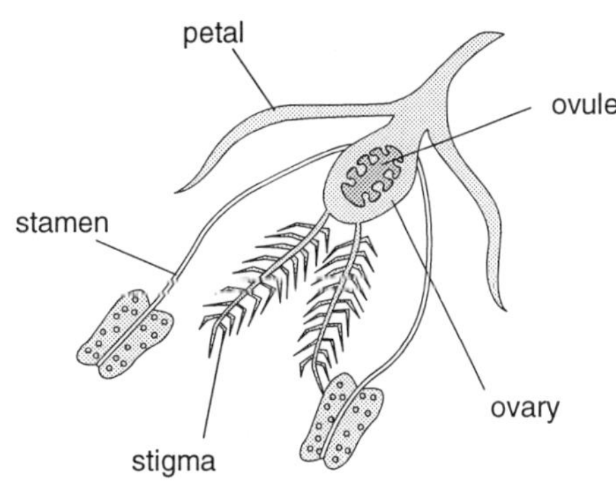

What do you know?

1 **a** Petal. **b** Ovary. **c** Stamen. **d** Pollen. **e** Stigma.

2 Pollination.

3 Lots of small, smooth pollen grains indicate wind pollination. Lots of pollen is made because not all of it will get to another flower – most is blown away. The pollen is small and smooth so it will be light and streamlined, to float easily.
A few big, spiky pollen grains indicate insect pollination. Fewer grains are made because they are more likely to be taken directly to another flower. They are big and spiky to make sure they stick to insect hairs.

Extras: The cycle of life

1 Puppy → mature dog → mating and birth of more puppies. The dog eventually ages and dies.
Baby → adolescent → adult, which has babies and rears them. A person lives a long time before reaching old age and death.
Seed → seedling → young tree → mature tree, which flowers and produces seeds for many years before ageing and death.
Frogspawn → tadpole → frog, which mates and lays more eggs for several years before death.

Assignments

4 Some flowers need to attract insects to carry their pollen away, so they are big and bright so the insects notice them. Other flowers use the wind to carry their pollen. It makes no difference to the wind what the flower looks like, so the plant doesn't need to use energy to make big, bright flowers.

5 Appropriate flower designs, with correct labels.

6 Pollination happens when pollen from one flower lands on the stigma of another flower. Fertilisation happens when the male sex cell from the pollen joins with the female sex cell (ovule).

The secrets of a flower

Running the activity

Use flowers that are in season, ideally enough for one between two pupils. Surprisingly few flowers are simple enough to be suitable for this exercise, and it is worth taking care in choosing which flowers to use. They need to be large, colourful and preferably have patterns on the petals (these are pollen guides, which attract the insects into the flower). The petals need to separate easily. Flowers made up of lots of tiny flowers, such as daisies, dandelions, chrysanthemums, etc. are not suitable. Wallflowers and forsythia are useful in the early part of the year.

Drawings should be big and bold, in pencil. If the petals are to be coloured, crayon rather than felt-tip pen should be used.

If any pupils have problems with the flower parts, suggest they use spread 9b of the pupils' book to help.

Differentiation

Sheets available are:

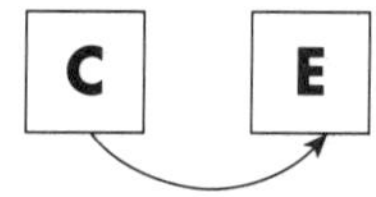

The Core sheet looks at insect-pollinated flowers. The Extension sheet looks at wind-pollinated flowers. Pupils work through a comparison with the insect-pollinated flower, which they organise for themselves with a few questions for guidance.

AT1 programme of study

Core: Pupils use dissection instruments in this activity. They have to manipulate material carefully and observe closely (**AT1.2a, b**).

Extension: Pupils make comparisons, look for patterns relating structure to function and draw conclusions (**AT1.3d, g**).

Risk assessment

A sharp scalpel is required to cut the carpels in half, and it is suggested on the sheet that the teacher is asked to do this. Avoid hyacinth, which is a common cause of allergy. Be aware of problems with hay fever.

Equipment needed

For each group (Core)
- insect-pollinated flower – tulip, daffodil, freesia, etc.
- white tile
- sheet of white or black paper
- scalpel or sharp knife
- mounted needle
- hand lens

For each group (Extension)
- equipment as for Core above
- wind-pollinated flower – grass, catkin, etc.

Scattering the seeds

Book coverage

Key ideas	PoS
Seeds develop inside fruits which form from the ovaries of plants. The fruits help the seeds to disperse using the wind or animals. Seeds are dispersed away from parent plants to find their own light, water and minerals.	AT2.3f

Answers

Activities in text

a If 'seedless' fruits are mentioned, comment that special sprays can stop seeds developing, as people like to eat fruits without seeds. Close examination of seedless grapes or bananas, for example, reveals the tiny undeveloped seeds.

b1 Light, water, minerals.

b2 So that they are not competing for the same light, water and minerals and have space in which to grow.

What do you know?

1 Once fertilisation has taken place in plants, the **flower** dies away. The seeds continue to develop inside the **ovary**. The ovary itself also grows to form the **fruit**. Once the seeds have formed, they are **dispersed**, which means they are carried away from the **parent plant**.

2 To protect the seeds, and to help make sure that they are carried as far as possible from the parent plant.

3 The seed with a parachute is dispersed by the wind, as it is small and light. The seed with a sticky coat is dispersed by animals, sticking to their fur.
The seed with the hard, woody outside is buried by animals as a food store, then forgotten and after a time the seed grows.

Extras: More ways of getting around

1 The coconut floats on water.
The liana flies, using wings.
The devils' thorn sticks into animals (it actually ensnares the foot of an ostrich!).
The proteus floats using parachutes, following a forest fire which opens the fruit to release the seeds.

Assignments

7 Plants produce fruits to protect the seeds and to help make sure that they are spread (dispersed) as far as possible from the parent plant.

8 Pupils will find that as well as the fresh, dried and tinned fruit which we use for food, fruit extracts are often used in cosmetics, bath products, shampoos and even bathroom and kitchen cleaners.

9 In most fruits pupils should see the seeds, the ovary and sometimes the ovary wall. In the apple, which will be a common choice, the ovary wall is the tough bit round the seeds in the middle, and another part of the flower (the receptacle) swells up to form the flesh. Some fruits, like bananas and seedless grapes, are specially bred to be seedless, but as mentioned above, if you look carefully you can see the tiny undeveloped seeds.

Make your own flying fruit

Running the activity

Pupils design and make a flying fruit capable of carrying a sunflower seed. They are given a limited time (15 minutes) to design, make, test and modify this fruit. If they are obviously pressed for time, extend it. For a fair comparison between the fruits, they must be dropped from the same height by the same person in the same air conditions.

Differentiation

Sheets available are: **C**

Differentiation will occur with pupils' abilities to modify their fruits, understand the fair test and draw out the important features for a flying fruit (differentiation by outcome).

AT1 programme of study

This is an investigation allowing pupils to make a range of decisions and look for patterns in the data (**AT1.1d, e**). They will need to select and control variables and make decisions about numbers of observations, and they could carry out trial runs (**AT1.1b, 1.2d, e, a**). They could have the opportunity to consider anomalous observations (**AT1.4b, c**) and evaluate their experiment (**AT1.4a, c**).

Equipment needed

For each group

- sunflower seed
- stiff card
- paper
- Blu-Tack
- Sellotape

From boy to man

Book coverage

Key ideas	PoS
Puberty is the stage when your body changes from that of a child to that of a young adult capable of reproducing. Testes are the male sex organs which make sperm. Semen is the mixture of sperm and liquid which leaves the penis.	AT2.1a, 2.2i,j

Answers

Activities in text

a Clothes, hairstyles, body shape, size and shape of face.

What do you know?

1 Changes during puberty for boys:

- Facial hair starts to grow.
- Body hair starts to grow.
- The larynx gets larger and the voice deepens.
- The sex organs (penis and testes) get larger, and the testes begin making sperm.
- Pubic hair grows.
- The body shape changes with bone and muscle growth.
- A growth spurt happens – the boy gets much taller.

Assignments

10 Appropriate answers.

The male reproductive system

Running the activity

The main purpose of this activity is to make sure that pupils have an accurate diagram of the male reproductive system to stick in their books, with a paragraph explaining its functioning. Depending on the make-up of your group, you could ask pupils to colour in the diagram when they have finished the labels and paragraph.

Differentiation

Sheets available are:

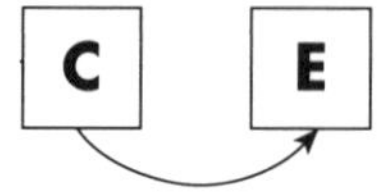

The Extension sheet gives an opportunity to transfer knowledge to a new situation, and to consider the advantages and disadvantages of two different ways of displaying the same information.

From girl to woman

Book coverage

Key ideas	PoS
During puberty a girl's ovaries start producing an ovum (egg) each month. Her uterus prepares for pregnancy each month and she has a monthly period. Her breasts develop to prepare for feeding a baby.	AT2.1a, 2.2i,j

Answers

What do you know?

1 Changes during puberty for girls:
- Breasts develop.
- The body shape changes as fat develops around the hips, bottom and thighs.
- Pubic hair grows.
- Periods start, as the ovaries begin to release an egg each month in preparation for pregnancy.
- A growth spurt happens – the girl gets taller.

Assignments

11 Appropriate answers.

12 In the early years of secondary school, everyone enters puberty at different times, and no one wants to be different or left out. By the age of 18 almost everyone has completed most of the changes of puberty, and so the pressure is much less.

The female reproductive system

Running the activity

The main purpose of this activity is to make sure that pupils have an accurate diagram of the female reproductive system to stick in their books, with a paragraph explaining its functioning. Depending on the make-up of your group, you could ask pupils to colour in the diagram when they have finished the labels and annotations.

Differentiation

Sheets available are:

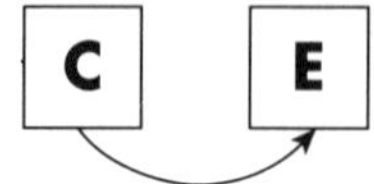

The Extension sheet gives an opportunity to transfer knowledge to a new situation, and to consider the advantages and disadvantages of two different ways of displaying the same information.

A new body, a new you?

Running the activity

This type of activity needs to be handled with tact if pupils are to be able to express their ideas happily. The groups might need to be managed to a certain degree. In mixed classes, it doesn't matter if groups are predominantly single sex. It is interesting to see if the male worries predicted by girls and the female worries predicted by boys reflect the results from the opposite sex groups, or mixed groups. If it does not happen naturally, deliberate mixing could be counterproductive as the pupils might then feel more inhibited. Much depends on the group.

Comparisons between the groups are best done as a class session led by the teacher, and the data for the bar chart collected at the same time. Make one table on the board for boys, and one for girls. Ask groups in turn for one area which they felt would be difficult to get used to, and why. Write this down, and then ask how many other pupils also identified that area. Record the total number (of groups or of pupils, depending on the class you are dealing with).

Task **6** could be done during the lesson or as homework. It is designed to help pupils see solutions to their own concerns.

Differentiation

Sheets available are:

C

More able pupils will probably complete the bar chart more rapidly, and so can be directed to task **6**.

Equipment needed

For each pupil

- graph paper

Meet the stork

Book coverage

Key ideas	PoS
During sexual intercourse, a man inserts his penis into the woman's vagina. When he ejaculates, sperm are released into the vagina. They swim up into the uterus and Fallopian tubes. If a sperm meets an ovum (egg), fertilisation takes place.	AT2.2j

Answers

What do you know?

1 To make a baby, an **ovum** and a **sperm** must join together. They are brought together during **sexual intercourse**. The man's **penis** becomes erect and is placed inside the **vagina** of the woman. When the man **ejaculates**, millions of sperm swim up through the **uterus** and on into the Fallopian tubes.
2 An ovum is only released once a month, so there isn't always one there in the Fallopian tubes. Many sperm die before they get to the Fallopian tubes.
3 The ovum is big and needs lots of resources from the mother to produce it. It stays in the body of the female, so it is safe. The sperm are very small and they have a long and difficult journey. Lots are needed to make sure at least one gets to the ovum.

Extras: The long and short of pregnancy

1 Because they have big babies which are well developed.
2 They can all see, breathe and feed from their mothers. Baby dolphins can swim and come up to the surface for air. Baby elephants can walk and run and very rapidly start to eat twigs. Baby humans are relatively helpless – it is about a year before they can walk.
3 They are very small animals so their babies don't have to grow so much. They produce babies which are blind with no fur, so they continue to develop after they are born.
4 Baby kangaroos can do nothing except pull themselves through the fur of their mother into her pouch. They then attach to a nipple and stay in the pouch for months before they are developed enough to survive on their own. The babies are small because the 'pregnancy' carries on while the baby is in the pouch.

Assignments

No homework assignments have been set on this sensitive topic. Teachers may choose to devise their own, with knowledge of the group they are working with.

What's happening inside?

Book coverage

Key ideas	PoS
The menstrual cycle lasts about 28 days. An egg ripens and leaves the ovaries each month. The lining of the uterus builds up. If the egg is fertilised it will implant into this lining. If not, the lining is lost as the period. Identical twins form when a single fertilised egg splits in two. Non-identical twins form from two eggs and two sperm.	AT2.2j,k

Answers

What do you know?

1 The menstrual cycle lasts for about **28** days. An **egg** ripens in the **ovary** and is released into the **Fallopian tube** after about two weeks. At the same time the lining of the **uterus** builds up. If the ovum is **fertilised** it will implant itself in the lining to **develop**. If the egg is not fertilised the lining is lost as the monthly **period**.

2 When a woman is pregnant, there is a baby growing in her uterus. No more eggs are produced while she is pregnant, and the lining of the uterus doesn't break down because it is helping to support the developing baby. All this means that the woman does not have periods.

3 a Some twins look exactly the same as each other because they are identical – they have formed from the same fertilised egg which split as it developed, and grew into two babies.

b Some twins happen because their mother released two eggs at about the same time, and each of these eggs was fertilised by a different sperm. The two babies are no more alike than other children of the same family, and they can be either the same sex or different sexes.

Assignments

13 A: Eggs begin to ripen in the ovary. The old lining of the uterus is lost for the first few days as the period, and a new lining starts to build up.
B: Ovulation takes place – a ripe egg is released from the ovary.
C: The egg dies. The thick lining remains for a few more days, then begins to break down.

14 a Both babies cannot be born at the same time – one always comes out first, and that baby is the oldest. It may be by only a few minutes, or by hours.

b Hannah and Charlotte are identical twins – they are the same sex, and are so similar that it is very difficult to tell them apart.

c The mother releases two eggs at about the same time, and each of these eggs is fertilised by a different sperm. The two babies are no more alike than any other children of the same family, and they can be either the same sex or different sexes.

The end and the beginning

Book coverage

Key ideas	PoS
A fetus is a developing baby. When a woman is pregnant, she has a growing fetus in her uterus. Pregnancy lasts for about 40 weeks. The process of giving birth is called labour.	AT2.1a, 2.2k

Answers

Activities in text

a Appropriate answers.

What do you know?

1 **a** A human pregnancy lasts for about **40** weeks.
b The developing baby is known as a **fetus**.
c The **placenta** supplies food and oxygen for the developing baby, and also removes all the waste products.
d The fetus is supported in a **bag of fluid** which protects it from knocks and bumps.
e The fetus is joined to the placenta by the **umbilical cord**.
f The baby is born at the end of several hours of **labour**.
g After the baby is born the **umbilical cord** is cut and the **placenta** comes away.

2 Inside the uterus: pupils need to think about the sounds of the mother's heart and gut, sounds from the outside world, any light that might come in when the mother undresses, the feeling of floating in fluid, feeling squeezed and being able to move less and less as the fetus gets bigger and the uterus reaches a size where it cannot grow any more.

Being born: Pupils need to think about the uterus squeezing and pushing the baby, travelling out of the body, the bright light and loud noises as it comes into the world, needing to breathe, being cuddled and held, having to feed for itself, etc.

Extras: It's cold outside

1 Remind the baby to breathe, or breathe for it; give it extra oxygen; provide food straight into the stomach so no sucking or swallowing is involved; keep the baby warm; keep the air moist so that the baby's skin does not dry out; keep the level of light and noise as low as possible; change nappies; stroke and talk to the baby so that it is aware of human contact and love; etc.

Assignments

15 **a** From the mother, through the placenta and then the umbilical cord.
b Waste is passed from the fetus to the mother through the umbilical cord and the placenta, and the mother gets rid of it.
c The fetus grows in a bag of fluid which protects it from knocks and bumps.
d Because it grows inside the body of the mother, the fetus is kept warm by the warmth of the mother.

16 At the end of pregnancy, the entrance to the uterus opens and the baby is pushed out through the vagina, usually head first. This is hard work for both mother and baby, and is called labour.

The human life support machine

Running the activity

The main purpose of this activity is to make sure that pupils have an accurate diagram of the fetus developing in the uterus to stick in their books. Depending on the make-up of your group, you could ask pupils to colour in the diagram when they have finished the labels and table.

Differentiation

Sheets available are: **C**

If any, by outcome of the tabulating exercise.

A fetus can't say no!

Running the activity

This activity can be used either as an individual task or as a group activity. The pupils need to be encouraged to put their message across clearly, with impact but also with plenty of real information. Some tact is needed as the mothers of some pupils will doubtless have smoked whilst pregnant. A useful point is that statistically, the birth weight is lower and the death rate is higher in the babies of smokers – but this does not mean that the baby of every smoker will be small or will die.

Differentiation

Sheets available are: C

Equipment needed

For each pupil/group
- paper
- colouring pencils/pens

Switching on

Book coverage

Key ideas	PoS
A switch is used to complete an electric circuit. An electric current flows around a complete circuit. A material that allows an electric current to flow is called an electrical conductor. A material that does not allow an electric current to flow is called an electrical insulator.	AT4.1

Notes: The need for a complete circuit before a current will flow and the use of switches is included in the Programme of Study for Key Stage 2.

Answers

Activities in text

a Light switch, alarm call button for an elderly person, switch on kettle.

b

Material being tested	Does the bulb light up?	Electrical conductor or insulator?
steel	yes	**conductor**
polythene	no	**insulator**
glass	no	**insulator**
aluminium	**yes**	**conductor**
wood	**no**	**insulator**

What do you know?

1 When a **switch** is closed, it makes a complete **circuit** so that an electric **current** can flow.

2 a The current flows through the metal wire, because it is an electrical conductor.

b The outside is made of plastic, an electrical insulator, so that the electric current stays in the metal wire.

Extras: Conductors or not?

1 Copper is the best conductor. Steel is a better conductor than pencil lead. Wood is an insulator.

2 All samples should be the same length and thickness.

Assignments

1

current	flows round a complete circuit
switch	close this to complete a circuit
electrical conductor	metal is an example
circuit	must be complete for a current to flow

2 a The bulb won't light because the switch is open.

b The bulb won't light because plastic is an insulator.

c The bulb will light because the switch is closed and aluminium is a conductor.

3

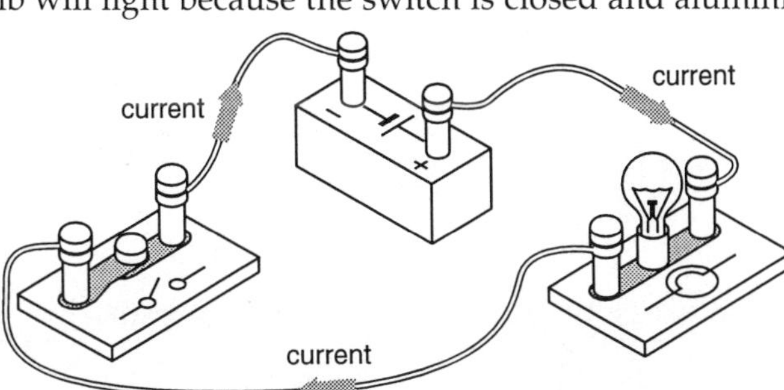

Burglars beware!

Running the activity

Pupils need not be given formal instructions on setting up circuits, as this is dealt with in the next activity. They may need to be shown the available batteries, connecting wires, crocodile clips and any other items which may be unfamiliar to them.

The activity involves sorting materials into conductors and insulators. It is likely that this will be familiar to them from Key Stage 2; however, you may feel it appropriate at this point to set the exercise of determining experimentally which materials are conductors. (This is dealt with on the spread in the pupils' book.)

This is a relatively informal activity which will give you an opportunity to gauge your pupils' experience in working with electrical components.

Differentiation

Sheets available are:

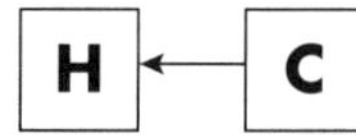

AT1 programme of study

Core: This sheet encourages pupils to approach the task in a systematic way, including the selection and use of equipment (**AT1.1h**). Since they have not yet learnt circuit symbols, they must record their circuit in the form of a drawing (**AT1.2e**). They are also required to write a description of how their circuit works (**AT1.3g**).

Help: This sheet provides ideas for constructing a switch, and for setting up the circuit.

Extension: Pupils could be asked to design a second switch which would allow the user to choose whether a bulb lights up or a buzzer sounds when the first switch is closed.

Equipment needed

For each group
- battery in holder
- bulb in holder
- buzzer
- 5 connecting wires
- 2 crocodile clips
- strips (approximately 5 cm × 1 cm) of the following:
  - copper sheet
  - aluminium foil
  - stiff card
  - plastic (from lemonade bottle)
- strip of wood, e.g. lollipop stick
- scissors
- block of wood
- drawing pins

Note
In all the activities in this unit, the batteries much be chosen to match the available bulbs.

Wiring up

Book coverage

Key ideas	PoS
Electric current flows from the positive end of a battery, around a circuit, to the negative end. Circuit symbols are used to represent electrical components. A circuit diagram shows how they are connected together.	AT4.1d implies understanding of current flow.

Answers

Activities in text

a Any battery-operated appliances, e.g. radio, personal stereo, remote controlled car, torch.
b One battery is the wrong way round.
c2 Close both switches.
c3 Close switch 1 for the buzzer, close switch 2 for the motor.
d Circuit B (with buzzer and motor).

What do you know?

1

Component	Symbol
battery	
bulb	
motor	M
switch	
buzzer	
connecting wire	

2 a **b**

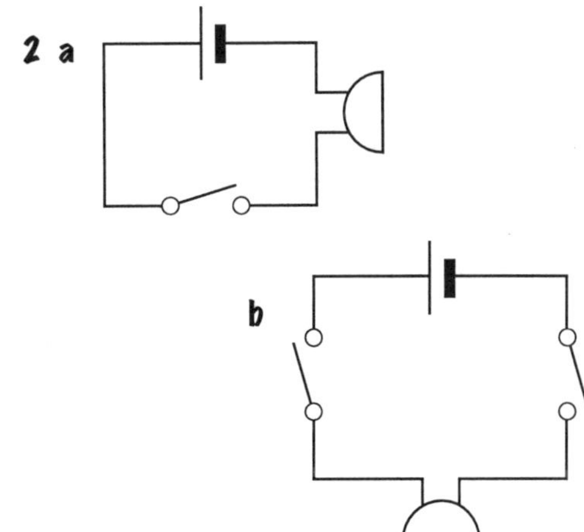

3 Switch 1: the buzzer sounds. Switch 2: the bulb lights. Switch 3: the motor turns.

Extras: Two-way switches

1 Bulb 1 will go off, and bulb 2 will come on.
2 The bulb is not lit until switch 1 is changed. Then changing switch 2 turns it off again.
3 This is used for a landing light with switches at the bottom and top of the stairs, for example.

Assignments

4 a The bulb stays off.
b The bulb stays off.
c The bulb comes on.

5 a,b

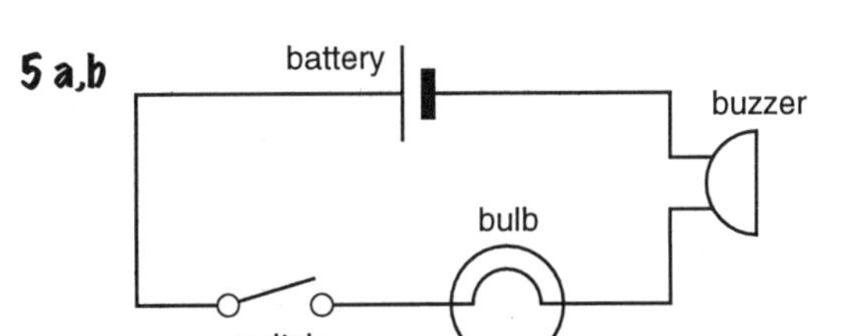

c When the switch is closed, the buzzer sounds and the bulb comes on.

6

Appliance	Used for	Could this be done without electricity?	How?
electric kettle	*boiling water*	*yes*	*use a gas cooker*
electric fire	**heating**	yes	**gas, coal fire, etc.**
radio	**making music, etc.**	no	—
torch	**light**	yes	**candle**
calculator	**doing sums**	yes	**brainpower**

b All are easier with electricity.

Circuits and symbols

Running the activity

This is a formal activity, in which pupils can learn how to set up a circuit from a drawing or from a diagram. They are likely to need some instruction in setting up circuits.

1 Set components out as they appear in the drawing or diagram.
2 Starting at the positive terminal of the battery, work systematically round the circuit, connecting components together. This should remind them of the way current flows round the circuit.
3 There is no significance in the colour of connecting wires.
4 Pupils should be encouraged to predict what will happen when the switches are closed before they complete the circuit. They could test each other on this.

Differentiation

Sheets available are:

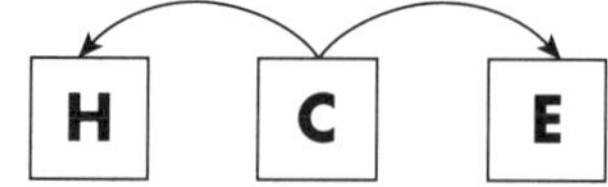

All three sheets relate to the same four circuits, with different degrees of detail provided. The circuits chosen include some of those shown in spreads 10a and 10b of the pupils' book. The Extension sheet has two extra questions, including an open-ended final question.

AT1 programme of study

Pupils should record their work in the form of circuit diagrams and sentences (**AT1.2e**) having used a range of equipment (**AT1.2a**).

Equipment needed

For each group
- battery in holder
- bulb in holder
- buzzer
- motor
- 2 switches
- 7 connecting wires

Note
Circuit boards are a good alternative to components joined by connecting wires.

Seeing the light

Book coverage

Key ideas	PoS
When batteries are connected end to end (in series), they give a bigger push and a bigger current flows. When two bulbs are connected in series, they shine less brightly than one bulb alone in a circuit. If you want them to continue to shine brightly, they should be connected in parallel (side by side).	AT4.1d

Answers

Activities in text

a Current flows in through one connection and out through the other.
b1 The more batteries, the brighter the bulb (unless you use too many batteries!).
b2 More batteries give more current.
c

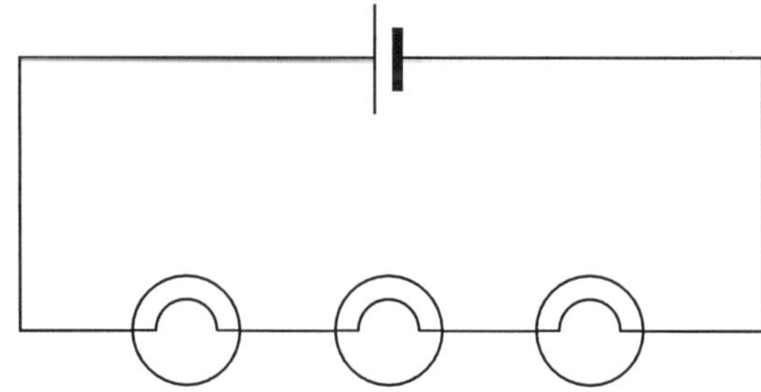

Three bulbs in series are even dimmer than two.
d If you unscrew bulb 2, it goes out, but bulb 1 stays lit.

What do you know?

1 **b** Two bulbs connected in parallel.
c Three bulbs connected in series.
d Three bulbs connected in parallel.
2 **a** These Christmas tree bulbs are connected in series (though this is not always the case).
b Car headlamps are connected in parallel.
3 **a** House lights are connected in parallel, so that they are all bright, and they can be switched on and off individually.
b There is a power supply, switch, then two lights in parallel.

Extras: Bright lights

1 Bulb 2 is brighter, because two batteries give more push.
2 Bulb 2 is brighter, because bulbs 3 and 4 are in series, so are dimmer.
3 Bulb 5 is not lit, because it is not part of a complete circuit.

Assignments

7

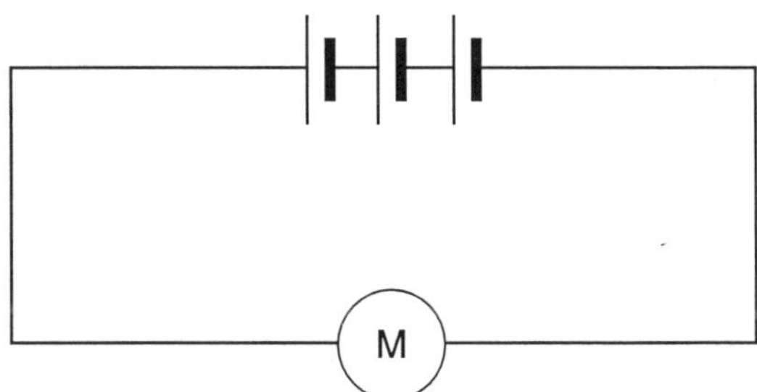

8 **a** Circuit B.
b Circuit A.
c In circuit A.
9 Connect the bulbs in parallel, or use more batteries in series.

Puzzlectric

Running the activity

The spread in the pupils' book includes a discussion of the construction of bulbs. For many pupils, this is unfamiliar, and you may wish to give them a chance to examine some bulbs to see the two connections and the filament. They could look at a torch bulb and at a 12 V bulb (and holder) whose construction is similar to a mains bulb. They could use two wires touching directly onto the connections to make the bulbs light up.

The activity sheet then reinforces the ideas of series and parallel connection, as discussed on the spread in the pupils' book. For the puzzle box, everything including the battery could be enclosed in the box, with just the two bulbs protruding.

Differentiation

Sheets available are: **C**

AT1 programme of study

Core: Pupils are asked to predict the outcome of completing each circuit (**AT1.1c**) and see if their data supports their prediction (**AT1.3f**).

Help: Some pupils may find it difficult to make predictions, but they should still be able to follow the diagrams on the sheet. Their attention could be drawn to the illustrations in the pupils' book.

Extension: Pupils could be asked to make a more complicated puzzle box, with three or more bulbs. They should be expected to draw the correct circuit diagram.

Equipment needed

For each group
- battery in holder
- 3 bulbs in holders
- 6 connecting wires
- small disposable cardboard box or packet

Resistance to change

Book coverage

Key ideas	PoS
You can change the current that flows in a circuit by changing the resistance of the components in the circuit. With more resistance, the current is smaller. A long, thin wire has more resistance than a short, fat wire.	AT4.1d

Answers

Activities in text

a Dimmer switch, volume control, heater control, etc.
b It is difficult to get through the long thin pipe – runners are queuing to get in.
c1 Dim – clips at ends; bright – clips close together.
c2 There is a greater resistance with the clips further apart.
c3 The current gets less with the clips further apart.

What do you know?

1 Every component in a circuit has electrical **resistance**. The more resistance there is in a circuit, the **smaller** the current that flows. A long wire has **more** resistance than a short wire.
2 Variable resistors: **b** brightness control, **c** volume control, **e** temperature control.

Extras: Slider circuits

1 Bulb 2.
2 There is an increasing resistance in bulb 2's circuit, but a decreasing resistance in the circuit of bulb 1.
3 In a theatre, for changing between two sets of lights, for example.
4

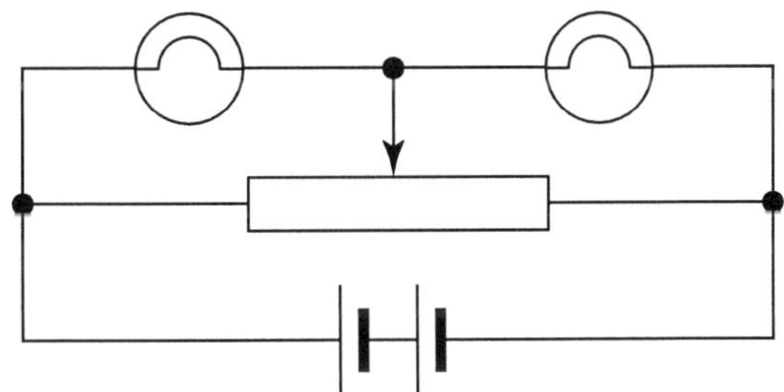

Assignments

10 When you turn the knob, you are turning a **variable resistor**. There is less **resistance** in the circuit, so a bigger **current** flows, and the ring gets **hotter**.
11 Radio volume control, VDU brightness control, car heater control, etc.
12 **a** B, because it is short.
b C, because it is long and thin.
c C has most resistance.
d A long thin wire has high resistance; a short fat wire has low resistance.

Using a variable resistor

Running the activity

Pupils should be able to set up the circuits shown in the spread of the pupils' book. It is likely that pupils will need to be shown how to make connections to a variable resistor (rheostat).

Differentiation

There are no photocopiable sheets for this activity.

Core: Pupils should set up a circuit in which a variable resistor controls the brightness of a bulb. They could be asked to change the circuit to control a motor.

They can then go on to use a pencil lead as the variable resistance in a circuit. Crocodile clips make connections on to the lead, and moving the clips can make a bulb become brighter or dimmer.

Extension: The illustration on the Extras page shows how to control two bulbs using one variable resistor.

AT1 programme of study

The activity allows discussion of controlling variables (**AT1.1d, e**), particularly with the pencil lead as a variable resistor. Careful observation should be encouraged (**AT1.2c, d**) and the identification of any pattern in the results (**AT1.3d**). Pupils should record their work in the form of circuit diagrams or drawings (**AT1.2e**).

Risk assessment

Pencil leads are made of graphite and clay. The graphite can become hot and burn if too high a current flows through the lead. In practice, this should not happen if batteries and bulbs are in the circuit. If pupils use a power supply, they may connect up a circuit which allows a large current to flow, and the pencil lead may overheat. Watch out for evidence of hot or smoking pencil leads.

Equipment needed

For each group

- battery in holder
- 2 bulbs in holders
- motor
- variable resistor (rheostat) less than 20 Ω
- 5 connecting wires
- pencil lead, e.g. 2B (as for propelling pencils)

How much current?

Book coverage

Key ideas	PoS
We use ammeters to measure electric current. We measure current in amperes (A). Current does not get used up around a circuit.	AT4.1c,e

Answers

Activities in text

a Dimmer bulbs mean there is less current flowing. You could measure the current with one bulb, and then with two.
b1 The cooker needs the biggest current.
b2 The light bulb needs the smallest current.
b3 1 A flows in, and 1 A flows out.

What do you know?

1 All ammeters read 0.5 A, because current is not used up in a circuit.
2 Because current has to flow all the way round a circuit, it doesn't matter whether the switch is before or after the bulb.
3 Put one ammeter in each of the parallel branches of the circuit. Could also include one next to the battery.

Extras: More obstacles

1 A short, fat wire has less resistance than a long, narrow wire.
2 Two wires connected side by side (in parallel) let through more current. (They have less resistance.)
3 Two wires end to end (in series) have more resistance, so less current flows.

Assignments

13 **a** 0.40 A. **b** 2 A. **c** 1.5 A.
14 2.5 A.
15

Component	Symbol
ammeter	—(A)—
motor	—(M)—
buzzer	[buzzer symbol]
connecting wire	——
lamp	[lamp symbol]
battery	—|⊢—

Ammeters at work

Running the activity

Pupils will need to be shown how to incorporate an ammeter in a circuit. They will also need to be shown how to read the value of current. It is best to avoid discussion of milliamps; stick to amps.

In part **4**, pupils should move the slider along at intervals of 2 cm. The brightness of the bulb can only be recorded as 'dim', 'very bright', etc.

Differentiation

Sheets available are:

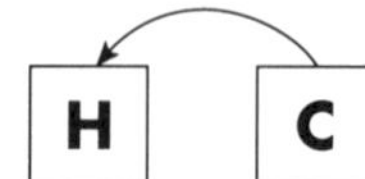

Core: This sheet presents circuits as circuit diagrams.

Help: This sheet has the same circuits as drawings, and a table for the results of part **4**.

Extension: Pupils could be asked to design their own circuit for part **4**.

AT1 programme of study

Pupils should conclude that the brightness of the bulb increases with the current flowing through it (**AT1.3d**). They should be encouraged to explain the variation of the current in terms of the resistance of the variable resistor, using their scientific knowledge (**AT1.3g**).

Alternative activity

Pupils might be asked to investigate the question: "Which has a greater resistance, a bulb, a buzzer or a motor?". They could investigate this by connecting the components in turn in a circuit with a battery and an ammeter. The greatest resistance gives the smallest current.

Equipment needed

For each group

- battery in holder
- 2 bulbs in holders
- variable resistor (rheostat)
- ammeter
- 5 connecting wires
- ruler or metre rule

Note

It is necessary to use robust ammeters in this activity. They should be chosen to have a range of up to 1 or 2 A. Auto-ranging digital ammeters are suitable.

Energy changes

Book coverage

Key ideas	PoS
An electric current flows all the way around a circuit. It gains electrical energy from a battery, and transfers it to devices which change it into different forms.	AT4.1e

Answers

Activities in text

a A hair drier: heat, kinetic energy, sound. B neon sign: light (and heat). C food whisk: kinetic energy, sound (and heat). D loudspeakers: sound (and heat).

b1 Current is not energy. Current carries energy. **b2** Correct.

b3 Current does lose energy, but that doesn't mean there is less current. **b4** Correct.

What do you know?

1

The obstacle race model	A real electric circuit
the obstacles	components in the circuit
start/finish line	the battery
the moving runners	**the current flowing**
the running track	**the wires of the circuit**

2

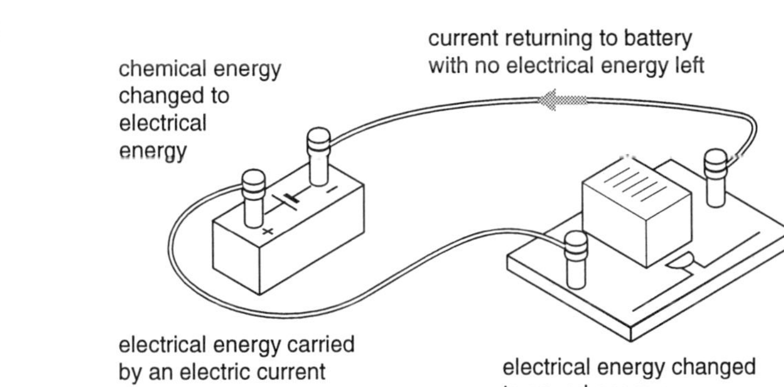

Extras: Central heating

1 Water flowing corresponds to the current.
You need a complete circuit.
Hot water has more heat energy, like current coming from the battery which has more electrical energy.
Colder water has less energy, like current returning to the battery.

Assignments

16 a Radio, television, loudspeaker, buzzer, etc.
b Heater, hair drier, etc.
c Motor, fan, etc.

17 a The electric current flowing out of a light bulb is **the same as** the current flowing into it.
b A light bulb changes **electrical energy** into light energy.
c You can measure the electrical energy carried by an electric current using **a joulemeter**.

18

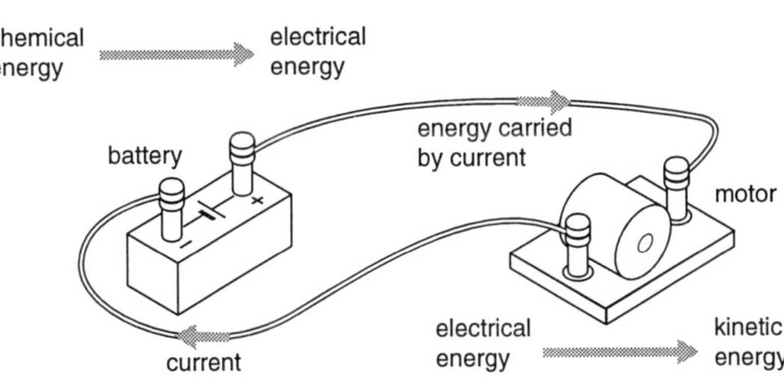

Revision demonstration

Running the activity

This demonstration is intended to help pupils to understand the distinction between electric current (which travels all the way round the circuit) and electrical energy (which does not). It gives you the opportunity to bring together all the ideas which have been developed in this unit.

Differentiation

There are no photocopiable sheets for this activity.

By appropriate questioning, you will be able to identify particular areas of uncertainty and help pupils to resolve some of their problems. Pupils may also be involved in setting up the circuits, and giving their predictions and explanations.

Sequence of demonstration

1 Set up a circuit with a battery or power supply and a 12 V, 30 W lamp. Ask pupils about their ideas of how current is flowing, and how energy is moving.
2 How can we find out the current? Use an ammeter. Add two ammeters to the circuit, to remind pupils that current is conserved around the circuit.
3 How can we find out about energy? Use a joulemeter. We can 'see' the energy coming from the bulb. Reduce the voltage; the bulb is dimmer, so less energy is being transferred. Add the joulemeter to measure the energy transferred. Observe the effect of changing the voltage.
4 Compare this circuit with the obstacle race described in the pupils' book.
5 The questions in the pupils' book give a suitable basis for pupils to record their understanding of this demonstration.

Equipment needed

For the demonstration

- batteries (up to 12V) or power supply
- 12V, 30W lamp in holder
- joulemeters
- 2 ammeters (preferably demonstration type)
- 8 connecting wires

Note

There are several types of joulemeter available, with different merits.

a Digital – gives a clear measurement in joules or kilojoules. May click to give an audible impression of rate of energy transfer.
b Flashing light – each flash equals 100J. Gives a visual impression of rate of energy transfer.
c Electricity meter type – like a domestic meter. The wheel spins faster as the bulb shines more brightly.

Sedimentary rocks

Running the activity

The activity involves identifying some common sedimentary rocks using a simple key. Start with a discussion of how to recognise sedimentary rocks, preferably showing samples with fragments of shells or fossils. Link back to Activity 11b 'Here comes the rain' to see how to identify limestone (fizzes with acid). Point out that the acid test only shows limestone if the rock itself fizzes – fossil shells are likely to fizz whatever rock they are in.

Pupils should use their answers to the key questions to build a brief description of each rock type. The Help sheet provides a text-marking method to help build this description.

Differentiation

Sheets available are:

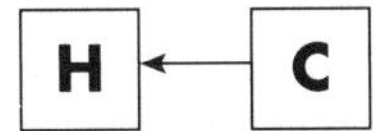

Differentiation is by outcome and the degree of help provided.

AT1 programme of study

The activity covers the use of keys and interpreting evidence (**AT1.2b, e**).

Risk assessment

Care is needed with acid in dropping pipettes. Pupils must act sensibly. Wear eye protection.

Equipment needed

For each group

- labelled samples of sedimentary rocks:
  - chalk (A)
  - shelly limestone (B)
  - *sandstone (C)
  - *shale (D)
  - *mudstone (clay rock without layers) (E)
  - conglomerate (pebble rock) (F)
- dilute hydrochloric acid (less than $2\,mol\,dm^{-3}$) with dropping pipette

Note

* test these samples first – some may have calcareous cement and so will fizz. Avoid these types.

Ideally, some of the samples should contain fossil seashells.

All change

Book coverage

Key ideas	PoS
Sedimentary rocks can be changed by heat and pressure into metamorphic rocks. Metamorphic rocks can melt, cool and set to form igneous rocks, which can break down to form sedimentary rocks. These changes make up the rock cycle.	AT3.2g

Answers

Activities in text

a1,2 Appropriate examples e.g. granite, sandstone, marble.
a3 Formed by weathering; carried by rivers; dumped in the sea; buried, squashed and heated; finally melted.

What do you know?

1 The rocks of the Earth are naturally **recycled** in the rock cycle. Igneous rocks break down on the surface of the Earth, forming **sediments**. Sedimentary rocks are changed into **metamorphic** rocks by heat and pressure. These rocks melt if they get too hot, turning back to **igneous** rock.
2 Bricks are baked in a kiln, slate was squashed by enormous pressures in the Earth.
3 The Earth's crust does not stay in position – it is lifted up in places.

Extras: Is it igneous or metamorphic?

1 Gneiss is metamorphic – the crystals are in layers.
Granite is igneous – the crystals are random.
2 Baking by hot lava is metamorphism by heat only. Pressure is needed to cause banding.

Assignments

16 a The high temperature bakes the rock.
b Metamorphic.
c Pressure is needed to form layers in a metamorphic rock.
d They are lost.
e Igneous.
17 a See diagram on page 147 of pupils' book.
b Clay = sediment; firing = metamorphic change; glaze melts, cools and sets = igneous rock; pottery crumbles = weathering to sediment again.
18 a The rocks have been uplifted, and the rocks above removed by weathering and erosion.
b 1 mm every 10 years.

▶ continued

Equipment needed

For each group ('Crystal size')
- selection of rock samples such as granite or gabbro (large crystals) and basalt or andesite (no crystals visible)

For the class ('Crystals from molten sulphur')
- flowers of sulphur
- test tube and holder
- Bunsen burner
- filter paper and funnel

Note
The trick is to melt the sulphur as gently as possible, without letting it start to brown. When poured into the filter paper, a crust will form fairly quickly. This should be broken open carefully before the sulphur has completely solidified. If the timing is right, long needle-shaped crystals will be exposed growing into the liquid centre. If the sulphur is kept in the filter paper, any remaining liquid will be held safely and will rapidly solidify.

For each group ('Test out your idea')
- microscope slides (hot and cold – see below)
- salol (premelted – see below)
- glass rods
- water bath

Note
Warm (not too hot) slides from an oven and cold slides from the fridge should be pre-prepared. It is probably easiest if tubes containing small amounts of molten salol (phenyl salicylate, m.p. 43 °C) and a glass rod are also prepared in advance and held in a water bath until required. The glass rod can be used to transfer the molten salol to the slides.

For each pupil (Extension)
assorted igneous rock samples, to include:
- vesicular (honeycomb) lava – pumice will do
- 'solid' lava, light and/or dark – avoid porphyritic lavas (those with a few large crystals)
- granite – crystals > 2 mm across, overall colour light grey or pink with few (if any) black crystals
- gabbro – crystals > 2 mm across, overall colour dark grey or black with more than 40% black crystals

Breaking the rocks

Book coverage

Key ideas	PoS
Rocks break down by both physical and chemical processes. Rocks can be broken by heating and cooling. Water freezing in cracks expands and pushes rocks apart. Rock fragments are worn down as they are rolled by rivers. Soil forms when rocks are weathered.	AT3.2f,g,h

Answers

Activities in text

a This is a stimulus for thought only, answered in the text below.
b Water and minerals (especially nitrogen, phosphorus and potassium).

What do you know?

1 Igneous rocks are slowly broken into pieces by **heating** and cooling. When water turns to **ice** it can push cracks open. Air and water then help to break the pieces down into **sand** and clay. This process is called **weathering**, and the rocks slowly turn into **soil**.
2 It shatters them into angular fragments.
3 Water expands as it freezes.

Assignments

4 The water in them expands as it freezes.
5 **a** A. **b** C. **c** B.
6 Plants cannot grow on bare rock waste. As the tips weather, soil forms and plants can grow. The older the tips, the more soil, so the more plants.
7 If the rock debris is covered with soil, plants will grow straight away so the area will soon look 'natural' again.

In the beginning . . .

Book coverage

Key ideas	PoS
There is hot liquid rock beneath the surface of the Earth. In places this forces its way to the surface and erupts as a volcano. Rocks that form when liquid rock cools and sets are called igneous rocks. Igneous rocks have crystals which form as the liquid rock cools. Lava rock has tiny crystals because it cools very quickly.	AT3.2g,h

Answers

Activities in text

a This is a stimulus for thought only – pupils are not expected to know about earthquake shock-wave analysis.
b A formed slowly, B quickly.
Pupils should have the chance to examine rock specimens if at all possible.

What do you know?

1 If liquid rock reaches the surface of the Earth, it will often build up a **volcano**. The liquid rock that comes out of a volcano is called **lava**. When this sets, it forms an **igneous** rock made up of **tiny** crystals. Gabbro is an igneous rock which has much **larger** crystals as it cooled more **slowly** within the Earth.
2 Volcanic eruptions.
3 The crater is open, so the lava spills out and flows down one side only. (The volcano does not explode catastrophically.)

Extras: Exploding volcanoes

1 Vesuvius suffered a catastrophic eruption, blowing its top off.
2 No – the pressure has had less time to build up in each case.
3 Yes – it's just about due for another eruption.

Assignments

1 **a,b** A volcano, B crater.
c Molten rock.
d 1000 °C.
e 15 °C.
fi Lava. **ii** It cools rapidly and solidifies to form rock.
gi It will have larger crystals. **ii** Because it cooled more slowly.
h Igneous rocks.
2 There was a volcano there, which erupted.
3 **a** It formed when molten rock cooled deep underground.
b The rock above has been removed in some way.

Igneous rocks

General note

Although it is not essential, the work undertaken in Unit 11 would clearly benefit from using any field work opportunities that are available nearby – investigation of coastal or inland outcrops of rocks, for example. As the opportunities for field work vary greatly from place to place, no details can be provided here. Science departments might like to liaise with geography departments, however, to maximise the opportunities for such work. Visits to local (or national) museums can also be helpful.

Running the activity

Core: The activity starts with a look at samples of igneous rock. If large specimens are available, a class discussion about the difference between lava, where no crystals can be seen, and granite (or gabbro), with large crystals, would be useful to lead on to question **2**.

'Crystals from molten sulphur' works very well as long as the safety instructions are followed carefully. The analogy will need to be drawn out. Close inspection of the fast-cooled crust shows tiny crystals (as in lava). The larger crystals inside grew as the liquid cooled more slowly, insulated by the crust (as in granite).

'Test out your idea' reinforces the idea that fast cooling gives small crystals, while slower cooling gives larger crystals.

Extension: Following the Core activity, pupils can use the key to identify samples of igneous rock. Make sure that the samples used are easily identifiable in the key. Do not confuse pupils with rocks such as dolerite (small crystals visible) or obsidian (volcanic glass). The key question here is still "Did the rocks cool fast on the surface or slowly underground?".

Differentiation

Sheets available are:

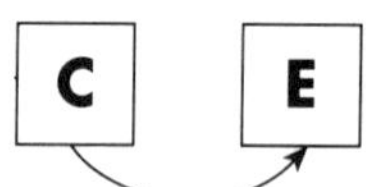

Differentiation is by outcome and level of teacher support.

AT1 programme of study

The activity covers the use of inference and analogy when working with 'uncontrollable' geological materials (**AT1.2a, b**).

Risk assessment

Core ('Crystals from molten sulphur'): Sulphur is flammable and should not be overheated – if it starts to burn it will give off toxic fumes of sulphur dioxide. For this reason, a fume cupboard should be available as a precaution. Burning sulphur can be extinguished by covering it with a damp cloth. Break open the cone of sulphur carefully – it will be hot. Wear eye protection.

Extension: Heavy rock specimens should be handled carefully and benchtops suitably protected with mats, boards or thick newspaper.

▶ continued

Weathering

Running the activity

Introduce the activity in the context of the weathering of rocks – rocks can be broken up by physical and chemical processes. Discuss 'Smash them up' and 'Here comes the rain' and then let groups of two or three pupils work at their own pace. The purpose is to give practical analogies for the weathering processes described in the pupils' book.

'How hard are the rocks?' can then be used as extension material as groups complete the first two experiments. This involves some simple planning and so can be used to develop AT1 skills. Alternatively, all pupils could plan this for a future practical session.

Differentiation

Sheets available are:

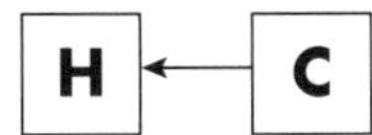

All pupils should try 'Smash them up' and 'Here comes the rain'. A fill-in Help sheet is available if needed. Some groups may be targeted to move on to 'How hard are the rocks?' rapidly, which extends the simple approach into an AT1 investigation.

AT1 programme of study

The first two experiments give practical experiences of processes occurring over longer time periods in nature (**AT1.1g**). 'How hard are the rocks?' gives an opportunity for simple planning (**AT1.1a**).

Risk assessment

'Smash them up': The broken bottle should remain sealed in its plastic bag. Eye protection must be worn during the experiment and when using acid. Warnings should be given concerning the dangers of touching red-hot glass.

For the other activities, general risks concerning the use of glassware and chemicals (including acids) apply.

Equipment needed

For the class ('Smash them up')
The day before the lesson, a glass, screw-topped bottle completely full of water should be sealed in a thick, transparent plastic bag and left in the freezer overnight. The glass should break as the water freezes and expands. This should be brought in, still in its bag for safety, at the start of the lesson so that the pupils can see it. If the introduction to the lesson is long, it may be kept in an insulated box or a bowl of crushed ice, to stop the ice in the bottle melting.

For each group ('Smash them up')
- glass rod
- tongs
- Bunsen burner
- beaker

For each group ('Here comes the rain')
- small chippings of marble, limestone, chalk, *sandstone, *shale, granite, lava
- watch-glasses
- dilute hydrochloric acid (less than 2 mol dm^{-3}) with dropping pipette

Note
* Test these samples first – some may have calcareous cement and so will fizz. Avoid these types.

For each group ('How hard are the rocks?')
- sandpaper
- access to a balance (2 decimal places)
- rock samples as above

(Other equipment may be needed.)

Move it!

Book coverage

Key ideas	PoS
Rivers carry sediment along as they flow. The faster they are flowing, the more sediment they carry. This sediment helps the river to wear away more rocks. Sediment is dropped as a river enters the sea. This can build up to form new land called a delta.	AT3.2g,h

Answers

Activities in text

a This is a stimulus for thought ('to the sea').
b1 Three layers.
b2 Pebbles at the bottom, then sand, and clay at the top.
b3 The larger pieces settle out first, while the water is still moving.
b4 Pebbles first, then sand, then clay. Clay moves the furthest in the water.

A screw-topped jar of 'soil' (pebbles, sand and clay) in water can be shaken and allowed to settle to demonstrate this effect.

What do you know?

1 Rivers carry material along as they **flow**. When they reach the **sea**, the water slows down so their material **settles** out. New layers of **sediment** are formed on the sea bed.
2 River A flows faster, as the pebbles and sand are not moving in river B.
3 Sand and mud from rivers or the sea settle out in the calm water and clog up the harbour.

Extras: What happened to the river?

1 The dam has stopped the flooding, but flooding brought in natural fertiliser, so this has stopped too.
2 The lake behind the dam has still water, so the river drops its load of mud and sand there.
3 The river built the delta into the sea by dumping its sand and mud very rapidly.
Now that this is no longer coming in, wave action is beginning to wash the delta away.

Assignments

8 They were moved when the river was in flood.
9 **a** It is carried by the river when it floods.
b The flood waters slow down once they have spread out, so even the finer material settles out.
10 The layer of sediment left behind carried fresh nutrients for the plants. No flooding means no natural fertiliser.
11 Waves move fast as they crash up the beach, moving even the big pebbles. The backwash is slower, sucking finer material back into the sea, where it settles in deeper, calmer water.

Rivers in action

Running the activity

Introduce the activity with the idea that the large-scale features in rivers can be modelled in the laboratory. The arrangement of the lesson then depends on how many stream tables are available – groups should not exceed five or six pupils. If there is only one stream table, it would be better to demonstrate its action to groups in rotation, while the rest of the class does written work from the pupils' book or assignments.

Alternatively, if the school has a suitable patch of sloping earth, a larger scale model can be demonstrated out of doors using a hose with some additional buckets of soil or sand.

Features to look for include: eroding channels with fine materials carried in the flow and coarser materials rolled along the bottom; meanders forming; ripples in the moving sand; rapids where larger pieces cause obstructions; split or abandoned channels; deltas building out into the tray of water at the base. The effects of flooding can be demonstrated by increasing the rate of flow.

Differentiation

Sheets available are: C

Differentiation is by outcome – quality of observation and the degree to which the analogy is seen between the model and real rivers.

AT1 programme of study

The activity covers the use of inference and analogy when working with uncontrollable geological materials (**AT1.1g, 1.2e, 1.3a**).

Equipment needed

For each group
- stream table made from a plank with 1 cm beading along side each (alternatives could include long tin trays, wide plastic guttering, etc.)
- large tray
- supporting blocks or clamp
- water supply
- soil: an artificial mix of sand, a little gravel and powdered chalk or clay works well, but any sandy soil will do
- bucket or bowl for waste disposal

New rocks from old

Book coverage

Key ideas	PoS
Sedimentary rocks are made from the broken fragments of other rocks or the remains of living creatures. The remains of living creatures found in sedimentary rocks are called fossils.	AT3.2g,h

Answers

Activities in text

a Air, from the gaps between the grains.
b Soft sandstone has larger air gaps. These gaps get filled in as the rock is cemented.
Pupils should have the chance to examine rock specimens if at all possible.

What do you know?

1 Sediments slowly harden to form **sedimentary rock**. Clay turns to **shale** when the water is squeezed out. Sand turns to **sandstone** when the grains are stuck together by **crystals** that grow from solution.
2 It is squeezed out by the weight of sediments above.
3 Accept suitable diagrams of crystals growing into the gaps.
4 They are made from fragments of other rocks.
5 In the sea.

Extras: A slice through time

1 They are younger, so less well cemented.
2 On land.
3 Trilobite fossils – trilobites lived in the sea.
4 100 million years.

Assignments

12 a Fossils.
bi In the sea. **ii** The rocks have been lifted up somehow.
13 It is quite young (as far as rocks go!) and so is not very hard in general.
14 a There is water in the gaps between the grains.
b Soft sandstones have little cement so the gaps are larger, giving lots of room for water. Harder sandstones have the gaps progressively filled with cement, leaving less and less room for the water.
15 a Clay gets squashed when it turns to shale, so the fossils get squashed too.
b Before – the nodules are hard and so don't get squashed.

Types of rocks

Running the activity

In the context of the rock cycle, pupils should get used to using the simple key to identify igneous, sedimentary and metamorphic rocks. Start with a collection of clear-cut examples that fit the key well, and then give less clear cut (but 'do-able') samples to those moving through the samples rapidly. The anomalous examples of slate, lava and marble should be introduced later, as they do not fit the key at all.

The local survey of the rocks in use can then be set up. This could either be done individually for homework, or as a class trail in lesson time. If the latter, a simple map of the area could be prepared for the pupils to mark up. A suitable route should be chosen to given the widest range of rock types and usages.

Differentiation

Sheets available are: C C

Differentiation is by outcome. The more able could be asked to write a report on the usage of natural stone in the area.

AT1 programme of study

The activity covers the use of keys and observations in the field (**AT1.1g, 1.2e**).

IT potential

Carry out a CD–ROM search for information on rocks and their uses, if a suitable disk is available.

Risk assessment

General rules for work in the field apply – act sensibly, politely and safely at all times. Beware of traffic. Do not damage any property, and keep to public areas. Parental permission and extra supervision may be needed – check local guidelines.

Equipment needed

For each group

- samples of obviously igneous rocks (random crystals), metamorphic rocks (banded crystals) and sedimentary rocks (fragmental) plus lava, slate and marble

Experimental and Investigative Science in the 1995 National Curriculum

The National Curriculum in science published in 1995 identifies four areas in experimental and investigative science:

1 planning experimental procedures
2 obtaining evidence
3 analysing evidence and drawing conclusions
4 considering the strength of the evidence.

These four areas form a model describing whole investigations:

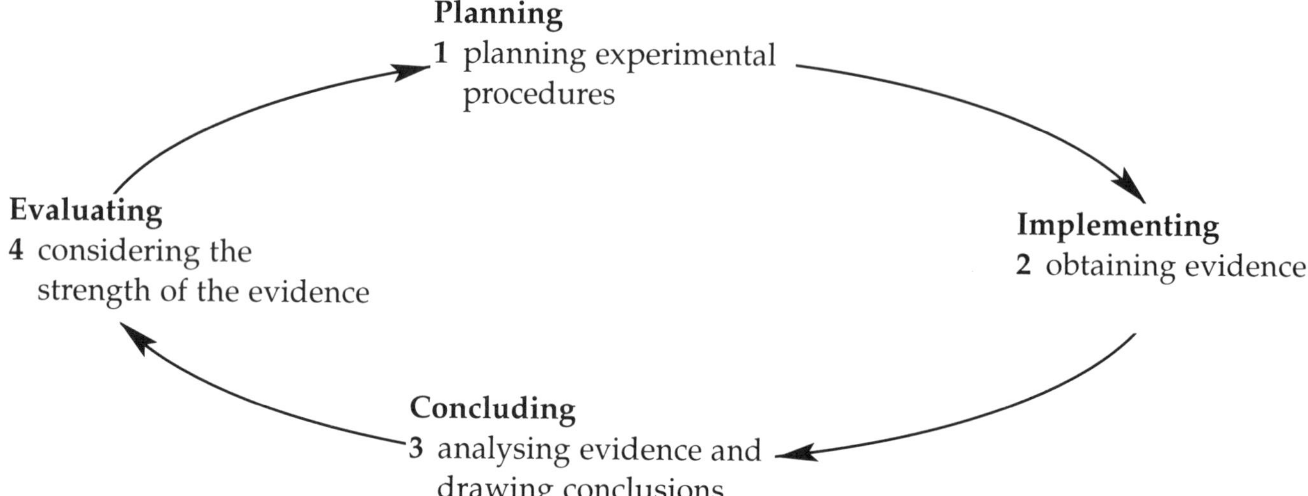

All four areas should be taught in contexts derived from the programmes of study for Life and Living Things, Materials and their Properties, and Physical Processes. There should be some opportunities for pupils to carry out the whole process, but this is not the sole requirement for the teaching or assessment of experimental and investigative science.

While teachers have always been free to develop programmes of work focused on individual aspects and skills of AT1, this is a new departure for assessment at Key Stage 3. Under the 1991 version of the National Curriculum, assessment was required to be based on whole investigations carried out by pupils. Now, although some assessment within the whole process of investigations is still seen to be desirable, a range of contexts for assessment are possible:

- assessing attainment in whole investigations
- assessing attainment in particular aspects of an investigation
- assessing attainment through activities which are not whole investigations.

In the previous Order (1991), the level of much AT1 work was defined by the nature and number of the variables included. Pupils were engaged in a great deal of quantitative work which in many teachers' experience was at the expense of qualitative work. The balance between types of practical work was found to be unsatisfactory.

In the new programme of study for Experimental and Investigative Science there is a greater emphasis on the variety of ways scientific evidence is obtained and on the quality of data obtained. An increased prominence is given to qualitative work and the programme of study reflects a broader range of experimental and investigative work. For example, detailed field work which involves thorough study of habitats and detailed observations can now more easily contribute to the assessment of AT1 compared with the difficulty of assessing such work using the 1991 statements of attainment.

The awarding of levels by using level descriptions in the new Order prevents particular combinations of variables dominating the assessment of science skills. In the new Order, teachers have a broader and more flexible framework for the planning of teaching and assessing Experimental and Investigative Science.

This section provides guidelines on constructing and implementing a teaching programme for AT1 and on assessing it based on the practical activity work in the *Science Now!* course.

Teaching AT1

The skills, knowledge and understanding to be developed are not always best taught and learned through investigations. It is certainly not necessary for the pupils to form hypotheses, plan experiments, carry them out, draw conclusions and evaluate the outcomes and procedures each time practical work is encountered.

If particular skills are to be developed, then focused activities that stress those skills are a less diffuse and more manageable way of teaching them. For example, pupils can be presented with data derived from an experiment that they have not carried out themselves, or with the outcomes of a class practical activity. In this way they can still be taught about drawing conclusions or can demonstrate their skills in analysing evidence and drawing conclusions.

On some occasions, pupils should attempt to carry out all steps of the process. Investigations help to consolidate skills and knowledge and, if well chosen and manageable, can also motivate pupils.

The activities in the *Science Now! Activity and Assessment Pack* provide a range of experiences based on the programme of study for Experimental and Investigative Science. They support the development of science skills, knowledge and understanding in pupils. The coverage of the programme of study related to AT1 which each offers is shown on the accompanying sheet of activity notes.

Putting experimental and investigative skills into context

All the activities in the pack are linked to double-page spreads in the book, so fitting easily into the overall teaching programme. This also makes it easy to integrate AT1 development and activities with the teaching of the programmes of study relating to ATs 2, 3 and 4. The book focuses on developing concepts within the three content areas of the Science curriculum and linked activities in this pack deepen and enhance this with experimental work.

Not every activity, of course, has to be used, nor do they have to be carried out exactly when the double-page spread to which they are linked is being studied. But the link enables you to see the relationship between concept and activity and where best to plan activities in a teaching programme based on the units in the book.

Planning an AT1 teaching programme

Science Now! gives a basis for planning a course with the units in the book providing both coverage and progression throughout the three years of Key Stage 3. When putting together schemes of work, schools may sequence work differently for different classes and so require experimental and investigational skills to be developed in different units.

The sequence of units will provide the basis for a scheme of work which includes a strategy for teaching and assessing AT1. The components of such a scheme could be:

- what is to be taught (skills, knowledge, understanding)
- the teaching sequence of the contents
- how much time is to be spent on a unit
- what resources are needed and are available
- what assessments are to be made and how.

Once a sequence of teaching units has been drawn up, the activities available for each unit can readily be identified. By looking at the reference chart for the activities on pages 309–11, suitable practical work can be selected to provide for progression and balance between the four areas of the programme of study related to AT1.

Information about progression in AT1 is found in the level descriptions for the attainment target. The programme of study describes what pupils should be taught as a series of teaching points identified by letters of the alphabet. The letters are for referencing purposes only and do not necessarily indicate a particular teaching sequence or hierarchy of knowledge, understanding and skills.

Using *Science Now!* activities in an AT1 programme

The chart on pages 309–11 shows the teaching points that can readily be covered in each activity in *Science Now!* Not all points need to be brought out during the activities, just those that are planned in the AT1 teaching programme or appropriate to particular pupils' needs at any one time. A couple of examples follow showing different approaches to particular activities.

Activity 4a
Capturing kinetic energy

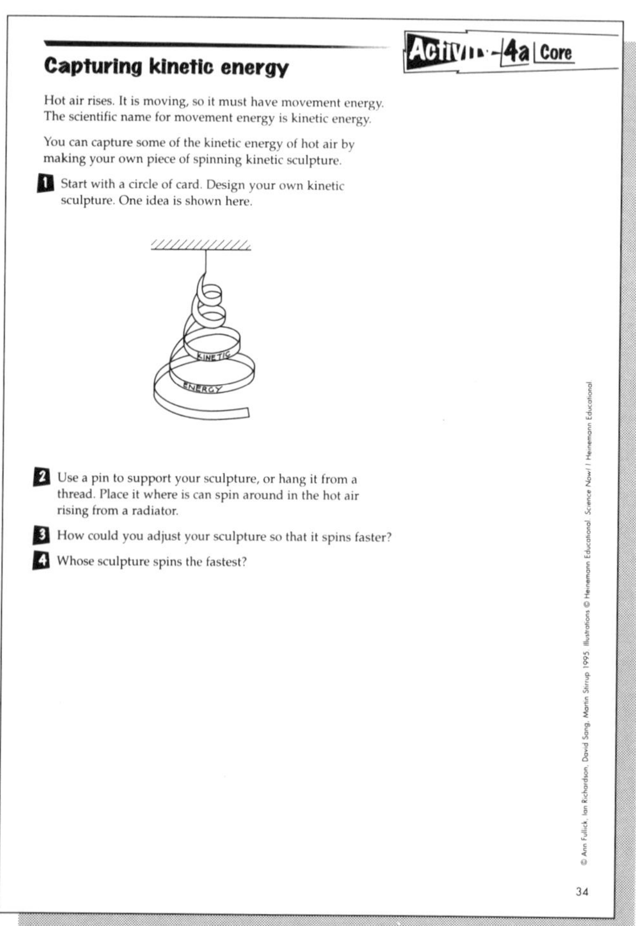

Activity 4a Core

Capturing kinetic energy

Hot air rises. It is moving, so it must have movement energy. The scientific name for movement energy is kinetic energy.

You can capture some of the kinetic energy of hot air by making your own piece of spinning kinetic sculpture.

1. Start with a circle of card. Design your own kinetic sculpture. One idea is shown here.

2. Use a pin to support your sculpture, or hang it from a thread. Place it where is can spin around in the hot air rising from a radiator.
3. How could you adjust your sculpture so that it spins faster?
4. Whose sculpture spins the fastest?

34

This activity involves pupils in carrying out trial runs (1.1b), selecting apparatus and techniques (1.1h), making observations (1.2b), looking for patterns in results (1.3d) and drawing conclusions (1.3e). These are the main teaching points which have been drawn out in the chart on page 309.

However, the same activity using the same equipment can be taught in such a way to demand repetition of measurements (1.2d), appropriate recording of evidence (1.2e), presentation of data (1.3a), predictions (1.1c) and deciding whether results support predictions (1.3f).

Activities can be made more open ended, requiring more decisions and contributions from the pupils. They can be given more structure, more detailed instructions or help in order to increase pupil confidence and reduce the risk of unsuccessful practical work.

Activity 3b
Which bag is best?

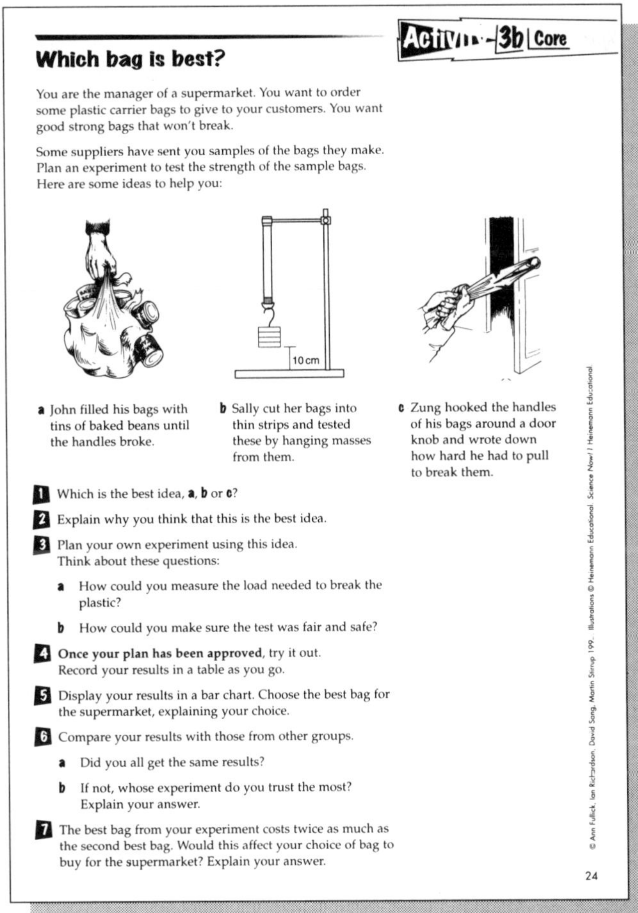

Activity 3b Core

Which bag is best?

You are the manager of a supermarket. You want to order some plastic carrier bags to give to your customers. You want good strong bags that won't break.

Some suppliers have sent you samples of the bags they make. Plan an experiment to test the strength of the sample bags. Here are some ideas to help you:

a John filled his bags with tins of baked beans until the handles broke.

b Sally cut her bags into thin strips and tested these by hanging masses from them.

c Zung hooked the handles of his bags around a door knob and wrote down how hard he had to pull to break them.

1 Which is the best idea, **a**, **b** or **c**?

2 Explain why you think that this is the best idea.

3 Plan your own experiment using this idea. Think about these questions:

- **a** How could you measure the load needed to break the plastic?
- **b** How could you make sure the test was fair and safe?

4 **Once your plan has been approved**, try it out. Record your results in a table as you go.

5 Display your results in a bar chart. Choose the best bag for the supermarket, explaining your choice.

6 Compare your results with those from other groups.

- **a** Did you all get the same results?
- **b** If not, whose experiment do you trust the most? Explain your answer.

7 The best bag from your experiment costs twice as much as the second best bag. Would this affect your choice of bag to buy for the supermarket? Explain your answer.

24

The programme of study reference for Materials (Unit 3) identifies the opportunity for pupils to carry out an investigation on the strength of carrier bags. If the teaching sequence for the science course has put the Materials unit first in Year 7, then the teacher may judge it too early for pupils to carry out an investigation for themselves.

There are two alternatives. The first is for the teacher to present the problem of: 'Which bag is strongest?', and discuss strategies, measurements needed, appropriateness of techniques, fair testing, etc. As a consequence of the discussion, an agreed procedure can be carried out, data gathered and analysed and conclusions drawn. The second alternative is to use the Help sheet provided which gives a step-by-step investigation of the strength of plastic bags.

Types of practical work

The programme of study related to AT1 describes what pupils should be taught, but it does not prescribe how they should be taught. Teachers have the scope to decide which strategy to use to teach the content and skills described in the programme of study. While the Order encourages that pupils should, on some occasions, carry out whole investigations, there is a range of practical work for teachers to choose from.

Forms of practical work can be grouped into a number of categories including:

- basic skills
- observations
- illustrations
- investigations.

The basic skills such as selecting and using apparatus, measuring, drawing graphs, using microscopes, etc. can be developed in the context of practical work. To support this development, *Science Now!* includes Skills sheets to use with pupils.

Observations are often the key to starting investigations and stimulating pupils' interest in the work. Accurate observation is fundamental to science work, and the gathering of relevant and precise data is dependent upon it. Observation is not simply seeing but involves the use of all senses in combination to describe situations and events.

When developing knowledge and understanding with a class of pupils an illustrative experiment can make learning outcomes more secure by having common procedures producing data that pupils can consider and compare as a class. Carrying out clear demonstrations of science work can be an effective way of promoting learning and understanding in pupils.

Investigations provide an opportunity for pupils to apply their knowledge and skills to solving a problem in a way that develops their curiosity and imagination.

Differentiation

The programme of study for each key stage is intended to be taught to the great majority of pupils. In Key Stage 3 this means pupils aged between 11 and 14 and in years 7–9 of their compulsory education.

Pupils who have particular special educational needs may need to work on areas of science drawn from earlier or later key stages to allow them to demonstrate progress in science. It would be helpful in these cases if such work was based in contexts that related well to the chronological age of the child, whether of higher or lower attainment.

The requirement for access for the majority of pupils is that the activities need to be differentiated for a range of ability. A curriculum that is differentiated for every pupil will build on past achievements, present challenges to allow for more achievements, provide opportunities for success and remove barriers to participation.

Strategies for differentiation include the following:

- grouping
- role
- task
- outcome
- resourcing
- extension and support.

Each strategy is explained in detail in the following pages.

Grouping

To help the management of learning, pupils can be grouped for work in a number of ways which could include pupils who have:

- similar needs for support
- comparable levels of attainment
- common experiences
- mixed gender
- mixed ability
- particular skills to be developed
- the same starting point in terms of knowledge and understanding.

It is necessary to use the grouping strategies as an active way of managing the learning situation for pupils. Most benefit will be gained by different groupings for different tasks as best suited to purpose. Because of the variety of demands made by successive activities, it is unlikely that one grouping policy will benefit all pupils in all activities.

Similar needs for support

The teacher or support staff can work with a group of pupils who have demonstrated lack of skill in displaying and handling data, to give a clear demonstration of appropriate techniques and the opportunity for pupils to ask questions to clarify their own thinking.

Comparable levels of attainment

A group of pupils who have completed a sequence of work and are in a position to tackle extension material could benefit from collaborating and discussing the extension materials.

Common experiences

In carrying out a field work survey, a number of pupils may have identified some factors of interest in common. They could be grouped together to plan and carry out an investigation of those factors.

Mixed gender

As part of a school policy on gender and in recognition of the need for pupils to experience working in a wide range of contexts, groups which have been consistently of one sex could be refreshed by mixing with members of the other sex.

Mixed ability

The range of abilities in a group can promote appropriate modification of working and allow pupils to adopt different roles. Lower attaining pupils, perhaps those with reading difficulties, can be helped to access the work by other pupils in the group. Given access, these pupils can demonstrate skills at which they have good attainment, for example, in observation or handling apparatus.

Particular skills to be developed

A number of pupils (for example, whose skills in using IT to collect and store scientific information need to be developed) could be gathered as a group. They can work together and be taught as a group the skills they require, which they can then practise, supporting each other as they work.

The same starting point in terms of knowledge and understanding

In a Year 7 class, different pupils coming from different primary schools may have had different science experiences. To avoid vain repetition of experiences, groups having common knowledge and understanding can be established to allow different work to be assigned to different groups.

Role

Within groups pupils can be given particular roles that give them the opportunity to work on different aspects of Experimental and Investigative Science, for example, selecting apparatus or handling data. As subsequent work is carried out, the role is changed to ensure a balanced experience and to develop other specific skills or understanding. If, for example, a pupil has had difficulty displaying and handling data, he or she can be given a partner to work with to develop the recording system while others in the group carry out the practical work.

Task

After establishing the teaching objectives for a class activity the next step is to develop tasks which help individual pupils to achieve these objectives, requiring different pupils to work on different tasks. There are many factors affecting the difficulty of a task:

- how familiar the pupils are with the materials;
- how familiar pupils are with the concept and vocabulary involved in the work;
- the required accuracy of measurements;
- the number and type of variable involved in the investigation;
- the extent to which the teacher prompts or helps the pupils

Outcome

This strategy for differentiation involves setting a class common tasks which are designed so that every pupil understands what is required of them. These activities will characteristically be open ended. Pupils use their knowledge, understanding and skills to achieve success at different levels. The higher attaining pupils may be expected to:

- use more difficult concepts in planning investigations
- plan and carry out more complex investigations
- complete more stages in an investigation
- complete more measurements
- measure more accurately
- record results more precisely
- describe findings using more sophisticated vocabulary or graphical techniques.

This technique is less likely to restrain the higher attaining pupils and can allow them to demonstrate their knowledge and skills at the highest level possible. Pupils who are lower attaining can still complete the task, but at a less demanding pitch.

Resourcing

The *Science Now!* activities aim to be self contained, either providing the resources needed or drawing on the pupil book for reference material. Other resources that can enrich the activity are suggested where appropriate.

The differentiated worksheets provided in this pack also assist in providing support for pupils' different needs in practical work. Some are teacher directed, that is, the teacher decides which sheet an individual pupil has before starting. Some Help sheets may be set out for pupils to use on a voluntary basis if they feel the need for more support.

Other reference books, CD-ROMS and other IT resources, while not necessary to use for the course, can also help pupils work at an appropriate level. IT opportunities arising from the activities in the pack are indicated in the activity notes where appropriate to give pupils practice in ways of using IT to collect, store, retrieve and present scientific information.

Extension and support

When a common task is set, pupils can be provided with materials that support lower attaining pupils and stretch the higher attainers.

The activities in the *Science Now!* pack provide differentiated levels of photocopiable worksheets for pupils where appropriate. The Core sheets describe the activity, the Help sheets provide support for pupils who need it and the Extension sheets stretch higher achieving pupils. These can be given out to pupils in accordance with their own level and needs, to guide them through the activities.

Assessing AT1

Using the level descriptions

The level descriptions in the Order describe the type and range of performance that pupils working at a level should characteristically demonstrate. The descriptions given are to be used in a best fit model of assessment.

This model works by estimating which level description a pupil's performance is closest to, but then considering the two adjacent descriptions to ensure the best fit for the pupil's performance. For example, a pupil might be placed at first assessment at level 5. This judgement is then consolidated by reading the level descriptions for level 4 and level 6, thus confirming that the description of level 5 best characterises the pupil's performance.

At the end of Key Stage 3, the performance of the great majority of pupils should be in the range 3–7, with level 8 being available for very able pupils. In addition to level 8, there is a description of exceptional performance that allows recognition of pupils' attainment beyond the highest number on the scale.

The level descriptions provide the basis of summative assessment at the end of the key stage. But the description of attainment that they provide can be used to inform assessment that takes place within the key stage and hence applies to the practical work of pupils carried out during the course. The level descriptions can still be used to guide judgements about attainment in individual activities.

Determining levels

Judgements required by the National Curriculum are summative and describe performance at the end of the key stage. Most schools of course also have an overall assessment policy to ensure consistency within department and school.

Within this, the best way to develop a common departmental assessment strategy with consistent standards is to generate a departmental portfolio of pupils' work. This is one portfolio per department, not per pupil. It should contain samples of pupils' work covering a range and balance of attainment targets and a range of levels. The levels related to evidence in pieces of work should be agreed amongst members of the department and used as benchmarks for teachers' assessment of pupils.

Planning assessment

Over the key stage, a suitable range of activities needs to be covered providing the evidence base for judgements about attainment as described in the level descriptions. It is not necessary for all four areas of Experimental and Investigative Science to be assessed at the same time. It may be that two or three whole investigations a year are carried out, but there will be many other opportunities to see pupils engaged in practical work and to make assessments of their attainment in those contexts.

Having decided on the frequency of assessment that can be managed effectively, teachers can set out in the schemes of work the sequence of activities to be assessed.

The coverage of the AT1 programmes of study by the activities in the pack is shown in the chart on page 309–11. These can be used for identifying the points of AT1 for assessment. An example for one activity is given here.

Activity 7c
Shadow challenge

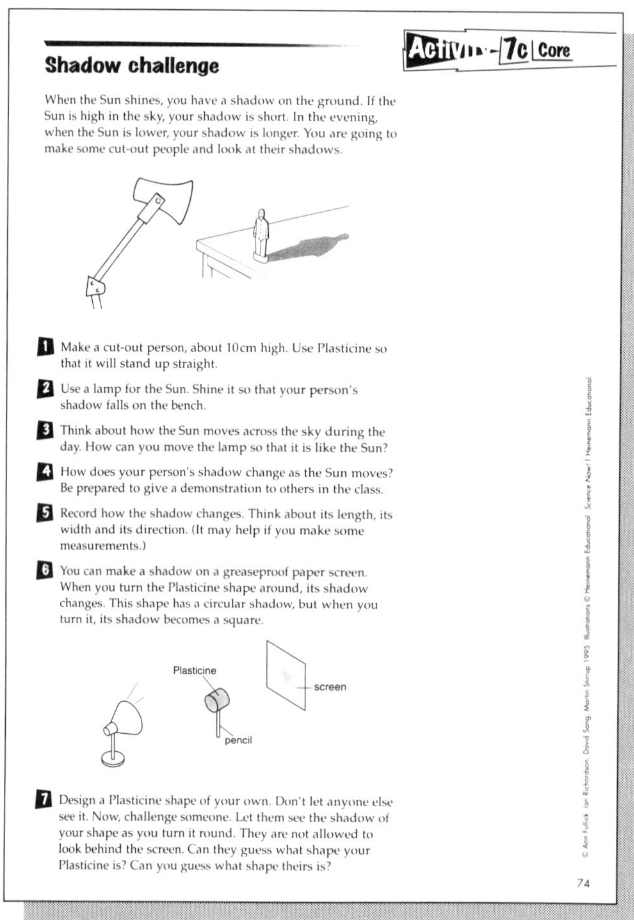

Activity 7c Core

Shadow challenge

When the Sun shines, you have a shadow on the ground. If the Sun is high in the sky, your shadow is short. In the evening, when the Sun is lower, your shadow is longer. You are going to make some cut-out people and look at their shadows.

1. Make a cut-out person, about 10cm high. Use Plasticine so that it will stand up straight.
2. Use a lamp for the Sun. Shine it so that your person's shadow falls on the bench.
3. Think about how the Sun moves across the sky during the day. How can you move the lamp so that it is like the Sun?
4. How does your person's shadow change as the Sun moves? Be prepared to give a demonstration to others in the class.
5. Record how the shadow changes. Think about its length, its width and its direction. (It may help if you make some measurements.)
6. You can make a shadow on a greaseproof paper screen. When you turn the Plasticine shape around, its shadow changes. This shape has a circular shadow, but when you turn it, its shadow becomes a square.

7. Design a Plasticine shape of your own. Don't let anyone else see it. Now, challenge someone. Let them see the shadow of your shape as you turn it round. They are not allowed to look behind the screen. Can they guess what shape your Plasticine is? Can you guess what shape theirs is?

74

Given the challenge, pupils continue to work on the practical and report their findings. The chart on page 310 shows teaching points 2b, 2e, 3a which could be addressed in this activity. Having observed the pupils working and having access to their written records of the experiment, teachers can read the level descriptions for Experimental and Investigative Science and make judgements about the attainment shown by the evidence.

Records of these judgements can be kept and used to inform summative judgements at the end of Key Stage 3. The data in the records is not used to calculate final levels via a formula, but it is used as evidence in applying the 'best fit' model to the assessment. In other words, it provides an overall picture of the pupils' attainment over the year and a 'feel' for their level.

Examples to consider

The purpose of this section is to give teachers the opportunity to think about the assessment strategy they may adopt with reference to the activities in the pack. A sample of activities has been selected to represent the range of work from different areas of science. For each of three example activities, a description of possible outcomes is described and these are related to the level descriptions given in AT1. The examples are given to attempt to illustrate the characteristic performance of pupils rather than to 'level' a piece of work. The activities could have been organised in a number of ways, modified by the teacher and used in a range of contexts for assessment:

- assessing attainment in whole investigations
- assessing attainment in particular aspects of an investigation
- assessing attainment through activities which are not whole investigations.

Activity 1c
Easy going

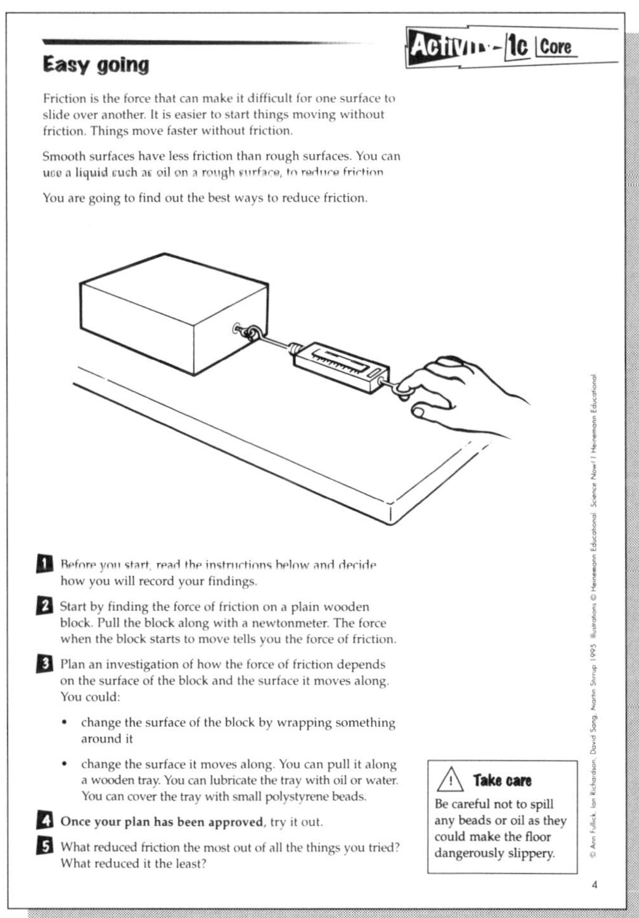

Activity 1c Core

Easy going

Friction is the force that can make it difficult for one surface to slide over another. It is easier to start things moving without friction. Things move faster without friction.

Smooth surfaces have less friction than rough surfaces. You can use a liquid such as oil on a rough surface, to reduce friction.

You are going to find out the best ways to reduce friction.

1. Before you start, read the instructions below and decide how you will record your findings.
2. Start by finding the force of friction on a plain wooden block. Pull the block along with a newtonmeter. The force when the block starts to move tells you the force of friction.
3. Plan an investigation of how the force of friction depends on the surface of the block and the surface it moves along. You could:
   - change the surface of the block by wrapping something around it
   - change the surface it moves along. You can pull it along a wooden tray. You can lubricate the tray with oil or water. You can cover the tray with small polystyrene beads.
4. **Once your plan has been approved**, try it out.
5. What reduced friction the most out of all the things you tried? What reduced it the least?

Take care
Be careful not to spill any beads or oil as they could make the floor dangerously slippery.

© Ann Fullick, Ian Richardson, David Sang, Martin Stirrup 1995 Illustrations © Heinemann Educational. Science Now! 1 Heinemann Educational

4

Pupils are introduced to the idea of friction and set the challenge of investigating how the force of friction depends upon the surface of the block and surface it moves along. If this activity is used for assessment purposes, it becomes more demanding and provides more scope for pupil contributions if less instruction and fewer suggestions are given.

At level 4 pupils will describe the importance of changing only one factor in their practical work and thus demonstrate the need for fair tests. They will therefore describe how the same block is pulled across different surfaces and will know that changing the block can only be done in a fair test if the other factors, i.e. materials and surfaces, stay the same. It is possible for them to make predictions because they know about the friction-reducing properties of oil, etc. They record their observations and measurement and present them in a table, and may also show the data on a bar chart. For pupils using the Help sheet, the tables are provided and thus the demand of the work is reduced. At level 4 they draw conclusions which can relate the reduction of friction to the properties of oil or the small polystyrene beads.

At level 5 the measurements they make are done with precision appropriate to the task and they will repeat measurements to attempt to spot and describe any reasons for differences in observations or measurements. In this activity, unintended variation in number of beads or amount of oil can produce variation in data.

At level 6 pupils may work with the extension sheet and consider a wider range of factors, deciding for themselves the number and accuracy of measurements to be made. They could make predictions as prompted on the Extension sheet and based on scientific knowledge and understanding. They may plot the data gathered on line graphs for which they choose the appropriate scale, and include data such as mass of the block, surface area in contact or force required to move the block. Their conclusions relate clearly to the evidence and are explained using scientific knowledge and understanding.

Activity 3b
Which bag is best?

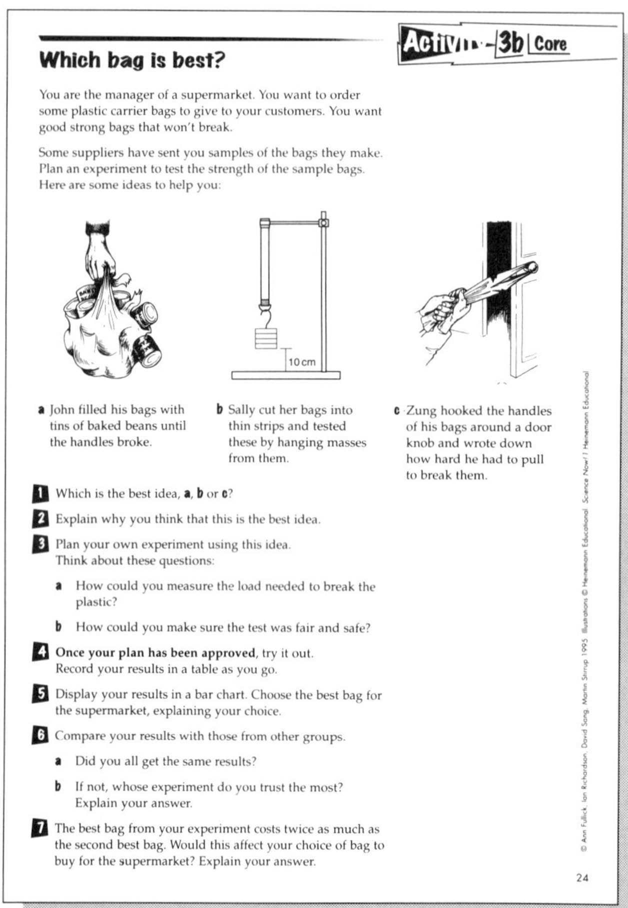
Activity 3b Core

Which bag is best?

You are the manager of a supermarket. You want to order some plastic carrier bags to give to your customers. You want good strong bags that won't break.

Some suppliers have sent you samples of the bags they make. Plan an experiment to test the strength of the sample bags. Here are some ideas to help you:

a John filled his bags with tins of baked beans until the handles broke.

b Sally cut her bags into thin strips and tested these by hanging masses from them.

c Zung hooked the handles of his bags around a door knob and wrote down how hard he had to pull to break them.

1. Which is the best idea, **a**, **b** or **c**?
2. Explain why you think that this is the best idea.
3. Plan your own experiment using this idea. Think about these questions:
   - **a** How could you measure the load needed to break the plastic?
   - **b** How could you make sure the test was fair and safe?
4. **Once your plan has been approved**, try it out. Record your results in a table as you go.
5. Display your results in a bar chart. Choose the best bag for the supermarket, explaining your choice.
6. Compare your results with those from other groups.
   - **a** Did you all get the same results?
   - **b** If not, whose experiment do you trust the most? Explain your answer.
7. The best bag from your experiment costs twice as much as the second best bag. Would this affect your choice of bag to buy for the supermarket? Explain your answer.

24

This activity is an opportunity for a whole investigation which could be used for assessment of all or some selected aspects of AT1. Pupils following the Core sheet are given a number of ideas from which they can develop their own plan. It would be possible to set the challenge without examples, but this would make it more difficult to come up with a strategy that would work. The degree of freedom given to pupils needs to be considered, as does the course of action to be taken by teachers when pupils are not progressing with the investigation. The Help sheet provides more structure to allow lower attaining pupils to carry out some practical work with an increased likelihood of success.

At level 4 pupils are testing the bags to see which is the strongest. There is no need to make predictions in every activity and in this case it is not appropriate. The equipment is used to make observations and measurements.

At level 5 they do not have the background knowledge needed to make predictions related to scientific knowledge and understanding. There is scope for pupils to generate their own list of apparatus required, or they may be allowed to choose from a range of equipment displayed. At level 5 the apparatus is used with precision appropriate to the task. Characteristic of level 5 also would be the repetition of observations and the identification of results that are anomalous. The data derived from the different tests could be displayed as block graphs, each block relating to a different bag.

A sophisticated treatment characteristic of level 5 would be the drawing of line graphs that showed the extension of the sample strips up to the point of failure.

At level 6 choice of line graphs would be made, as would the choice of scales to show the data effectively, the repetition of measurements, having a concern for accuracy of measurement and using instruments with fine divisions. At level 6 it would be characteristic to offer comment on anomalous data, draw conclusions that are consistent with the evidence and offer explanations based on scientific knowledge and understanding. At the start of the investigation, the identification of key factors out of a range of possible factors that need to be considered is supportive of level 6. In less complex contexts involving only a few factors this is less demanding and may support level 5 attainment.

Activity 9c
Make your own flying fruit

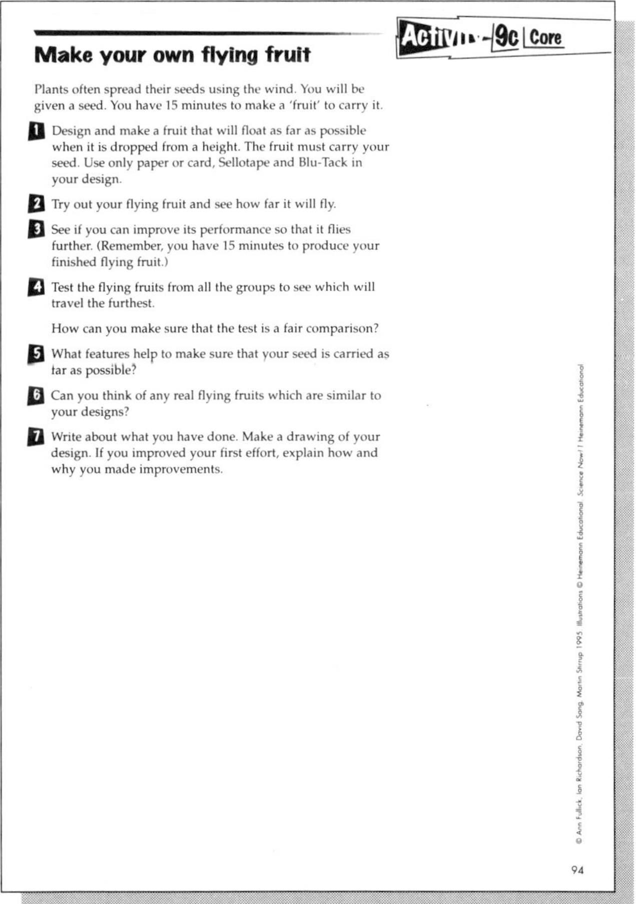

Activity 9c Core

Make your own flying fruit

Plants often spread their seeds using the wind. You will be given a seed. You have 15 minutes to make a 'fruit' to carry it.

1. Design and make a fruit that will float as far as possible when it is dropped from a height. The fruit must carry your seed. Use only paper or card, Sellotape and Blu-Tack in your design.
2. Try out your flying fruit and see how far it will fly.
3. See if you can improve its performance so that it flies further. (Remember, you have 15 minutes to produce your finished flying fruit.)
4. Test the flying fruits from all the groups to see which will travel the furthest.

   How can you make sure that the test is a fair comparison?
5. What features help to make sure that your seed is carried as far as possible?
6. Can you think of any real flying fruits which are similar to your designs?
7. Write about what you have done. Make a drawing of your design. If you improved your first effort, explain how and why you made improvements.

94

This activity as described on the activity sheet is a timed exercise as part of a learning experience. This time pressure in an assessment situation may not prove helpful to many pupils. If the assessment is aimed to give pupils the opportunity to show what they can do then it is inappropriate to limit their performance by inadequate time allocation. There will naturally be the constraint of lesson length and the need for the teacher to move pupils on to continue with the scheme of work.

There is the opportunity for pupils to have a wide range of ideas here, and to test out by trial and error. They could also develop shapes for their fruit based on theory, their own rationale and their understanding of science.

At level 4 pupils would recognise the importance of fair testing, a notion which is clear in the activity sheet. The use of pieces of material that have the same size is evidence that the pupil has a fundamental understanding of fair testing. This understanding is commensurate with level 4 attainment.

At level 5 more detailed descriptions of factors that need to be controlled such as surface area and uniformity of launching procedures would be expected.

At level 6 more than a few factors are recognised and key factors such as mass and surface area are identified.

Measurement in this practical and the calculation of speed can be carried out with varying degrees of accuracy. Data gathering can be increasingly systematic with more concern for repetition of measurements.

At level 4 a series of observations are made and could be displayed in a table.

At level 5 repetition of measurements concerning the travel of the fruit are characteristic.

At level 6 pupils are repeating the measurements of distance and considering results that do not fit the pattern of others. They will also have the notion of sufficiency of the data for the task undertaken.

When drawing conclusions, level 6 could involve ensuring a consistency with evidence and an explanation based on scientific understanding such as the size of mass and surface area and the consequent effect on air resistance. Conclusions should be supported by the data and begin to offer some relationship with science knowledge.

Science Now! Activities and AT1

Unit 1 Forces

		Experimental and Investigative Science: Teaching points																						
	AT1 code	1a	1b	1c	1d	1e	1f	1g	1h	2a	2b	2c	2d	2e	3a	3b	3c	3d	3e	3f	3g	4a	4b	4c
Activity	1a			Y					Y		Y			Y						Y				
	1b			Y							Y								Y					
	1c		Y		Y	Y	Y							Y					Y	Y				
	1d		Y	Y		Y							Y		Y						Y		Y	Y
	1e						Y								Y	Y	Y	Y	Y					

Unit 2 Living things

		Experimental and Investigative Science: Teaching points																						
	AT1 code	1a	1b	1c	1d	1e	1f	1g	1h	2a	2b	2c	2d	2e	3a	3b	3c	3d	3e	3f	3g	4a	4b	4c
Activity	2a					Y	Y				Y	Y	Y											
	2b					Y	Y	Y	Y										Y		Y			Y
	2c									Y	Y			Y										
	2d										Y			Y										
	2e										Y			Y										
	2f										Y			Y										
	2h										Y			Y										

Unit 3 Materials

		Experimental and Investigative Science: Teaching points																						
	AT1 code	1a	1b	1c	1d	1e	1f	1g	1h	2a	2b	2c	2d	2e	3a	3b	3c	3d	3e	3f	3g	4a	4b	4c
Activity	3a				Y	Y	Y			Y	Y	Y	Y	Y	Y					Y				
	3b	Y	Y	Y	Y	Y	Y		Y	Y	Y	Y	Y	Y	Y			Y	Y	Y	Y	Y	Y	Y
	3c										Y	Y						Y					Y	
	3d			Y	Y	Y									Y	Y		Y						
	3e	Y	Y	Y	Y	Y	Y		Y	Y	Y	Y	Y	Y	Y		Y	Y	Y	Y	Y	Y	Y	

Unit 4 Energy

		Experimental and Investigative Science: Teaching points																						
	AT1 code	1a	1b	1c	1d	1e	1f	1g	1h	2a	2b	2c	2d	2e	3a	3b	3c	3d	3e	3f	3g	4a	4b	4c
Activity	4a		Y						Y		Y							Y	Y					
	4b	Y		Y	Y	Y	Y			Y	Y		Y	Y	Y			Y			Y	Y		Y
	4c	not appropriate																						
	4d	not appropriate																						
	4e		Y										Y	Y			Y	Y			Y		Y	

Unit 5 Food for life

		Experimental and Investigative Science: Teaching points																						
AT1 code		1a	1b	1c	1d	1e	1f	1g	1h	2a	2b	2c	2d	2e	3a	3b	3c	3d	3e	3f	3g	4a	4b	4c
Activity	5a				Y	Y				Y														
	5b	Y						Y		Y					Y									
	5c														Y	Y		Y						
	5d														Y	Y		Y						
	5e	Y		Y																		Y		Y
	5f	Y		Y																		Y		Y

Unit 6 Mixtures

		Experimental and Investigative Science: Teaching points																						
AT1 code		1a	1b	1c	1d	1e	1f	1g	1h	2a	2b	2c	2d	2e	3a	3b	3c	3d	3e	3f	3g	4a	4b	4c
Activity	6a	Y			Y				Y	Y									Y					
	6b			Y	Y	Y	Y							Y	Y				Y	Y				
	6c1			Y	Y	Y	Y			Y	Y												Y	Y
	6c2			Y	Y	Y	Y												Y	Y				
	6d									Y														
	6e		Y							Y								Y	Y			Y		

Unit 7 Sound and light

		Experimental and Investigative Science: Teaching points																						
AT1 code		1a	1b	1c	1d	1e	1f	1g	1h	2a	2b	2c	2d	2e	3a	3b	3c	3d	3e	3f	3g	4a	4b	4c
Activity	7a										Y								Y					
	7b											Y		Y	Y	Y	Y							
	7c										Y			Y	Y									
	7d	not appropriate																						
	7e										Y	Y	Y											

Unit 8 Changing materials

		Experimental and Investigative Science: Teaching points																						
AT1 code		1a	1b	1c	1d	1e	1f	1g	1h	2a	2b	2c	2d	2e	3a	3b	3c	3d	3e	3f	3g	4a	4b	4c
Activity	8a									Y	Y	Y		Y	Y	Y	Y	Y	Y		Y			
	8b									Y	Y	Y		Y	Y	Y		Y	Y		Y			
	8c																Y		Y		Y			
	8d									Y	Y								Y		Y			
	8e	Y		Y				Y	Y					Y		Y	Y					Y		

Unit 9 Reproduction

		Experimental and Investigative Science: Teaching points																						
	AT1 code	**1a**	**1b**	**1c**	**1d**	**1e**	**1f**	**1g**	**1h**	**2a**	**2b**	**2c**	**2d**	**2e**	**3a**	**3b**	**3c**	**3d**	**3e**	**3f**	**3g**	**4a**	**4b**	**4c**
Activity	**9b**									Y	Y							Y			Y			
	9c		Y		Y	Y				Y			Y	Y				Y	Y			Y	Y	Y
	9d	not appropriate																						
	9e	not appropriate																						
	9d,e	not appropriate																						
	9h	not appropriate																						

Unit 10 Electricity

		Experimental and Investigative Science: Teaching points																						
	AT1 code	**1a**	**1b**	**1c**	**1d**	**1e**	**1f**	**1g**	**1h**	**2a**	**2b**	**2c**	**2d**	**2e**	**3a**	**3b**	**3c**	**3d**	**3e**	**3f**	**3g**	**4a**	**4b**	**4c**
Activity	**10a**								Y					Y							Y			
	10b									Y				Y										
	10c			Y																				
	10d				Y	Y						Y	Y	Y				Y						
	10e																	Y			Y			
	10f	demonstration: not applicable																						

Unit 11 Rocks

		Experimental and Investigative Science: Teaching points																						
	AT1 code	**1a**	**1b**	**1c**	**1d**	**1e**	**1f**	**1g**	**1h**	**2a**	**2b**	**2c**	**2d**	**2e**	**3a**	**3b**	**3c**	**3d**	**3e**	**3f**	**3g**	**4a**	**4b**	**4c**
Activity	**11a**									Y	Y											Y		
	11b	Y						Y																
	11c							Y						Y	Y									
	11d										Y			Y										
	11e							Y						Y										

Forces

You can find the answers on these pages of the book.

1 There is no movement in this tug-of-war. What do the forces need to be like for the rope to move? Underline the correct word.

red team white team

unbalanced **strong** **balanced** **pulling** 2–3

2 The red team wins. Draw an arrow on the diagram to show the direction of the bigger force. 2–3

3 What are forces measured in? Underline the correct word.

kilograms **newtons** **centimetres** 2–3

4 If you throw a ball up into the air, _ _ _ _ _ _ _ pulls it back down again. 4–5

5 **a** The _ _ _ _ _ _ of something is the pull of the Earth's gravity on it.

b What is this pull measured in? Underline the correct word.

kilograms **newtons** **centimetres** 4–5

6 **a** What is the amount of 'stuff' (matter) in an object called?

_ _ _ _

b What is it measured in? Underline the correct word.

kilograms **newtons** **centimetres** 4–5

7 If something is moving, _ _ _ _ _ _ _ _ acts to slow it down. 6–7

8 The skier is moving downhill. Draw an arrow to show the force of friction from the ground. 6–7

9 A _ _ _ _ _ _ _ _ _ _ _ shape meets less friction in air and in water. 8–9

10 Underline the correct word to complete the sentence.
A parachute helps to slow you down because of the force of:

gravity **streamlining** **air resistance** 8–9

11 Underline the correct word to complete the sentence.
A woman can just lift a weight of 500 N using a lever.
To lift a weight of 900 N, she would need a lever that was:

longer **shorter** **the same** 10–11

Living things

Test yourself Unit 2

You can find the answers on these pages of the book.

1 Complete this list of the seven features of life.

M _ _ _ _ _ _ _ _ R _ _ _ _ _ _ _ _ _ _ _ _

S _ _ _ _ _ _ _ _ _ _ _ _ _ G _ _ _ _ _ _ E _ _ _ _ _ _ _ _ _ _

R _ _ _ _ _ _ _ _ _ _ _ _ _ _ _ N _ _ _ _ _ _ _ _ _ _ 14–15

2 Complete these sentences using the words **animals** or **plants**.

________ make their own food, but ________ do not.

Most ________ can only move very slowly, but most ________ can move fast. 16–17

3 **a** This is a typical animal cell. Add the labels:

cytoplasm nucleus cell membrane

____

____

____

b Draw in two other parts that you would find in a plant cell, and label them. 18–19

4 Complete the sentence. When scientists put things into groups, they

_ _ _ _ _ _ _ _ them. 20–1

5 Write these words in the correct columns to complete the table.

human robin slug snail octopus toad

Vertebrates	**Invertebrates**

20–1

6 Which group of invertebrate animals can be split into centipedes and millipedes, crustaceans, insects and spiders? _ _ _ _ _ _ _ _ _ _ _ _ _ 22–3

7 There are five vertebrate groups. Two are given here. Add the other three.

amphibians reptiles _ 24–5

8 Underline the names of the four main plant groups.

flowering plants trees conifers ferns shrubs mosses 26–7

9 Why do scientists use keys? Underline the correct answer.

a to group organisms together **c** to identify organisms
b to join organisms **d** to examine organisms 28–9

Materials

You can find the answers on these pages of the book.

1 Write these words in the correct columns to complete the table.

pencil air water orange juice desk

Solids	Liquids	Gases

32–3

2 The space something fills is called its _ _ _ _ _ _. 32–3

3 Draw in lines to link the words together to make correct sentences.

a Gases — have a fixed shape and a fixed volume.
b Liquids — have no fixed shape and no fixed volume.
c Solids — have no fixed shape but have a fixed volume. 32–3

4 Which of these can easily be squashed? Underline the correct answer.

solid liquid gas

5 Tick the property that applies to each material.

	Heavy	Light
aluminium		
iron		

	Strong	Weak
straw		
brick		

	Hard	Soft
diamond		
pencil lead		

34–5

6 Underline the correct word to complete each sentence.

a Most metals are **shiny/dull**.
b Metals are good **insulators/conductors** of heat and electricity.
c **Iron/Aluminium** is attracted to a magnet. 36–7

7 If you want to compare the 'heaviness' of iron and aluminium, what size pieces would you need to make it a fair test? Underline the correct answer.

A a 2 cm cube of iron and a 1 cm cube of aluminium
B a 1 cm cube of iron and a 1 cm cube of aluminium
C a 3 cm cube of iron and a 4 cm cube of aluminium 38–9

8 The mass of a 1 cm cube of a material is its _ _ _ _ _ _ _. 38–9

9 Underline the two materials with the highest density.

water feathers lead aluminium gold air 38–9

10 Underline the correct number in this sentence. 38–9
The density of water is **1/10/100** grams per cubic centimetre.

11 Tick the boxes that describe the particles in solids, liquids and gases.

	Solids	Liquids	Gases
closely packed			
not free to move			
free to move			
far apart			

40–1

Energy

Test yourself Unit 4

You can find the answers on these pages of the book.

1 Complete the following sentences. Use the words below to fill the gaps.

movement turning transfers kinetic

A hamster _______ energy to its wheel by _______ it. The wheel has _______

energy. The scientific name for this kind of energy is _______ energy. 44–5

2 **a** A substance that burns to release energy is a fuel, for example, coal. Which type of stored energy is in the fuel before it is burnt? Underline the correct word.

elastic energy heat energy chemical energy

b Complete the energy transfer diagram for the burning of coal. 46–7

3 **a** What is the fuel used by the human body? _ _ _ _

b What is the energy value of food measured in? Underline the correct word.

kilograms kilojoules kilowatts

c Which of the foods given below would have the highest energy value per 100 g of food? Underline the correct word.

milk bread peanuts 48–9

4 Name four types of energy that electrical energy can be changed into.

a _______________ **c** _______________

b _______________ **d** _______________ 50–1

5 Complete the following sentences. Use the words below to fill the gaps.

coal power stations gas generators

Most electricity comes from _______ _____________. There they burn fuels

such as _______ or _______ to release energy to turn _______. 50–1

6 Draw in lines to link the words together to make correct sentences.

a Anything that is stretched or squashed has — chemical energy.

b A battery stores — gravitational energy.

c Something lifted off the ground has — elastic energy. 52–3

Food for life

You can find the answers on these pages of the book.

1 This diagram shows how a leaf makes food.
Use the words below to complete the labels. 56–7

oxygen water carbon dioxide sunlight starch chloroplasts

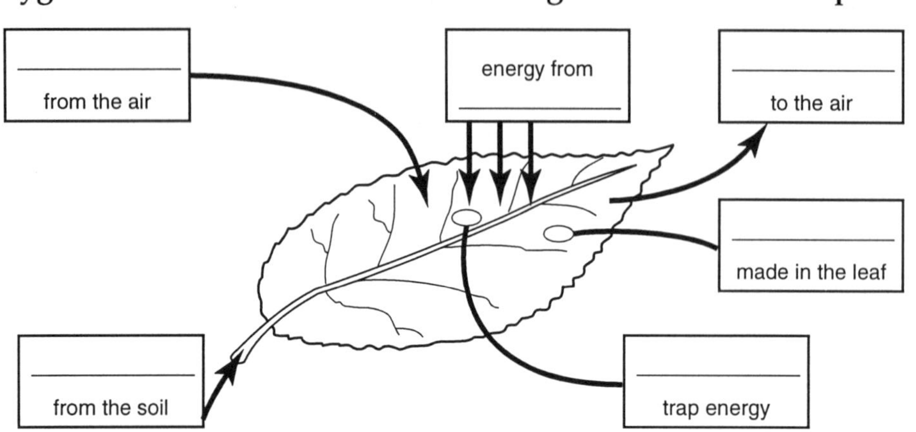

2 Underline the correct word to complete each sentence.

- **a** Plants need **minerals/food/strength** from the soil to be healthy.
- **b** Plants can be given nitrogen, phosphorus and **carbon/oxygen/potassium** in fertiliser.
- **c** Plants need roots for anchorage, and to get water and **minerals/food/carbon**. 58–9

3 Draw in lines to link the words together to make correct sentences describing parts of a balanced diet. One has been done for you.

a Fats	have a lot of energy and are easily digested.
b Carbohydrates	are needed for healthy bones and teeth.
c Proteins	have the highest energy content.
d Minerals	are essential for health.
e Vitamins	help to build muscle and repair your body. 60–1

4 Name a mineral that is needed for:

- **a** keeping bones hard and strong. _ _ _ _ _ _ _
- **b** the blood to carry oxygen. _ _ _ _ 62–3

5 Name the vitamins needed to prevent:

- **a** rickets. _ _ _ _ _ _ _ _ _
- **b** scurvy. _ _ _ _ _ _ _ _ _ 62–3

6 Underline the correct words to complete the sentences.
Plants make their own food and are called **consumers/producers**. Animals take in food by eating other organisms and are called **consumers/producers**. The place where a plant or animal lives is called a **den/habitat/food chain**. 64–5

7 Where many plants and animals live together in the same place, the food chains are connected in a _ _ _ _ _ _ _. 66–7

Mixtures

Test yourself Unit 6

You can find the answers on these pages of the book.

1 Draw in lines to link each mixture with the best way of separating it. One has been done for you.

Mixture	Method	
a different coloured Smarties	sieve	
b magnetic and non-magnetic screws	filter	
c flour and currants	decant	
d liquid with sediment	pick out by hand	
e suspension of chalk and water	use a magnet	70–1

2 Underline the correct word to complete each sentence.

a A solid dissolves in water to form a **solute/solvent/solution**.
b In a solution, water is called the **solute/solvent/solution**.
c In a solution, the solid that dissolves is called the **solute/solvent/solution**.
d Some solids do not dissolve. They are **soluble/insoluble**.
e Solids that dissolve are **soluble/insoluble**.
f When no more solid will dissolve, the solution is **saturated/unsaturated**. 72–3

3 What are the three main ways to speed up dissolving?

a ______________________________

b ______________________________

c ______________________________ 72–3

4 Underline the correct words to complete each sentence.

a Puddles of water disappear because water **evaporates/dissolves** into the air.
b Energy from the sun will **speed up/slow down** this process.
c You can get the solute back from a solution by **pouring off/evaporating** the water. 74–5

5 Underline the correct words to complete each sentence.

a Windows steam up because **water vapour/liquid water** in the air hits the cold glass and **evaporates/condenses** to form small drops of **liquid water/water vapour**.
b To get pure water back from a solution, you can **filter/evaporate** the water and **condense/dissolve it**. This is called **distillation/condensation**. 76–7

6 What is the name of the process for separating different dyes out of a mixture?

_ _ _ _ _ _ _ _ _ _ _ _ _ _ 78–9

Sound and light

Test yourself Unit

You can find the answers on these pages of the book.

1 A door bangs and you hear it. The table shows what things happen for you to hear the sound. Put them in the correct order by writing a number beside them (1 for the first thing to happen, 2 for the second, and so on). The first one has been done for you.

The three little bones vibrate.	
The eardrum vibrates.	
The air vibrates.	
You hear a sound.	
A message is sent to the brain.	
The fine hairs vibrate.	
The door vibrates.	1

82–3

2 John, aged eight, his dad and his grandad were outside at night. A bat flew past, squeaking.

a Who is most likely to hear the bat? ____________

b Who is least likely to hear the bat? ____________ 84–5

3 An echo is a sound which is ____________ from a hard, smooth surface. 84–5

4 In a car, there is always a lining of thick felt between the driver and the engine. The felt quietens the engine noise, because it is a soft, rough material which

_ _ _ _ _ _ _ _ the sound. 84–5

5 Underline the correct words to complete each sentence.

a Light rays travel **in straight lines/around corners**.

b When someone blocks rays of light, they form **an echo/a shadow/a reflection**. 86–7

c Light travels **faster than/slower than/at the same speed as** sound. 88–9

6 Complete the following sentences. Use the words below to fill the gaps.

smooth rough scattered reflected

Jo noticed that she could see her reflection in the shiny foil wrapper of her Kit-Kat. When she screwed the wrapper up and opened it out again, she could no longer see her reflection in it. This is because light is ____________ well from ____________ surfaces but it is ____________ by ____________ surfaces. 90–1

7 Write these words in the correct columns to complete the table.

Sun star Moon Earth torch headlight cat's eye mirror

Produce light (sources)	**Only reflect light**

90–1

Changing materials

You can find the answers on these pages of the book.

1 Underline the correct words to complete the sentences.

a Ice is **rock/solid water**. The gas that water forms when it evaporates is called **water vapour/air**.

b A solid changes into a liquid at its **melting point/boiling point**.

c A liquid changes into a gas at its **melting point/boiling point**. 94–5

2 Underline the correct word to show whether the following statements are **true** or **false**.

a Pure water freezes at 0 °C. **true/false**

b All substances boil at 100 °C. **true/false**

c Mixtures melt and boil at one temperature. **true/false**

d Evaporation and melting are reversible. **true/false** 94–5

e When salt is added to water, it makes the freezing point lower. **true/false** 96–7

3 The table shows whether substances are solid, liquid or gas at room temperature. Are their melting points and boiling points higher or lower than room temperature? Fill in the blanks in the table using the words **higher** or **lower**.

State at room temperature	Melting point	Boiling point
solid	higher	
liquid		
gas		lower

96–7

4 Underline the correct words to complete each sentence.

a When most solids are heated they **expand/contract**, and when they are cooled they **expand/contract**.

b Thermometers work because most liquids **expand/contract** when heated and **expand/contract** when cooled. 98–9

5 Underline the correct word to show whether the following are **physical** or **chemical** changes.

a melting wax **physical/chemical**

b boiling water **physical/chemical**

c burning a candle **physical/chemical** 100–1

6 Underline the correct word to show whether the statements are **true** or **false**.

a Changes of state are physical changes. **true/false**

b Chemical changes are easily reversible. **true/false**

c Chemical changes may give off light and heat energy. **true/false**

d Physical changes are not reversible. **true/false**

e Chemical changes form new substances. **true/false** 100–1

7 When something burns:

a What gas does it use from the air? ____________

b Is it a physical or chemical change? ____________ 102–3

8 When iron rusts:

a What two things does it need to rust? ____________ and ____________ 102–3

b Is it a physical or chemical change? ____________

Reproduction

You can find the answers on these pages of the book.

1 Underline the correct word to show whether the following statements are **true** or **false**.

a Only some living organisms need to reproduce. **true/false**
b When male and female sex cells join, it is called fertilisation. **true/false**
c In animals, the female sex cell is called the ovule. **true/false**
d The male sex cell in plants is carried in the pollen. **true/false** 106–7

2 Complete the following sentences. Choose from the words below to fill the gaps.

ovary stem seeds fertilisation flower pollination stamens

The part of a plant that contains the sexual organs is the __________. When pollen from one plant lands on the stigma of another, it is called __________. The pollen then fertilises the ovules in the __________. The fertilised ovules become __________. 108–9

3 It is important for the plant to scatter its seeds around. This is called __________. 110–11

4 Complete the following sentences. Choose from the words below to fill the gaps.

urethra testes hair puberty penis voice sperm

The name given to the time when your body becomes capable of reproducing is __________. For boys, this is the time when they grow more __________ on their faces and bodies, their __________ gets deeper and their __________ and __________ get larger. 112–13

5 Tick the things that happen to a girl's body at puberty.

A Her breasts develop. ☐
B Her monthly periods begin. ☐
C Her uterus becomes much smaller. ☐
D Her ovaries start producing an ovum each month. ☐ 114–15

6 Underline the correct word to show whether the statements are **true** or **false**.

a During sexual intercourse, sperm are released into the woman's vagina and swim up into the uterus and Fallopian tubes. **true/false**
b A fertilised egg implants at the top of the Fallopian tube, near to the ovary. **true/false**
c If a sperm meets an egg (ovum), the egg is fertilised. **true/false** 116–7

7 Tick those things that happen **every month** in a woman's menstrual cycle.

A An egg ripens and leaves the ovaries. ☐
B An egg is fertilised. ☐
C The cycle lasts about 28 days. ☐
D The lining of the uterus builds up during the cycle. ☐ 118–19

8 Underline the correct answer to complete each sentence.

a A fetus is **an unfertilised egg/a developing baby/a ripening ovary**.
b A developing baby in the uterus receives food and oxygen from its mother through the **Fallopian tubes/placenta/vagina**. 120–1

Electricity

You can find the answers on these pages of the book.

1 Underline the correct words to complete each sentence.
- **a** For electric current to flow, you need a **bulb/buzzer/complete circuit**.
- **b** An insulator **does/does not** allow an electric current to pass through it.
- **c** A conductor **does/does not** allow an electric current to pass through it. 124–5

2 Complete the following sentence.

Electric current flows from the ___________ end of a battery, round a circuit to the ___________ end of the battery. 124–5

3 Write in the correct component for each symbol to complete the table.

Symbol	Component
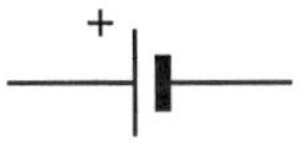	
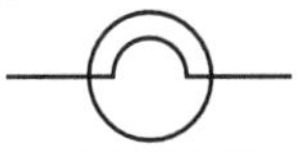	
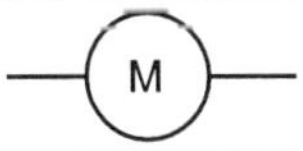	

126–7

4 In circuit 1, bulbs A and B are connected in series. In circuit 2, bulbs X and Y are connected in parallel. All the bulbs have the same resistance. Circuits 1 and 2 each have just one battery.

- **a** Which bulbs shine more brightly, A and B or X and Y? ___________
- **b** In circuit 1, bulb B is unscrewed. What happens to bulb A? ___________
- **c** In circuit 2, bulb X is unscrewed. What happens to bulb Y? ___________ 128–9

5 Complete the following sentences. Choose from the words below to fill the gaps.

resistance smaller bigger energy

Current flows around a circuit. All the components in the circuit have electrical ___________. The more resistance there is in the circuit, the ___________ the current that flows. A long wire has a ___________ resistance than a short wire. 130–1

6 Underline the correct words to complete each sentence.
- **a** Current **does/does not** get used up around a circuit. 132–3
- **b** A battery is a store of chemical **heat/electricity/energy**.
- **c** The energy stored in the battery is transferred along a circuit in the form of **light/electrical/current** energy.
- **d** A lamp changes **battery/chemical/electrical** energy into **kinetic/elastic/light** energy and **heat/stored/sound** energy. 134–5

Rocks

You can find the answers on these pages of the book.

1 Complete the following sentences. Use the words below to fill the gaps.

igneous volcano crystals lava

When molten rock reaches the Earth's surface, a __________ is often formed.

The liquid rock is called __________. When it cools, a rock with

__________ forms. This is called __________ rock. 138-9

2 The table shows what things happen to cause the physical weathering of rock to form small pebbles and soil. Put them in the correct order by writing a number beside them (1 for the first thing to happen, 2 for the second, and so on). The first one has been done for you.

Pieces of rock drop off.	
Water enters cracks in the rock.	1
Pieces of rock are carried away in rivers.	
Water freezes and expands.	
Pieces of rock crash against each other and become smaller.	

140-1

3 Sophie shakes a sample of soil with water and leaves it to stand. In which order do the particles of sand, clay and pebbles settle? Tick the beaker below that shows the correct order of layers.

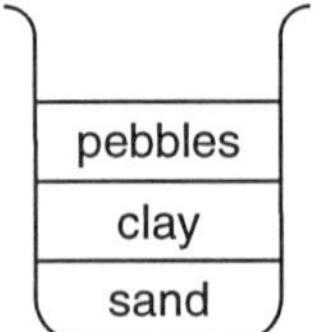

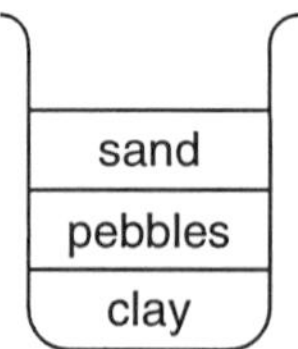

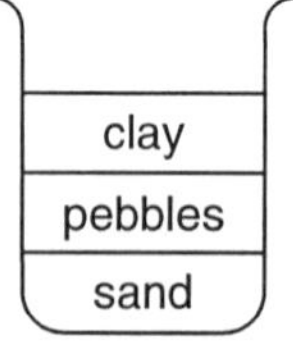

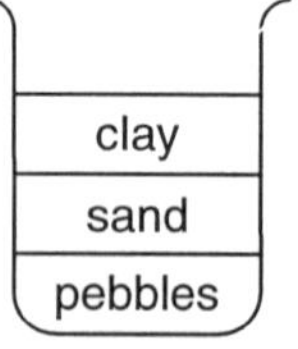

142–3

4 One of these statements is **not** true about sedimentary rocks. Which one is it? Cross it out of the list.

A They are formed from small particles like sand, clay or shells.
B They often contain fossils.
C They are made entirely of crystals.
D Some are formed by small particles being squeezed together hard.
E In some, small particles are stuck together by crystals growing between them. 144–5

5 Add the correct labels from the list below at A, B and C on the diagram.

weathering melting changing by pressure and heat

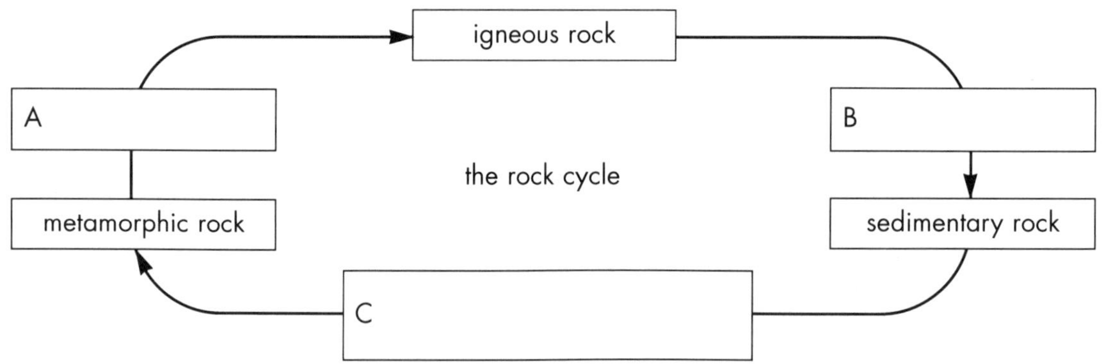

146–7

Forces

1 There is no movement in this tug-of-war. What do the forces need to be like for the rope to move? Underline the correct word.

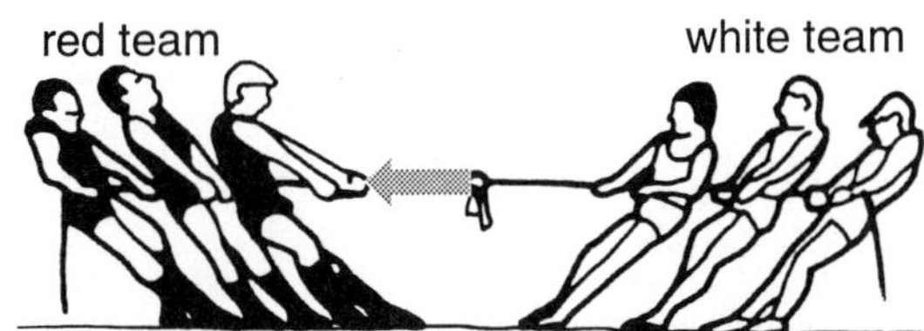

unbalanced (underlined) **strong** **balanced** **pulling**

2 The red team wins. Draw an arrow on the diagram to show the direction of the bigger force.

3 What are forces measured in? Underline the correct word.

kilograms **newtons** (underlined) **centimetres**

4 If you throw a ball up into the air, g r a v i t y pulls it back down again.

5 **a** The w e i g h t of something is the pull of the Earth's gravity on it.

b What is this pull measured in? Underline the correct word.

kilograms **newtons** (underlined) **centimetres**

6 **a** What is the amount of 'stuff' (matter) in an object called?
m a s s

b What is it measured in? Underline the correct word.

kilograms (underlined) **newtons** **centimetres**

7 If something is moving, f r i c t i o n acts to slow it down.

8 The skier is moving downhill. Draw an arrow to show the force of friction from the ground.

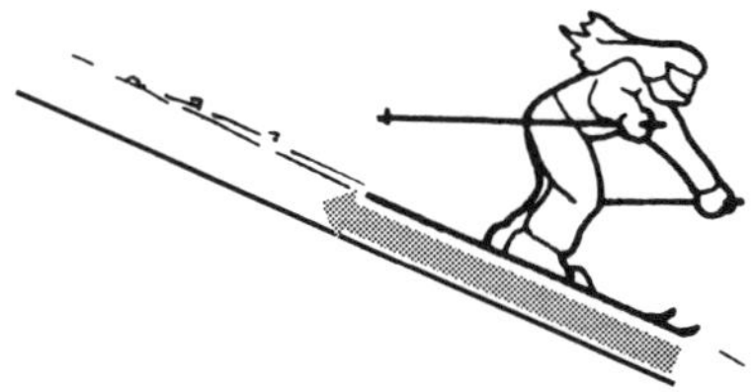

9 A s t r e a m l i n e d shape meets less friction in air and in water.

10 Underline the correct word to complete the sentence.
A parachute helps to slow you down because of the force of:

gravity **streamlining** **air resistance** (underlined)

11 Underline the correct word to complete the sentence.
A woman can just lift a weight of 500N using a lever.
To lift a weight of 900N, she would need a lever that was:

longer (underlined) **shorter** **the same**

Living things

1 Complete this list of the seven features of life.

M ovement **R** espiration

S ensitivity **G** rowth **E** xcretion

R eproduction **N** utrition

2 Complete these sentences using the words **animals** or **plants**.

Plants make their own food, but animals do not.

Most plants can only move very slowly, but most animals can move fast.

3 **a** This is a typical animal cell. Add the labels:

cytoplasm **nucleus** **cell membrane**

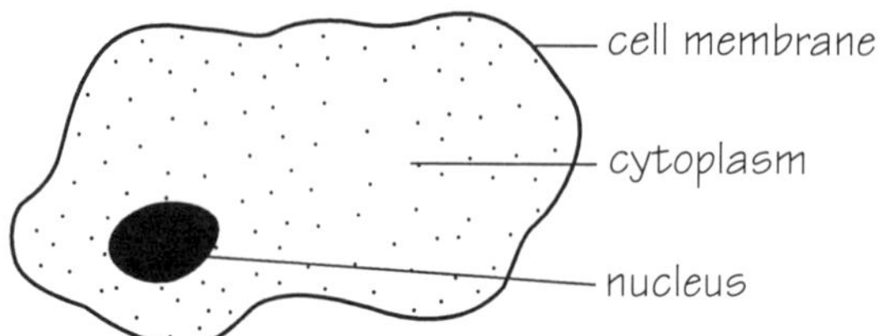

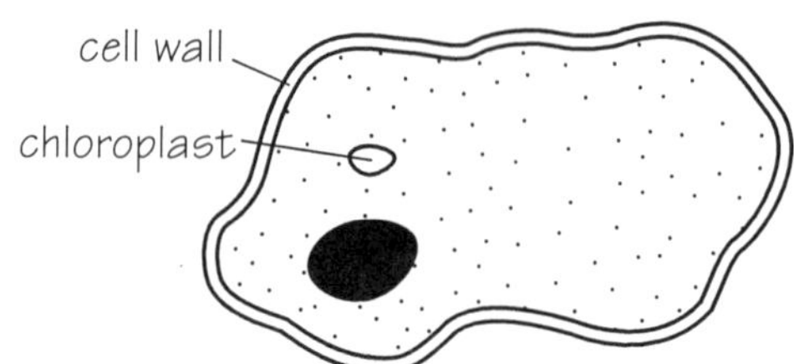

b Draw in two other parts that you would find in a plant cell, and label them.

4 Complete the sentence. When scientists put things into groups, they classify them.

5 Write these words in the correct columns to complete the table.

human **robin** **slug** **snail** **octopus** **toad**

Vertebrates	**Invertebrates**
robin	snail
human	slug
toad	octopus

6 Which group of invertebrate animals can be split into centipedes and millipedes, crustaceans, insects and spiders? arthropods

7 There are five vertebrate groups. Two are given here. Add the other three.

amphibians **reptiles** fish birds mammals

8 Underline the names of the four main plant groups.

flowering plants **trees** **conifers** **ferns** **shrubs** **mosses**

9 Why do scientists use keys? Underline the correct answer.

a to group organisms together **c** to identify organisms
b to join organisms **d** to examine organisms

Materials

1 Write these words in the correct columns to complete the table.

pencil air water orange juice desk

Solids	Liquids	Gases
desk	water	air
pencil	orange juice	

2 The space something fills is called its v o l u m e.

3 Draw in lines to link the words together to make correct sentences.

a Gases — have no fixed shape and no fixed volume.
b Liquids — have no fixed shape but have a fixed volume.
c Solids — have a fixed shape and a fixed volume.

4 Which of these can easily be squashed? Underline the correct answer.

solid liquid <u>gas</u>

5 Tick the property that applies to each material.

	Heavy	Light
aluminium		✔
iron	✔	

	Strong	Weak
straw		✔
brick	✔	

	Hard	Soft
diamond	✔	
pencil lead		✔

6 Underline the correct word to complete each sentence.

a Most metals are **<u>shiny</u>/dull**.
b Metals are good **insulators/<u>conductors</u>** of heat and electricity.
c **<u>Iron</u>/Aluminium** is attracted to a magnet.

7 If you want to compare the 'heaviness' of iron and aluminium, what size pieces would you need to make it a fair test? Underline the correct answer.

A a 2 cm cube of iron and a 1 cm cube of aluminium
B <u>a 1 cm cube of iron and a 1 cm cube of aluminium</u>
C a 3 cm cube of iron and a 4 cm cube of aluminium

8 The mass of a 1 cm cube of a material is its d e n s i t y.

9 Underline the two materials with the highest density.

water feathers <u>lead</u> aluminium <u>gold</u> air

10 Underline the correct number in this sentence.
The density of water is **<u>1</u>/10/100** grams per cubic centimetre.

11 Tick the boxes that describe the particles in solids, liquids and gases.

	Solids	Liquids	Gases
closely packed	✔	✔	
not free to move	✔		
free to move		✔	✔
far apart			✔

Energy

1 Complete the following sentences. Use the words below to fill the gaps.

movement turning transfers kinetic

A hamster transfers energy to its wheel by turning it. The wheel has movement energy. The scientific name for this kind of energy is kinetic energy.

2 **a** A substance that burns to release energy is a fuel, for example, coal. Which type of stored energy is in the fuel before it is burnt? Underline the correct word.

elastic energy heat energy <u>chemical energy</u>

b Complete the energy transfer diagram for the burning of coal.

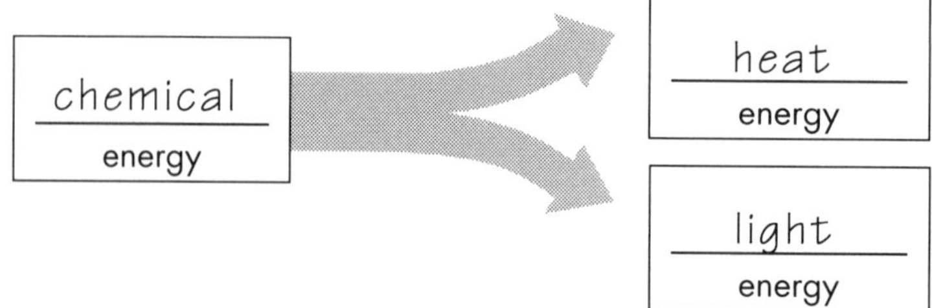

3 **a** What is the fuel used by the human body? f o o d

b What is the energy value of food measured in? Underline the correct word.

kilograms <u>kilojoules</u> kilowatts

c Which of the foods given below would have the highest energy value per 100g of food? Underline the correct word.

milk bread <u>peanuts</u>

4 Name four types of energy that electrical energy can be changed into.

a sound

b light

c heat

d movement/kinetic

5 Complete the following sentences. Use the words below to fill the gaps.

coal power stations gas generators

Most electricity comes from power stations. There they burn fuels such as coal or gas to release energy to turn generators.

6 Draw in lines to link the words together to make correct sentences.

a Anything that is stretched or squashed has — elastic energy.

b A battery stores — chemical energy.

c Something lifted off the ground has — gravitational energy.

Food for life

1 This diagram shows how a leaf makes food.
Use the words below to complete the labels.

oxygen water carbon dioxide sunlight starch chloroplasts

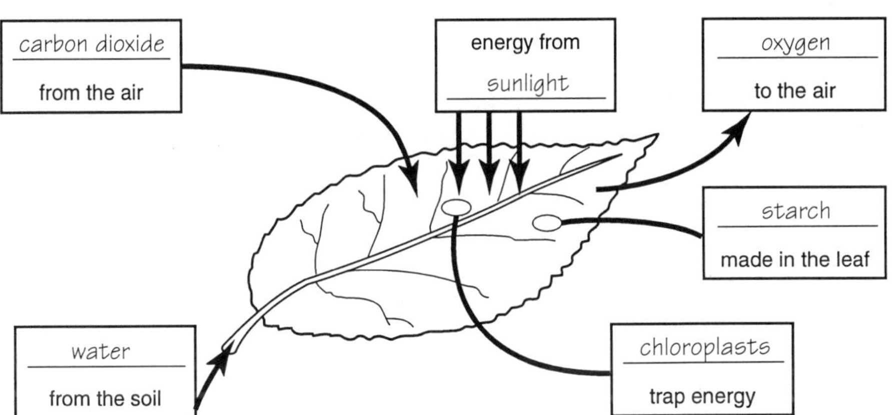

2 Underline the correct word to complete each sentence.

a Plants need **minerals/food/strength** from the soil to be healthy.

b Plants can be given nitrogen, phosphorus and **carbon/oxygen/potassium** in fertiliser.

c Plants need roots for anchorage, and to get water and **minerals/food/carbon**.

3 Draw in lines to link the words together to make correct sentences describing parts of a balanced diet. One has been done for you.

a Fats	have a lot of energy and are easily digested.
b Carbohydrates	are needed for healthy bones and teeth.
c Proteins	have the highest energy content.
d Minerals	are essential for health.
e Vitamins	help to build muscle and repair your body.

4 Name a mineral that is needed for:

a keeping bones hard and strong. c a l c i u m

b the blood to carry oxygen. i r o n

5 Name the vitamins needed to prevent:

a rickets. v i t a m i n D

b scurvy. v i t a m i n C

6 Underline the correct words to complete the sentences.
Plants make their own food and are called **consumers/producers**. Animals take in food by eating other organisms and are called **consumers/producers**. The place where a plant or animal lives is called a **den/habitat/food chain**.

7 Where many plants and animals live together in the same place, the food chains are connected in a f o o d w e b.

Mixtures

1 Draw in lines to link each mixture with the best way of separating it. One has been done for you.

Mixture / **Method**

- **a** different coloured Smarties — pick out by hand
- **b** magnetic and non-magnetic screws — use a magnet
- **c** flour and currants — sieve
- **d** liquid with sediment — decant
- **e** suspension of chalk and water — filter

2 Underline the correct word to complete each sentence.

- **a** A solid dissolves in water to form a **solute/solvent/<u>solution</u>**.
- **b** In a solution, water is called the **solute/<u>solvent</u>/solution**.
- **c** In a solution, the solid that dissolves is called the **<u>solute</u>/solvent/solution**.
- **d** Some solids do not dissolve. They are **soluble/<u>insoluble</u>**.
- **e** Solids that dissolve are **<u>soluble</u>/insoluble**.
- **f** When no more solid will dissolve, the solution is **<u>saturated</u>/unsaturated**.

3 What are the three main ways to speed up dissolving?

- **a** warming or heating the solution
- **b** stirring or shaking the solution
- **c** breaking up the solid to make smaller particles

4 Underline the correct words to complete each sentence.

- **a** Puddles of water disappear because water **<u>evaporates</u>/dissolves** into the air.
- **b** Energy from the sun will **<u>speed up</u>/slow down** this process.
- **c** You can get the solute back from a solution by **pouring off/<u>evaporating</u>** the water.

5 Underline the correct words to complete each sentence.

- **a** Windows steam up because **<u>water vapour</u>/liquid water** in the air hits the cold glass and **evaporates/<u>condenses</u>** to form small drops of **<u>liquid water</u>/water vapour**.
- **b** To get pure water back from a solution, you can **filter/<u>evaporate</u>** the water and **<u>condense</u>/dissolve** it. This is called **<u>distillation</u>/condensation**.

6 What is the name of the process for separating different dyes out of a mixture?

c h r o m a t o g r a p h y

Sound and light

1 A door bangs and you hear it. The table shows what things happen for you to hear the sound. Put them in the correct order by writing a number beside them (1 for the first thing to happen, 2 for the second, and so on). The first one has been done for you.

The three little bones vibrate.	4
The eardrum vibrates.	3
The air vibrates.	2
You hear a sound.	7
A message is sent to the brain.	6
The fine hairs vibrate.	5
The door vibrates.	1

2 John, aged eight, his dad and his grandad were outside at night. A bat flew past, squeaking.

a Who is most likely to hear the bat? John

b Who is least likely to hear the bat? grandad

3 An echo is a sound which is reflected from a hard, smooth surface.

4 In a car, there is always a lining of thick felt between the driver and the engine. The felt quietens the engine noise, because it is a soft, rough material which a b s o r b s the sound.

5 Underline the correct words to complete each sentence.

a Light rays travel **in straight lines/around corners**.

b When someone blocks rays of light, they form **an echo/a shadow/a reflection**.

c Light travels **faster than/slower than/at the same speed as** sound.

6 Complete the following sentences. Use the words below to fill the gaps.

smooth rough scattered reflected

Jo noticed that she could see her reflection in the shiny foil wrapper of her Kit-Kat. When she screwed the wrapper up and opened it out again, she could no longer see her reflection in it. This is because light is reflected well from smooth surfaces but it is scattered by rough surfaces.

7 Write these words in the correct columns to complete the table.

Sun star Moon Earth torch headlight cat's eye mirror

Produce light (sources)	Only reflect light
Sun	cat's eye
star	Moon
torch	Earth
headlight	mirror

Changing materials

1 Underline the correct words to complete the sentences.

a Ice is **rock/solid water**. The gas that water forms when it evaporates is called **water vapour/air**.

b A solid changes into a liquid at its **melting point/boiling point**.

c A liquid changes into a gas at its **melting point/boiling point**.

2 Underline the correct word to show whether the following statements are **true** or **false**.

a Pure water freezes at 0 °C. **true/false**

b All substances boil at 100 °C. **true/false**

c Mixtures melt and boil at one temperature. **true/false**

d Evaporation and melting are reversible. **true/false**

e When salt is added to water, it makes the freezing point lower. **true/false**

3 The table shows whether substances are solid, liquid or gas at room temperature. Are their melting points and boiling points higher or lower than room temperature? Fill in the blanks in the table using the words **higher** or **lower**.

State at room temperature	Melting point	Boiling point
solid	higher	higher
liquid	lower	higher
gas	lower	lower

4 Underline the correct words to complete each sentence.

a When most solids are heated they **expand/contract**, and when they are cooled they **expand/contract**.

b Thermometers work because most liquids **expand/contract** when heated and **expand/contract** when cooled.

5 Underline the correct word to show whether the following are **physical** or **chemical** changes.

a melting wax **physical/chemical**

b boiling water **physical/chemical**

c burning a candle **physical/chemical**

6 Underline the correct word to show whether the statements are **true** or **false**.

a Changes of state are physical changes. **true/false**

b Chemical changes are easily reversible. **true/false**

c Chemical changes may give off light and heat energy. **true/false**

d Physical changes are not reversible. **true/false**

e Chemical changes form new substances. **true/false**

7 When something burns:

a What gas does it use from the air? oxygen

b Is it a physical or chemical change? chemical

8 When iron rusts:

a What two things does it need to rust? water/moisture and oxygen/air

b Is it a physical or chemical change? chemical

Reproduction

1 Underline the correct word to show whether the following statements are **true** or **false**.

a Only some living organisms need to reproduce. **true/<u>false</u>**
b When male and female sex cells join, it is called fertilisation. **<u>true</u>/false**
c In animals, the female sex cell is called the ovule. **true/<u>false</u>**
d The male sex cell in plants is carried in the pollen. **<u>true</u>/false**

2 Complete the following sentences. Choose from the words below to fill the gaps.

ovary stem seeds fertilisation flower pollination stamens

The part of a plant that contains the sexual organs is the _flower_. When pollen from one plant lands on the stigma of another, it is called _pollination_. The pollen then fertilises the ovules in the _ovary_. The fertilised ovules become _seeds_.

3 It is important for the plant to scatter its seeds around. This is called _dispersal_.

4 Complete the following sentences. Choose from the words below to fill the gaps.

urethra testes hair puberty penis voice sperm

The name given to the time when your body becomes capable of reproducing is _puberty_. For boys, this is the time when they grow more _hair_ on their faces and bodies, their _voice_ gets deeper and their _penis_ and _testes_ get larger.

5 Tick the things that happen to a girl's body at puberty.

A Her breasts develop. ☑
B Her monthly periods begin. ☑
C Her uterus becomes much smaller. ☐
D Her ovaries start producing an ovum each month. ☑

6 Underline the correct word to show whether the statements are **true** or **false**.

a During sexual intercourse, sperm are released into the woman's vagina and swim up into the uterus and Fallopian tubes. **<u>true</u>/false**
b A fertilised egg implants at the top of the Fallopian tube, near to the ovary. **true/<u>false</u>**
c If a sperm meets an egg (ovum), the egg is fertilised. **<u>true</u>/false**

7 Tick those things that happen **every month** in a woman's menstrual cycle.

A An egg ripens and leaves the ovaries. ☑
B An egg is fertilised. ☐
C The cycle lasts about 28 days. ☑
D The lining of the uterus builds up during the cycle. ☑

8 Underline the correct answer to complete each sentence.

a A fetus is **an unfertilised egg/<u>a developing baby</u>/a ripening ovary**.
b A developing baby in the uterus receives food and oxygen from its mother through the **Fallopian tubes/<u>placenta</u>/vagina**.

Electricity

1 Underline the correct words to complete each sentence.

a For electric current to flow, you need a **bulb/buzzer/complete circuit**.

b An insulator **does/does not** allow an electric current to pass through it.

c A conductor **does/does not** allow an electric current to pass through it.

2 Complete the following sentence.

Electric current flows from the positive end of the battery, round a circuit to the negative end of the battery.

3 Write in the correct component for each symbol to complete the table.

Symbol	Component
+ ─┤▌─	battery
─(◠)─	bulb
─(M)─	motor
─o╱ o─	switch
buzzer symbol	buzzer
────	connecting wire

4 In circuit 1, bulbs A and B are connected in series. In circuit 2, bulbs X and Y are connected in parallel. All the bulbs have the same resistance. Circuits 1 and 2 each have just one battery.

a Which bulbs shine more brightly, A and B or X and Y? X and Y

b In circuit 1, bulb B is unscrewed. What happens to bulb A? It goes out.

c In circuit 2, bulb X is unscrewed. What happens to bulb Y? It stays lit.

5 Complete the following sentences. Choose from the words below to fill the gaps.

resistance smaller bigger energy

Current flows around a circuit. All the components in the circuit have electrical resistance. The more resistance there is in the circuit, the smaller the current that flows. A long wire has a bigger resistance than a short wire.

6 Underline the correct words to complete each sentence.

a Current **does/does not** get used up around a circuit.

b A battery is a store of chemical **heat/electricity/energy**.

c The energy stored in the battery is transferred along a circuit in the form of **light/electrical/current** energy.

d A lamp changes **battery/chemical/electrical** energy into **kinetic/elastic/light** energy and **heat/stored/sound** energy.

Rocks

1 Complete the following sentences. Use the words below to fill the gaps.

igneous volcano crystals lava

When molten rock reaches the Earth's surface, a volcano is often formed.

The liquid rock is called lava. When it cools, a rock with crystals forms. This is called igneous rock.

2 The table shows what things happen to cause the physical weathering of rock to form small pebbles and soil. Put them in the correct order by writing a number beside them (1 for the first thing to happen, 2 for the second, and so on). The first one has been done for you.

Pieces of rock drop off.	3
Water enters cracks in the rock.	1
Pieces of rock are carried away in rivers.	4
Water freezes and expands.	2
Pieces of rock crash against each other and become smaller.	5

3 Sophie shakes a sample of soil with water and leaves it to stand. In which order do the particles of sand, clay and pebbles settle? Tick the beaker below that shows the correct order of layers.

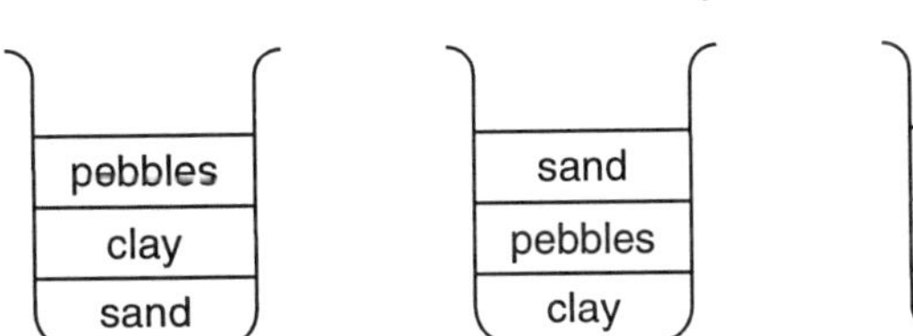

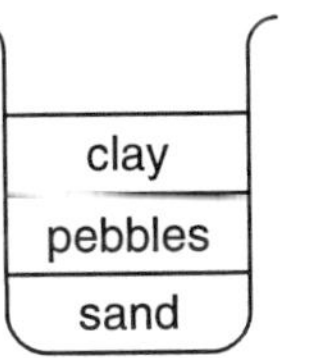

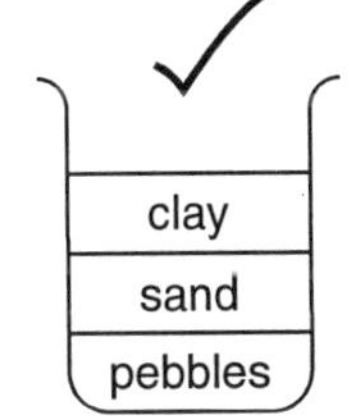

4 One of these statements is **not** true about sedimentary rocks. Which one is it? Cross it out of the list.

A They are formed from small particles like sand, clay or shells.
B They often contain fossils.
~~**C** They are made entirely of crystals.~~
D Some are formed by small particles being squeezed together hard.
E In some, small particles are stuck together by crystals growing between them.

5 Add the correct labels from the list below at A, B and C on the diagram.

weathering melting changing by pressure and heat

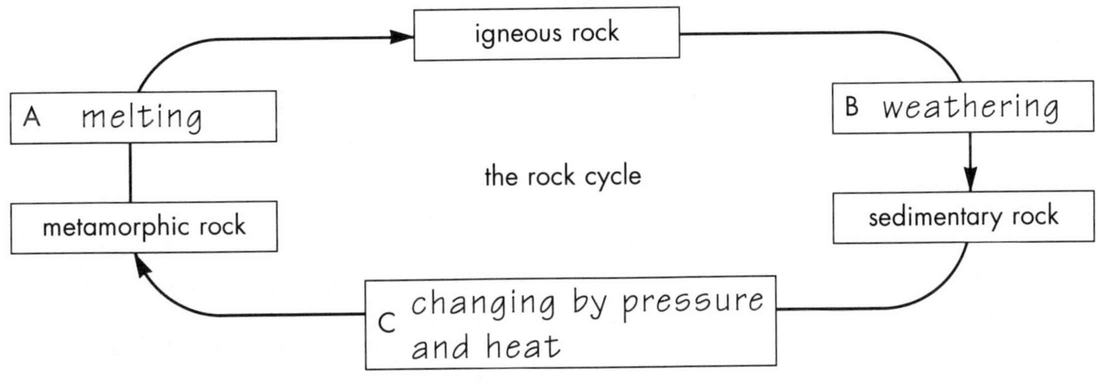

Core

1 Copy and complete the following sentences.
Choose from the words below to fill the gaps.

balanced forces kilograms newtons unbalanced friction newtonmeters

Pushes and pulls are ________. They are measured in

______________. An object will not start to move if the forces

on it are ______________.

2 Write down the correct letter. The force you get when one surface rubs against another is called:

A pulling
B rubbing
C friction
D balanced.

3 Write down the correct letter. Sameena uses a force of 300 N to push a chair across a smooth wooden floor. To push it across a rough carpet she would need:

A a smaller force
B a bigger force
C the same force.

4 All these animals can swim.
Copy and complete the sentence.

The ________ can move fastest

under water because its shape is

________________________.

5 An astronaut has a mass of 75 kg, measured on Earth.
Copy and complete the following table. Choose from the words below to fill the gaps.

75 kg 750 N zero
less than 750 N
more than 750 N
75 N 750 kg

	Earth	Moon	Outer space
Weight			
Mass	75 kg		

6 Copy these pictures.
Draw in arrows to show the forces in each case.
(There is more than one force acting on each object.)

a

falling ball

b

moving train

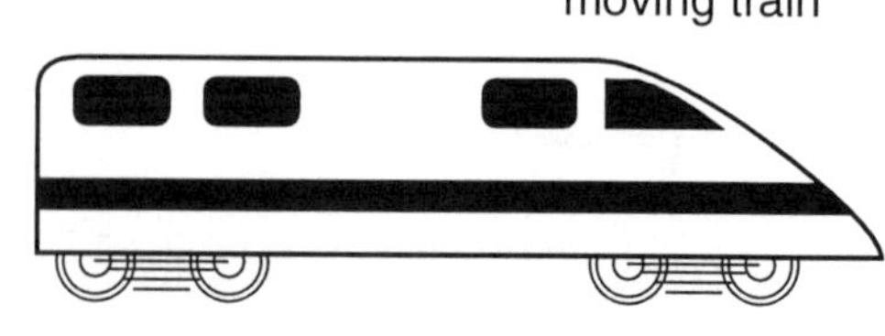

7 If you find it difficult to get the lid off a tin of paint, you could use a screwdriver to help you. What sort of machine is a screwdriver when it is used like this?

8 The see-saw is balanced. Sameena weighs 500 N. How much does John weigh?

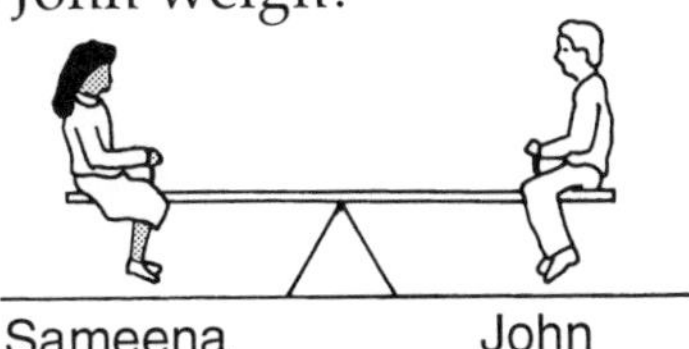

9 John is replaced by his dad. Where will his dad need to sit to keep the see-saw balanced?

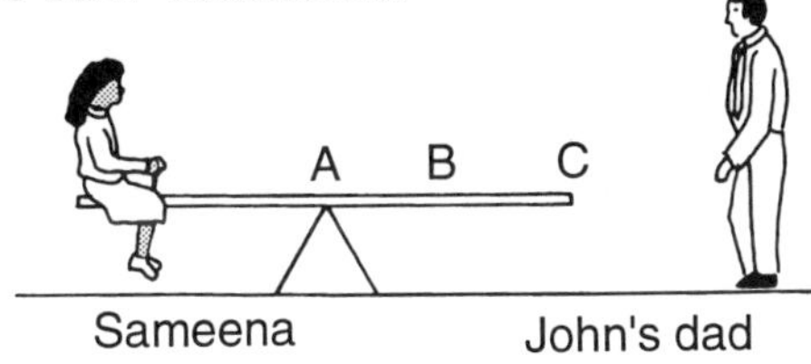

Extension

10 These pictures show a racing car and a family saloon. How does the racing car's shape allow it to go faster?

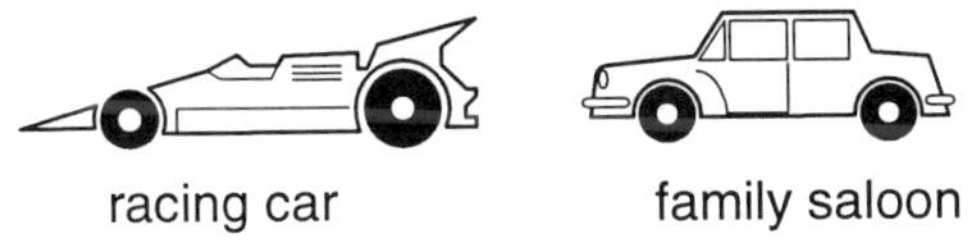

11 **a** The racing car sets off and speeds up. The arrows show two forces on the car. Say what the forces are.

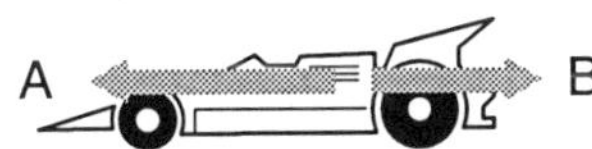

b The car moves at a steady speed of 300 km every hour. Draw a diagram to show the forces on the car.

c This picture shows the car slowing down. Explain why this is happening.

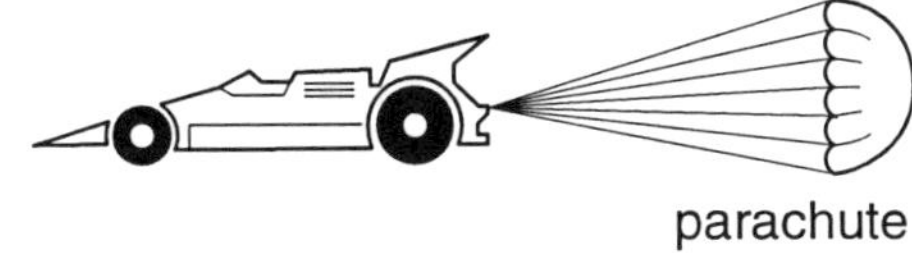

12 **a** Emma weighs 350 N. What is Laura's mass? Show how you worked out your answer.

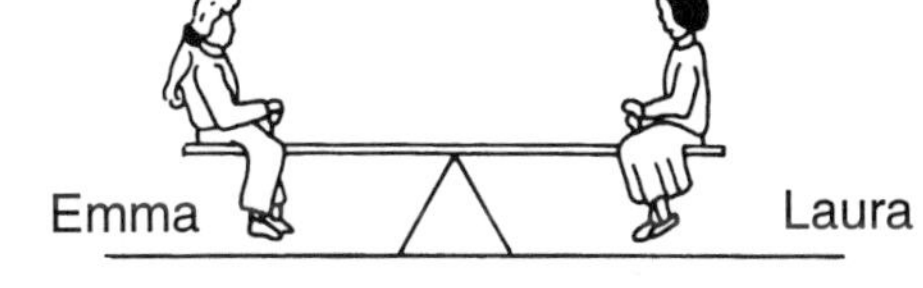

b Laura is replaced by her mum. Copy the picture and show where you think the pivot should be so that the see-saw balances.

Core

1 Here are some pairs of groups of living organisms.
Which pair shows the two largest groups?
Write down the correct letter.

A vertebrates and invertebrates
B mammals and reptiles
C invertebrates and fish
D plants and animals

2 How do plants normally move? Write down the correct letter.

A slowly
B not at all
C quickly
D very quickly

3 Why do some animals have skeletons?
Write down the correct letter.

A nutrition
B reproduction
C support
D excretion

4 Draw a table with these headings to show what dogs can do and what motor bikes can do. Use the words below to complete the table. You can use words more than once.

feed grow move reproduce

Dogs can	Motor bikes can

5 Copy and complete this diagram.
Use the words below to fill the gaps.

crocodile amphibian reptile bird
mammal penguin dog toad

vertebrates

fish
carp

6 The diagram shows a human cheek cell.

a Name part A.
b Plant cells have some parts that animal cells do not have.
Name two of these parts.

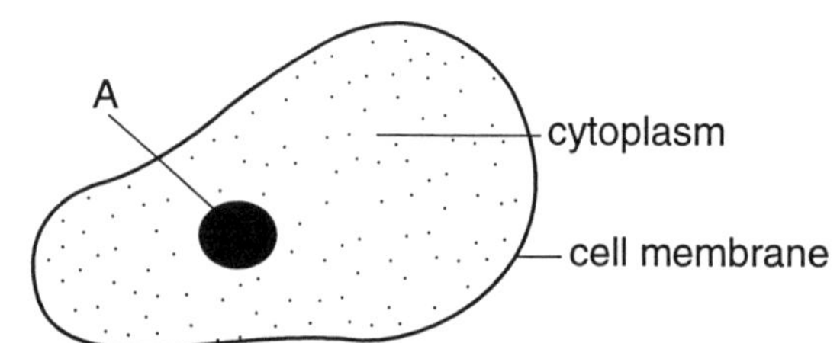

7 The plant world is divided into four main groups. These are mosses, ferns, conifers and flowering plants.
Write down the correct letter. Which of the plants A to D is:

a a fern
b a moss
c a conifer
d a flowering plant?

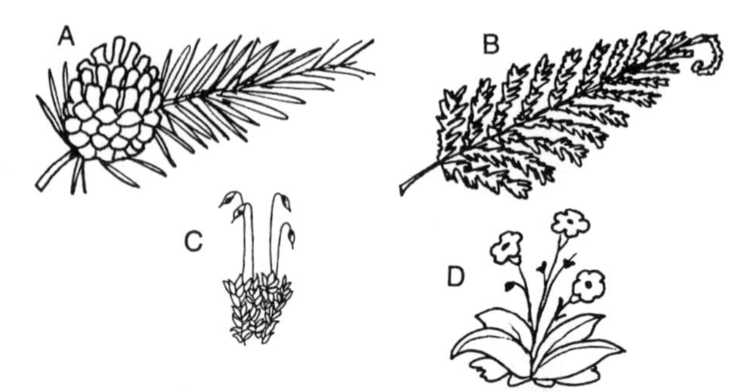

8 Here are four types of tree.
Copy and complete the following sentences.
Use the key below to find the names of the trees.

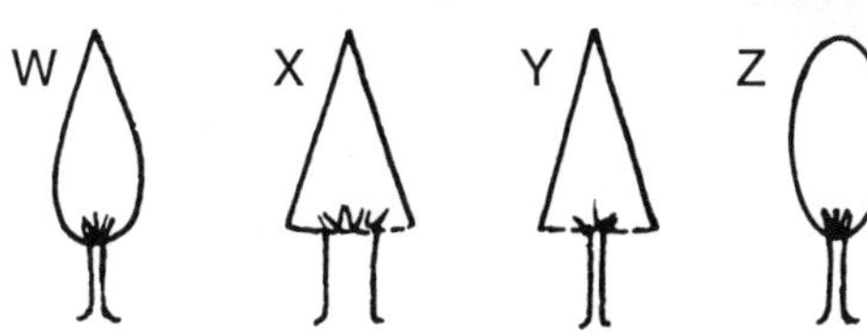

a Tree W is . . . **c** Tree Y is . . .
b Tree X is . . . **d** Tree Z is . . .

1 Is it triangular in shape?	Yes: go to question **2**	No: go to question **3**
2 Does it have a wide trunk?	Yes: December fir	No: lap fir
3 Is it pointed at the top?	Yes: Wem poplar	No: trench poplar

Extension

9 The picture shows six leaves.
Copy and complete the following sentences.
Use the key below to identify the leaves.
The first two have been done for you.

a Drawing P is leaf E. **d** Drawing S is leaf __.
b Drawing Q is leaf B. **e** Drawing T is leaf __.
c Drawing R is leaf __. **f** Drawing U is leaf __.

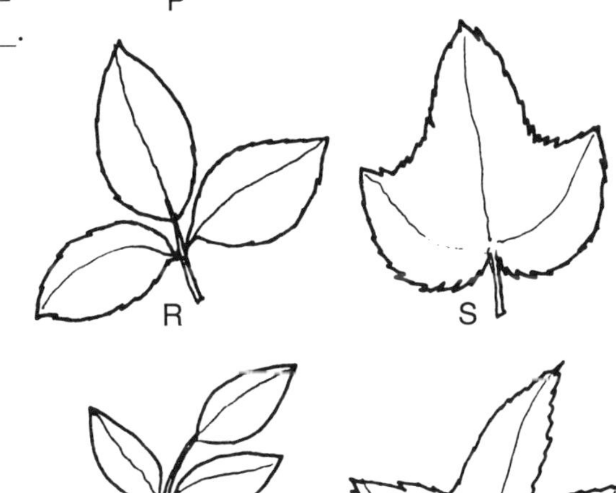

T

U

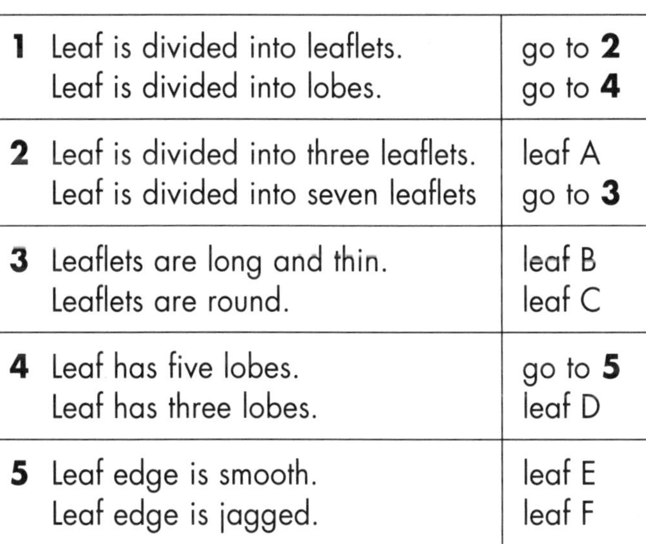

1 Leaf is divided into leaflets. Leaf is divided into lobes.	go to **2** go to **4**
2 Leaf is divided into three leaflets. Leaf is divided into seven leaflets	leaf A go to **3**
3 Leaflets are long and thin. Leaflets are round.	leaf B leaf C
4 Leaf has five lobes. Leaf has three lobes.	go to **5** leaf D
5 Leaf edge is smooth. Leaf edge is jagged.	leaf E leaf F

10 Green plants are classified into four groups – mosses, ferns, conifers and flowering plants. Below are descriptions of four plants. To which group does each plant belong?

a has thin needle-shaped leaves
b has non-waterproof leaves
c has seeds
d has spores and waterproof leaves

11 **a** An animal has scales.
Which of these groups could it belong to?

fish amphibians reptiles birds mammals

b Classify this scaly animal into one correct group.
Choose one group. What else would you look for on its body to help you decide if it belongs to this group?

Core

1 Copy the table. Tick the boxes that best describe the properties of solids, liquids and gases.

	Solids	Liquids	Gases
fixed shape			
no fixed shape			
fixed volume			
no fixed volume			

2 Diagrams **a**, **b** and **c** show different arrangements of particles. For each diagram, write whether it represents a solid, a liquid or a gas.

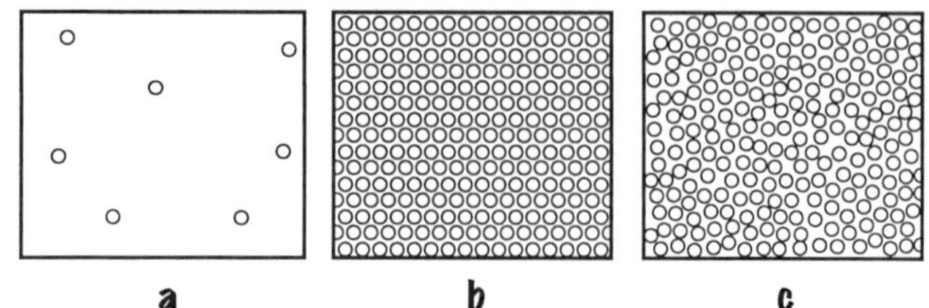

3 Coal is carried in lorries. Oil and gas are often transported by pipeline. What property of oil and gas makes this possible?

4 Solids can be **heavy** or **light**, **weak** or **strong**, **hard** or **soft**. For each of the following sentences, choose one of these words to describe the property being used.

- **a** Diamond is used to make rock-cutting drills.
- **b** Aluminium is used to make the body of an aeroplane.
- **c** Steel is used to make the bars of a prison window.

5 Quartz has a hardness value of H7. A penny has a hardness value of H3.5. Quartz will scratch a penknife. A penknife will scratch a penny. Suggest a hardness value for the penknife.

6 You discover a new material. When you polish it, it looks shiny. It is not magnetic.

- **a** What suggests that it is a metal?
- **b** Write down two other properties you would expect it to have if it is a metal.
- **c** Tom says "It is not magnetic, therefore it can't be a metal." Give an example of a metal that is not magnetic.

7 The picture shows three materials.

- **a** Which material has the highest density?
- **b** Why is it a fair test?

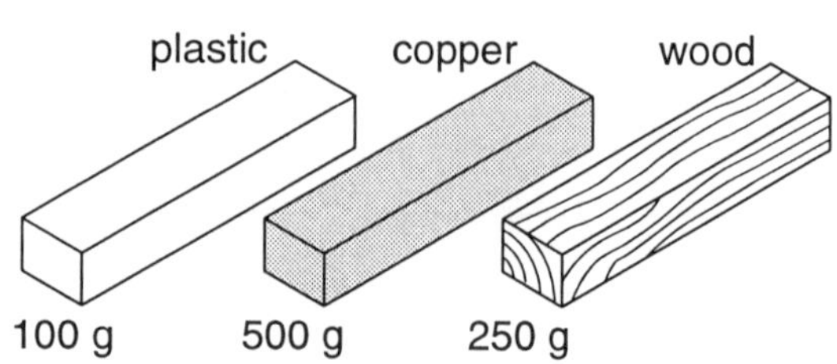

8 The picture shows two balloons.
Which gas, helium or carbon dioxide, has the higher density?

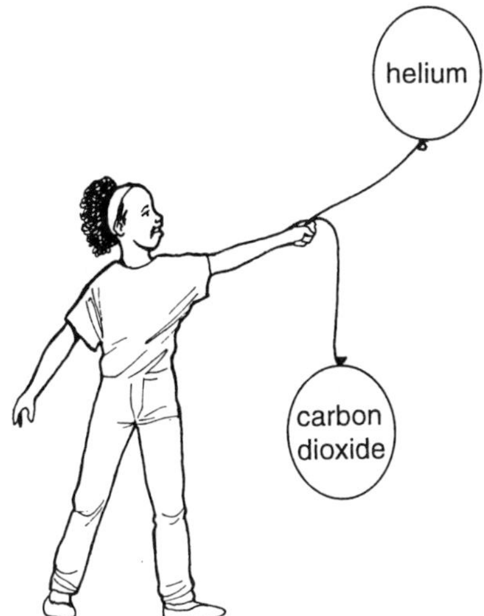

9 In a solid, the particles are closely packed and stuck together.
Write a similar sentence to describe the particles in:
a a liquid **b** a gas.

10 Use the idea of particles to explain why:
a you can squash a gas, but not a liquid or a solid
b you can pour a liquid, but not a solid.

Extension

11 The picture shows samples of different materials.
Their densities are given in grams per cubic centimetre.
a Which two of these materials are gases?
b A $2\,cm^3$ block of aluminium has a mass of 5.4 g. What is the density of aluminium in grams per cubic centimetre?

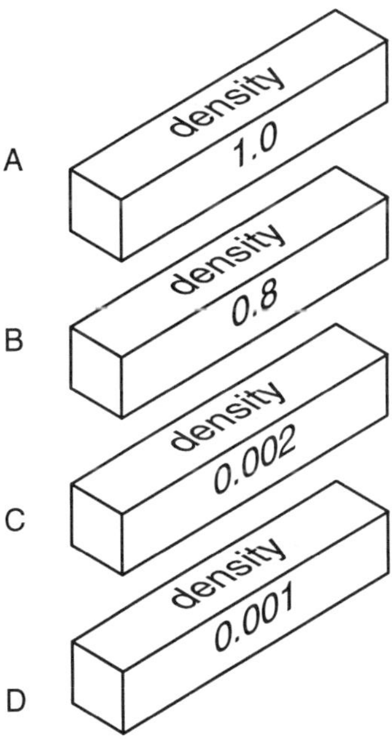

12 Butane (camping gas) is stored under pressure in cylinders. It is stored as a liquid. If the pressure is released, it changes from liquid to gas.
a What happens to the particles when the pressure is released?
b What is the advantage of storing butane as a liquid?
c You are asked to make a container to keep the liquefied gas under pressure. The following materials are available.

Material	Strength	Flexibility
X	low	medium
Y	high	low
Z	medium	high

i Which material would you use?
ii Why did you choose this one?

13 Use the particle model to explain what happens when ice changes into water. You can use a diagram if you wish.

Core

1 Match the descriptions in column A with the energy types in column B. Write down the matched pairs.

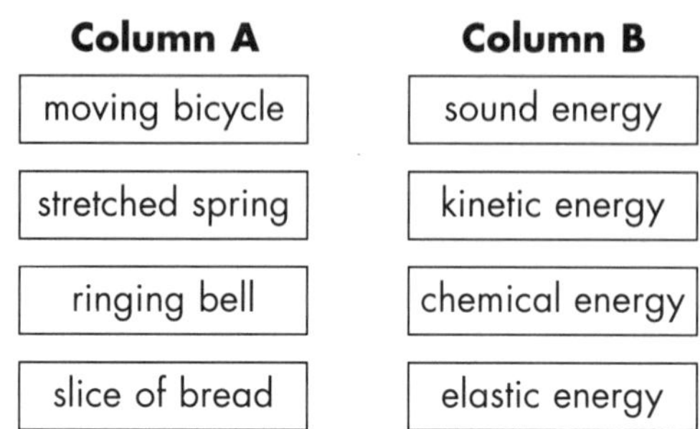

2 Copy and complete these diagrams showing energy transfers. Choose from the words below to fill the gaps.

kinetic energy **light energy** **sound energy** **electrical energy**
elastic energy **heat energy** **chemical energy**

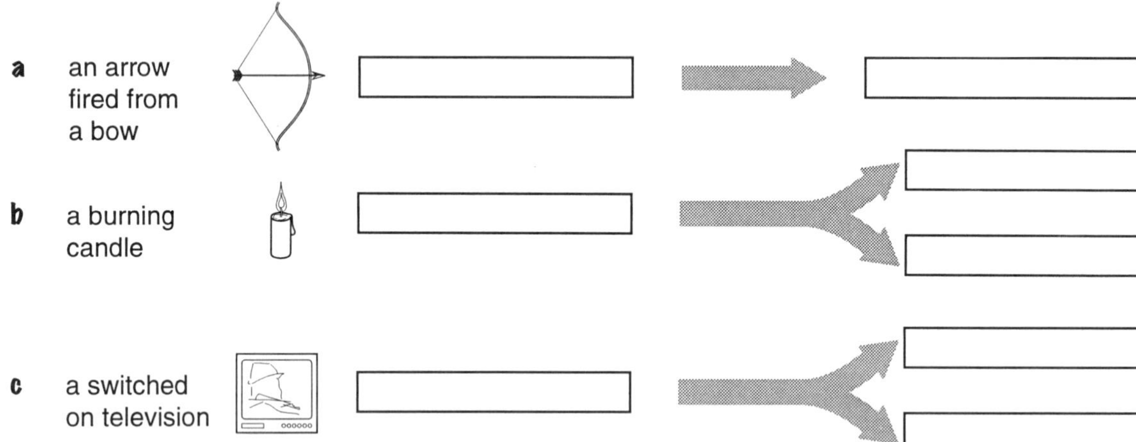

3 Fuels can be burnt to release energy.

a Name a fuel that can be burnt in an open fire.

b Name a fuel that is stored in tanks and can be burnt in a boiler to heat a large building like a school.

c Which one of the following is not a fuel?

gas **electricity** **wood** **paraffin**

4 Here are three food labels.

A

per 100 g	
energy	569
protein	2.0 g
carbohydrate	27.3 g
fat	0.5 g

B

per 100 g	
energy	1457
protein	10.7 g
carbohydrate	71.5 g
fat	1.6 g

C

per 100 g	
energy	486
protein	26.6 g
carbohydrate	0.1 g
fat	1.1 g

a What is the unit of energy missing from the labels?

b **i** Which food would be the best for someone who is very active?

ii Why have you made this choice?

5 Dynamos can be used to generate electricity. The diagram shows a dynamo connected to a lamp on a bicycle. The movement of the bicycle is lighting the lamp.

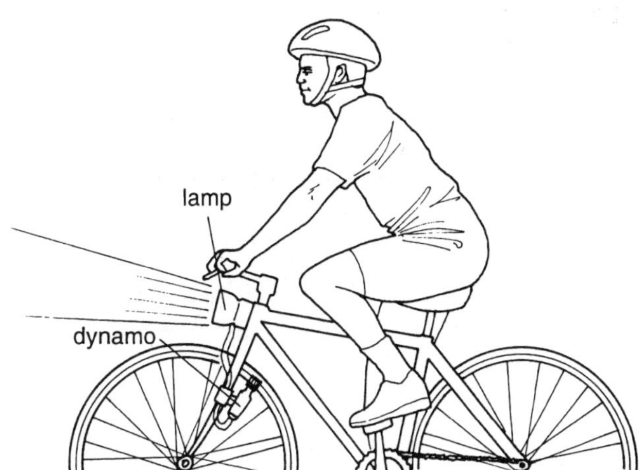

a What useful energy change is taking place in the bulb?

b What main energy change is taking place in the dynamo?

c What main energy change is taking place in the person?

d Going uphill, the bicycle slows down.

- **i** How does the brightness of the bulb change?
- **ii** Why does the brightness of the bulb change?

6 Steam trains burn coal as a fuel to boil water. When the water boils, it produces steam. This is used to push a piston, which drives the wheels.

What forms of energy are in the energy transfer diagram? Write down your answers for **a** and **b**.

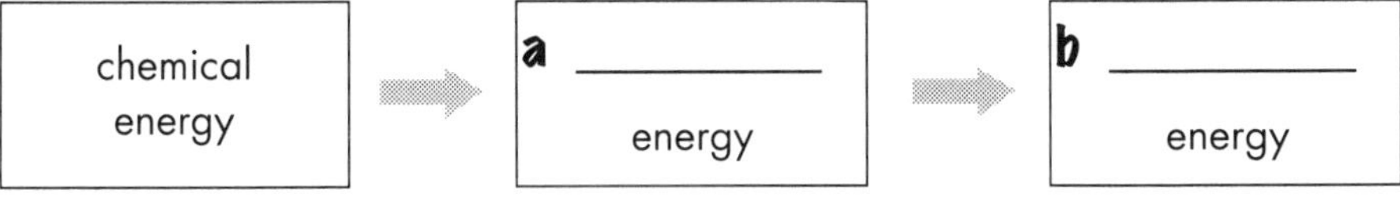

Extension

7 This energy transfer diagram shows how a steam engine can be used to light a bulb.

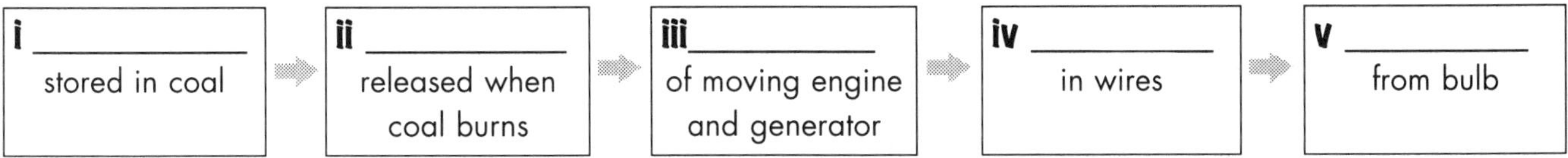

a What forms of energy are in the energy transfer diagram? Write down your answers for **i** to **v**.

b Energy is escaping from the system, mostly in one form.

- **i** In what form is this energy?
- **ii** From where is this energy escaping?
- **iii** Where does this escaping energy go?

8 The main task of an electric motor is to convert electrical energy into kinetic energy.

a If 100 kJ of electricity are supplied to the motor, how much kinetic energy would you expect to be given out? Choose from the values below.

120 kJ   100 kJ   90 kJ

b Why did you choose this value?

Core

1 What we eat can affect our health. Our diets should contain the right amounts of the following substances:

carbohydrate fat fibre protein vitamins minerals

a Copy and complete the following table. Choose from the words above to fill the gaps.

Food substance	Needed for
	energy that the body can use easily
	makes sure food moves along the gut
	for strong bones and teeth

b Copy and complete the following table to show why each food substance is needed by the body.

Food substance	Needed for
protein	
fat	
vitamin C	

2 Plants make food in their leaves by a process called photosynthesis.

a Use the words below to answer the following questions about photosynthesis.

carbohydrates carbon dioxide oxygen sunlight water

i Which substances are made (produced)?
ii Which gas from the air is needed for photosynthesis?
iii Where does the energy for this process come from?

b Why are chloroplasts needed for photosynthesis?

3 The diagram shows a food web.

a Name a producer in this food web.
b Name a consumer that eats plants.
c Write a food chain from the food web.
d What does any food web show?

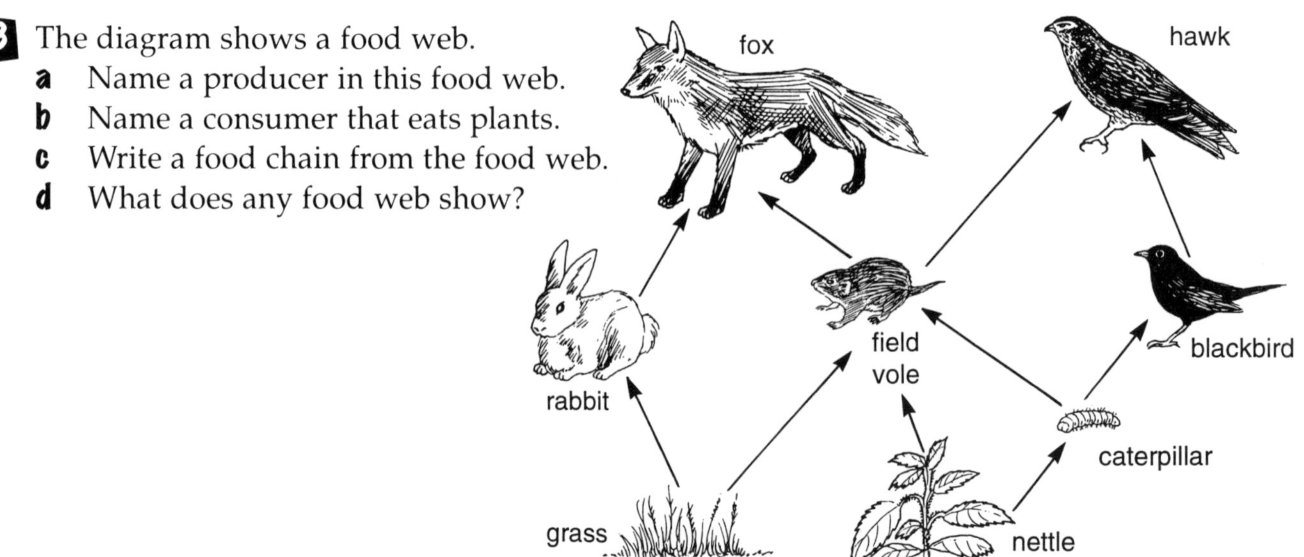

4 What are the three main jobs of plant roots?

Extension

5 SuperGro is a chemical fertiliser which can be added to the soil. This label shows some ingredients in SuperGro.

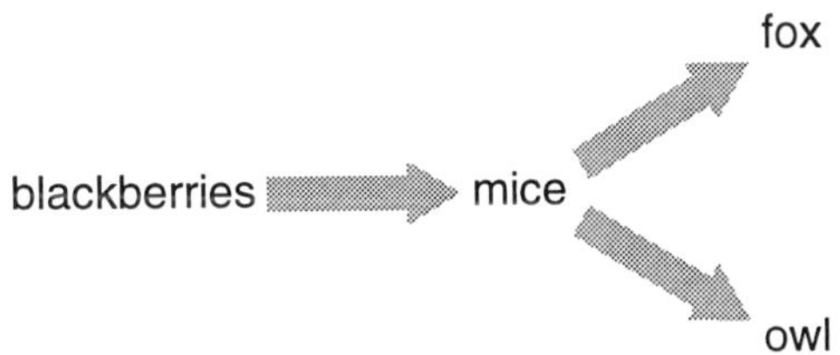

a What is X likely to be?
b Why is fertiliser added to soil?
c Why is phosphorus important to plants?
d Name a substance that is used as a natural fertiliser.

6 The diagram below shows part of a food web.

One summer, a disease killed off most of the mice.

a **i** What do you think happened to the numbers of foxes and owls?
ii Explain your answer.

b What do you think happened to the number of blackberries?

7 Copy and complete the diagram showing the process of photosynthesis. Use the words below to fill the gaps.

carbohydrate carbon dioxide chloroplasts oxygen sunlight water

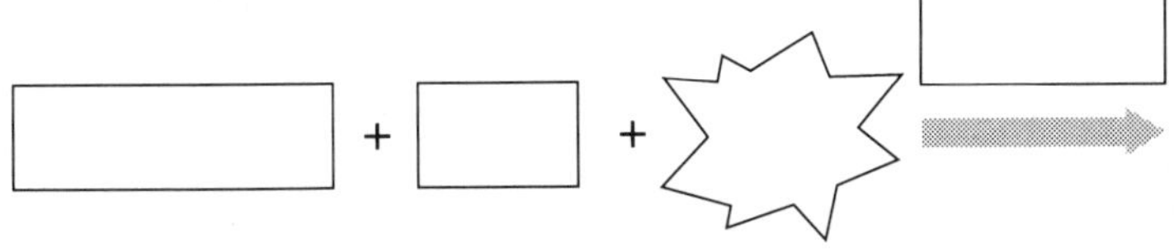

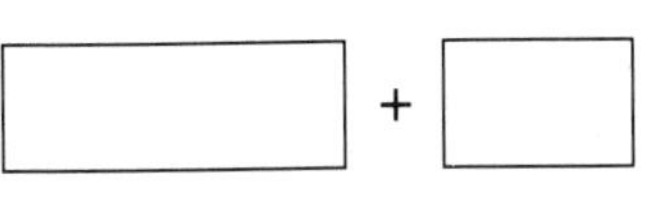

Core

Jane mixes sand and salt and water. The salt disappears.

sand, salt and water mixture
funnel
beaker

1 **a** Copy and complete the following sentence. Choose from the words below to fill the gap.

solvent soluble saturated insoluble

Salt is _______________ in water.

b She separates the sand from the liquid using this apparatus. What is this method of separation called?

2 The table shows different methods of separation and how they are used.
Write down the missing words for **a** to **d**.

Method	Used to separate	Example
a	solid sediment from a liquid	soil and water
sieve	**b**	sand and peas
c	different coloured dyes	black ink
magnet	**d**	copper tacks and iron nails

3 Look at the diagram. Write down the missing words for **a** to **d**, choosing from the words below.

solute solution saturated insoluble solvent unsaturated

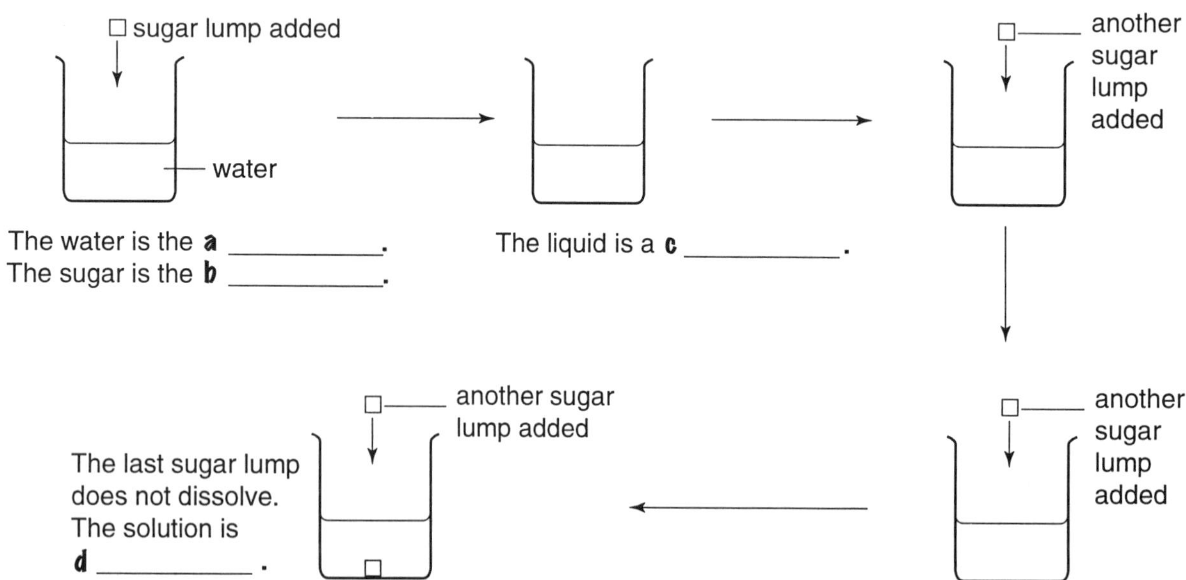

e Describe two ways you could make the sugar dissolve faster in this experiment.

4 It is Sam's birthday. Sam's dad is decorating his cake with brown Smarties. He sorts out the sweets on white paper. Water spills on the paper, and colour from the sweets runs into the paper.
The brown colour is not brown any more.
Three colours appear on the paper.
Sam's mum did an experiment at work to find out which dyes are in brown Smarties.
Here are the results.

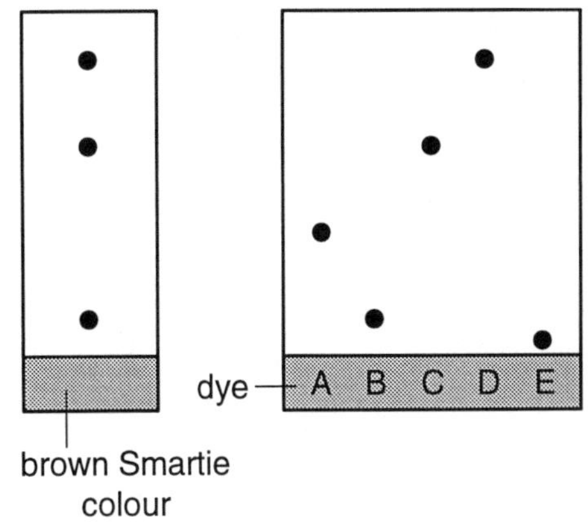

a Write down the letters of the dyes in the brown Smartie.
b What solvent do you think Sam's mum used in the experiment?
c Which dye travelled fastest?

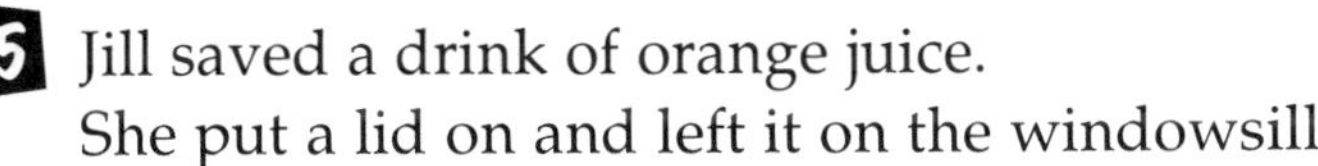

5 Jill saved a drink of orange juice.
She put a lid on and left it on the windowsill.

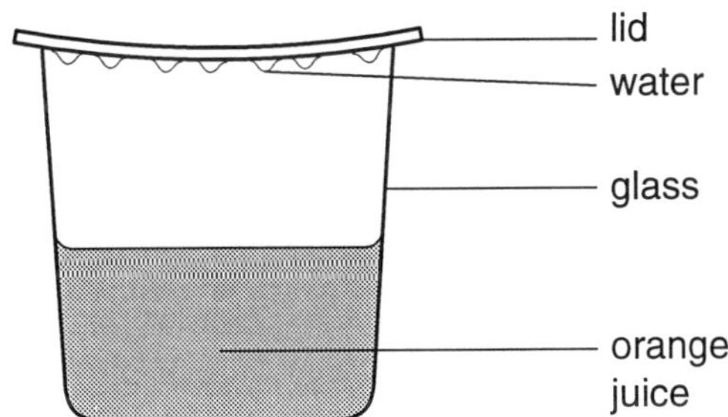

Drops of water formed underneath the lid.
This happens in two stages.
a How does the water leave the orange juice?
b Why do drops of water form on the lid?

The apparatus in the diagram can be used to obtain pure water from orange juice.

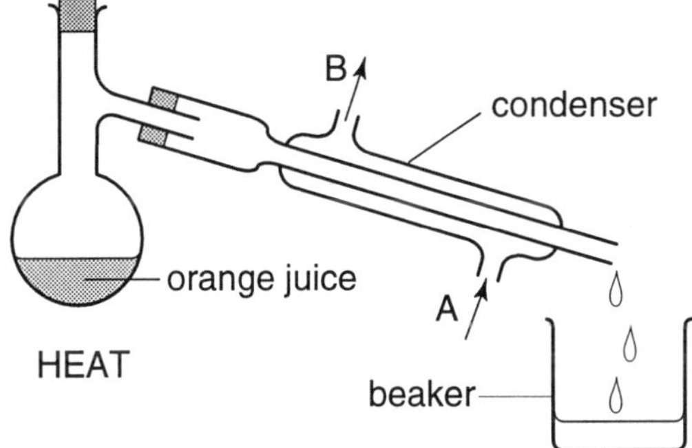

c What is this method of separation called?
d Name the part of the apparatus that is heated.
e What flows into the condenser at A (and out at B)?

Extension

6 The diagram shows the water cycle. Describe what is happening at each of the points A to E.
Use all the following words at least once.

condenses **cools** **energy** **evaporates**
moist air **rain** **water vapour**

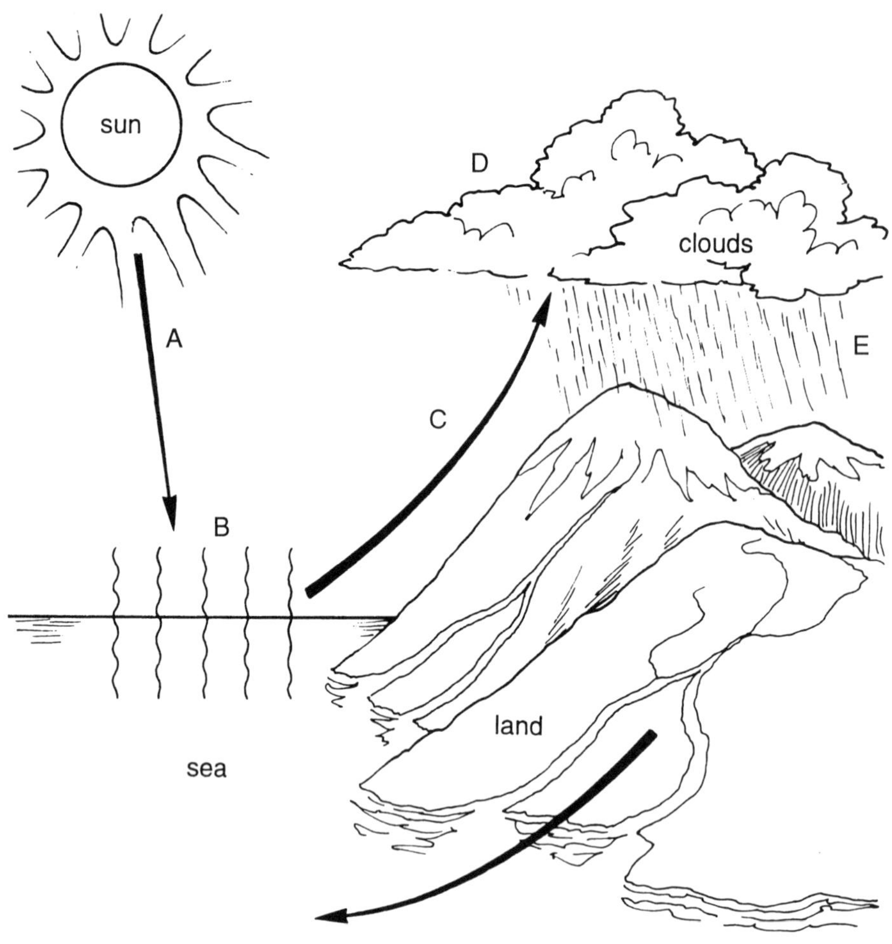

7 Fred says the green colour in plants is really a mixture of two different substances, one green and one orange.
He did the following experiment to prove this to his friends.

a What separation method will he use at stage 3?

b What separation method will he use at stage 4?

c Draw a diagram to show the results of stage 4 if Fred's idea is correct.

d Why do you think propanone was used instead of water in this experiment?

e It is dangerous and expensive to throw away propanone. How could Fred get **pure** propanone from the green mixture he has used? (Propanone boils at 57 °C.)

Cabbage leaves are ground up with sand in a mortar

stage 2

Propanone (a solvent) is added to the mixture

stage 3
The sand and leaf bits are removed from the liquid

stage 4
The colours in the liquid are separated

Core

1 Which thing below is happening at positions **a**, **b**, **c** and **d** on the diagram? Write down the correct letter.

A The air carries the vibrations.
B An object vibrates, producing sound.
C The sound is heard.
D The eardrum vibrates.

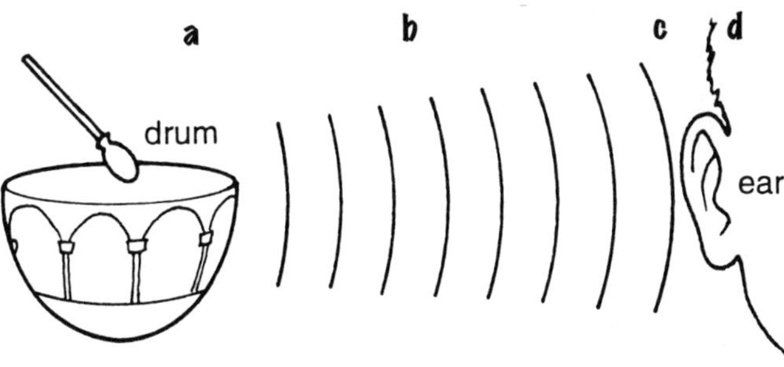

2 The boat sounds its hooter. Moments later the ship's crew hear the hoot again.

a What happens to the sound when it hits the cliff?

b What is the second sound called?

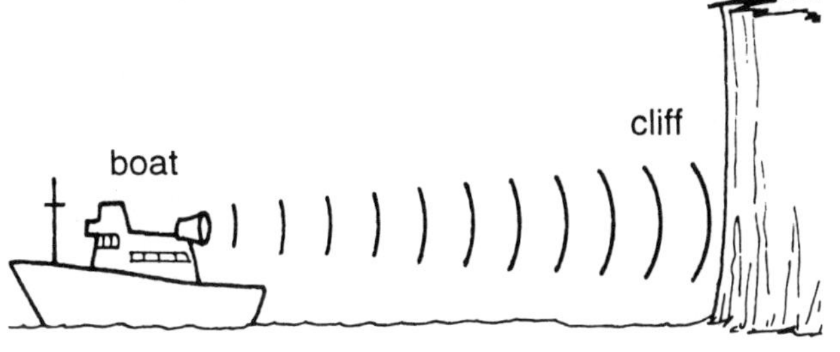

c The boat again sounds its hooter, but no second sound is heard. Why is this? Write down the correct letter.

A The trees have soft, rough surfaces which absorb sound.
B The trees have soft, rough surfaces which reflect sound.
C The trees have hard, smooth surfaces which absorb sound.
D The trees have hard, smooth surfaces which reflect sound.

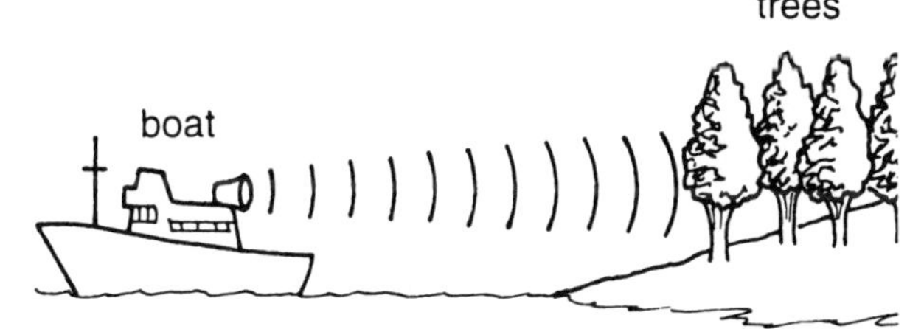

3 Write down the correct letter. When a volcano erupts, you see the flames, then hear the sound, then you are hit by the ash. This is because:

A light is fast, sound is slower, ash is very slow
B ash is slow, light is very slow and sound is even slower
C sound is fast, light is slow and ash is fast
D ash is fast, sound and light are slow.

4 Which of these pairs of diagrams correctly shows light reflecting from a mirror and from sandpaper? Write down the correct letter.

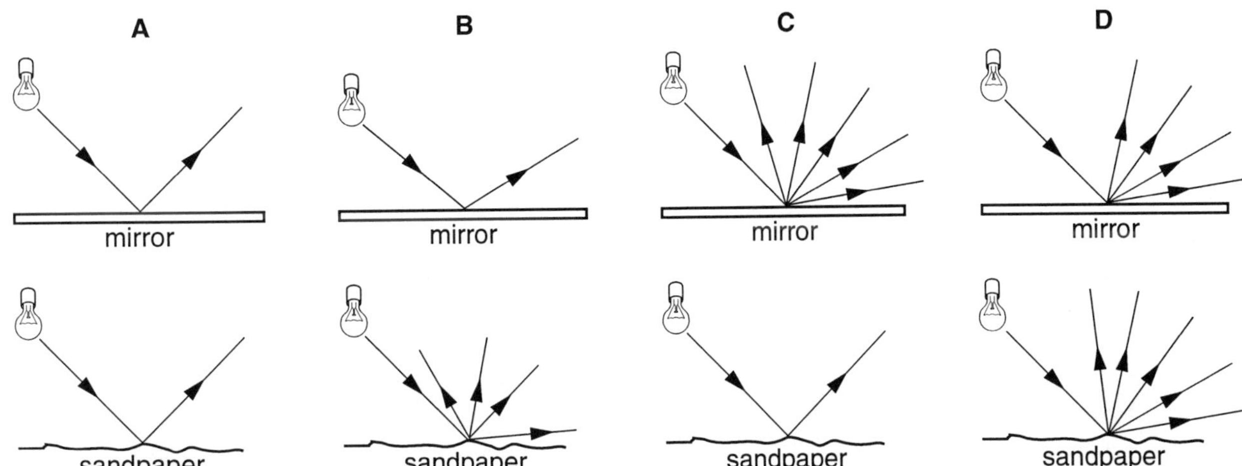

5 Sam is a schoolgirl, and her older sister works in a noisy night club. Her mother is a teacher, and her grandad is a gardener.

a Who is most likely to have damaged ears?
b What could this person do to reduce the damage to their ears?
c Who is most likely to be able to hear a bat squeaking?

6 Copy and complete this table. Write the underlined words in the correct columns.

On a moonlit night, Mark and Mary have a bonfire in their garden. They use torches to set up fireworks. A frightened cat runs across the grass and jumps over the fence.

Produces light	Only reflects light

7 The diagram shows a factory machine on a hard workbench, with a cover over it.

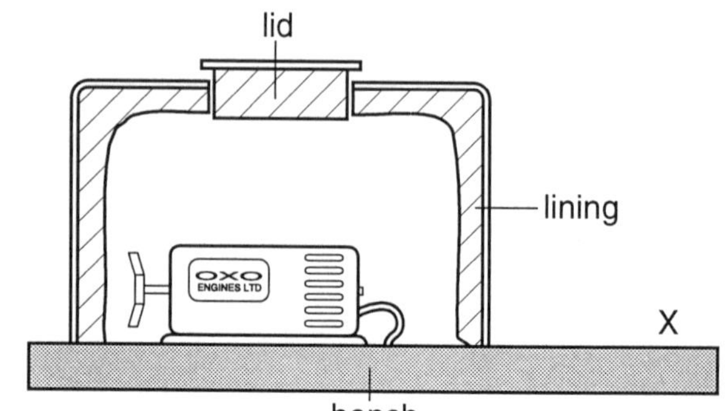

a Which of the following linings would you use to reduce the noise in the factory? Write down the correct letter.

A concrete   **C** plastic foam
B steel   **D** wood

b If you placed your hand on the bench at X, what would you feel?
c If the covering was removed and the machine used, which two parts of your ear could be damaged?

8 Copy this diagram.

a Draw the light rays from the light box to the screen.
b Clearly draw the position of the shadow on the screen.

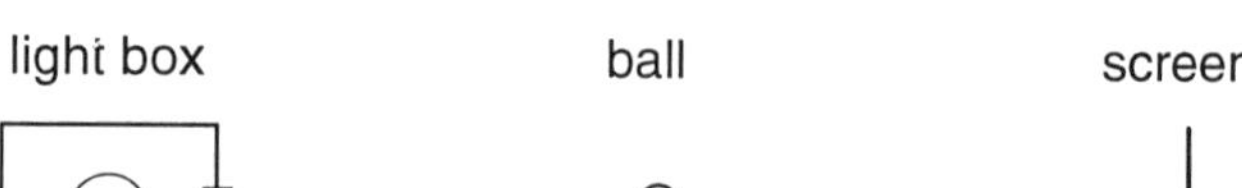

Extension

9 **a** Write the names of the parts of the ear labelled **A**, **B** and **C**.

b Write down the correct letter. Which part:
i picks up sound vibrations from the air
ii vibrates, then sends messages to the brain
iii carries the vibration from the outer ear to the inner ear?

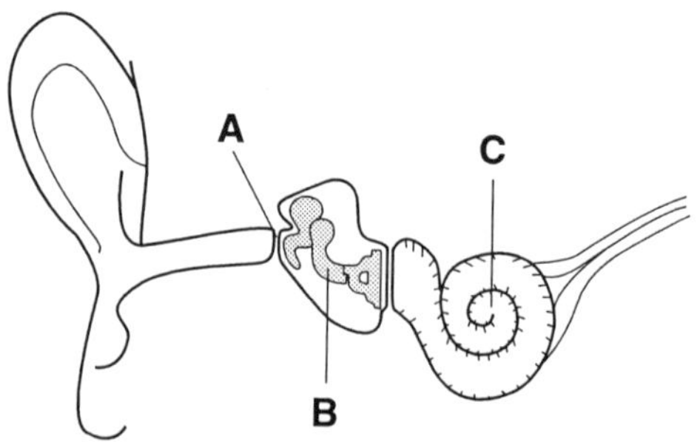

10 The captain of the ship wants to find the wreck. He uses an echo sounder. This produces a sound that travels down through the water, reflects off the bottom and is heard (detected) back on the ship. This is the echo.

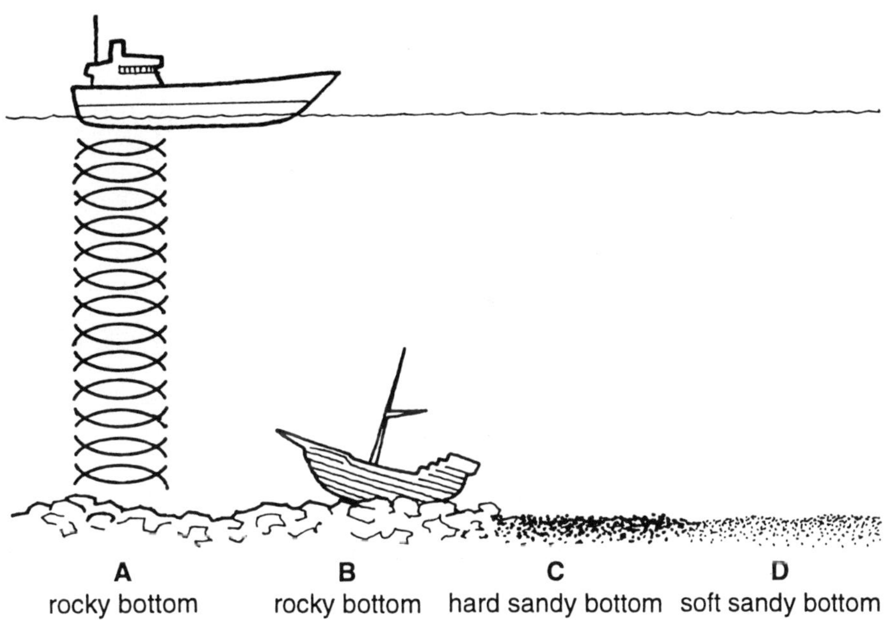

a At which position would the echo return the quickest? Write down the correct letter.

b At which position would the echo be very unclear? Write down the correct letter.

11 This is a periscope.
The light ray from the object is shown (A).

a Copy the diagram and draw on the path of the light to the eye.

b Where might a periscope be used?

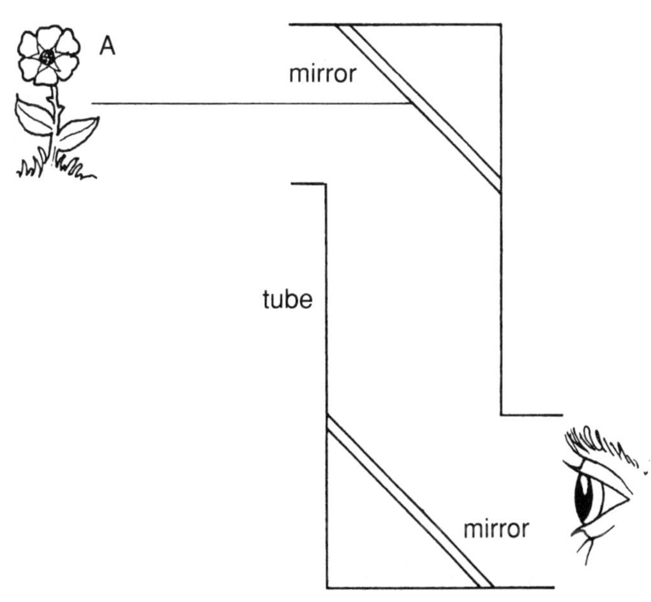

Core

1 Copy and complete the following sentence.

When you heat ice, the ice melts at __________°C and the water then boils at __________°C.

2 Choose a word from those below to describe each of the following changes.

melts freezes boils condenses

a An orange ice lolly changes into orange juice.
b Steam from a kettle touches a cold window.
c Ice forms on puddles in water.

3 Coal is a fuel. When it burns, a gas from the air is used, energy is released and new substances are formed.
a Name the gas used when coal burns.
b Name the two types of energy released.
c When coal burns, is this a physical change or a chemical change?
d Give two reasons for your answer to **c**.

4 Fred says painted nails don't go rusty. What two things that cause rusting are prevented from reaching the nail by the paint?

5 Are the following changes physical or chemical?
a a rusting nail
b steam condensing on the window
c burning a match
d toasting a teacake

a road bridge seen from above

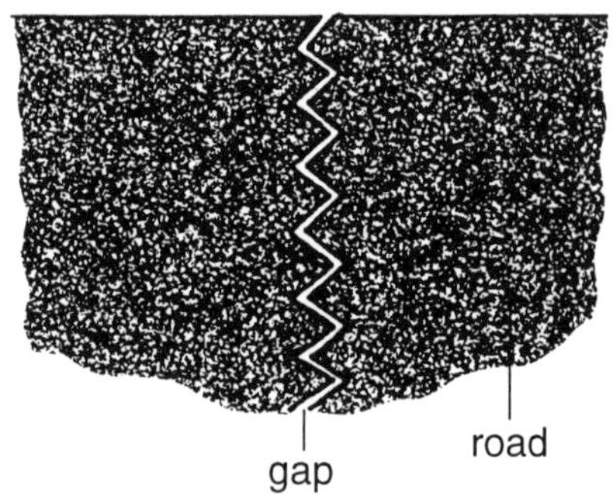

6 These sentences are about the bridge in the diagrams. Copy and complete the following table. Write the sentences in the correct columns.
a It is hotter.
b The gap gets smaller.
c It is cooler.
d The road expands.
e The road contracts
f The gap gets bigger.

Summer	Winter

7 What would happen in very hot weather if the roadway of a suspension bridge had no gaps?

8 Professor Harwood discovers three new substances in rock from Mars. She called them misto, fristo and wisto. The table shows the melting points and boiling points of the substances.
Copy and complete the following table to say if they are solid, liquid or gas at room temperature (20 °C).

Substance	Melting point	Boiling point	State at room temperature
fristo	–50 °C	5 °C	**a**
wisto	90 °C	150 °C	**b**
misto	10 °C	80 °C	**c**

Extension

9 This question is about hailstones forming in clouds where the air temperature is –5 °C.

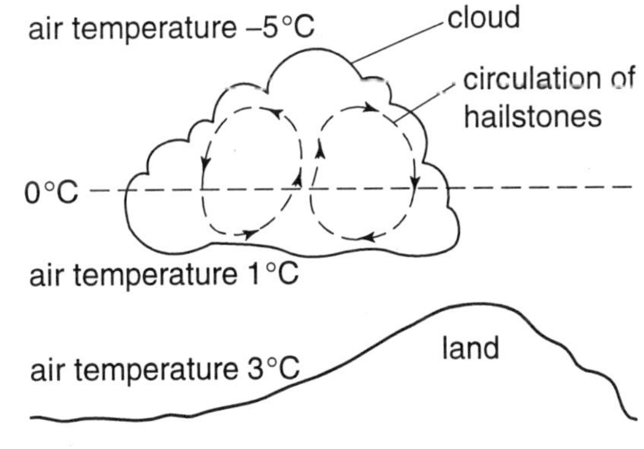

a Explain how the water droplets in the cloud form hailstones.

b This is a picture of a hailstone cut in half.

How many times has the hailstone been round in the cloud?

c Explain your answer to **b**.

10 This diagram shows how a bimetallic strip is used as part of a thermostat in a central heating system.
Brass expands much more than steel.

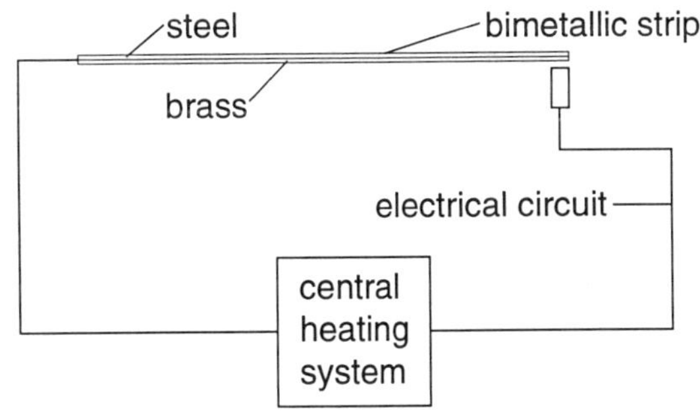

a Draw a diagram to show how the bimetallic strip changes as the temperature rises.

b How does this control the temperature of the house on a cold day?

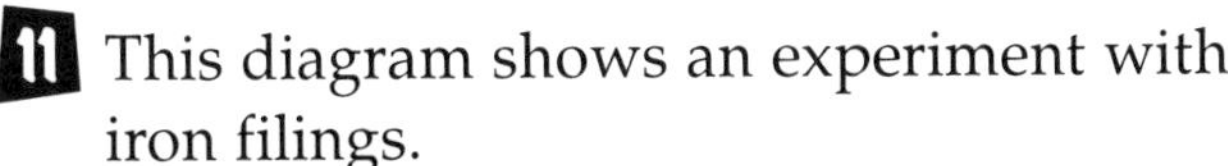

11 This diagram shows an experiment with iron filings.

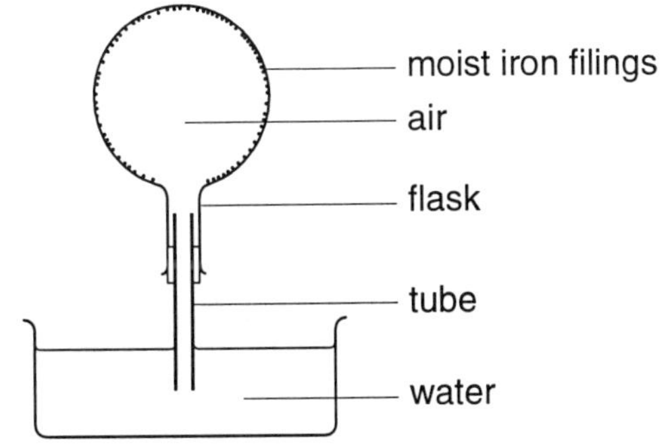

a How will you know if the iron filings undergo a chemical change?

b Why will the water rise up the tube?

c At the end of the experiment, what fraction of the flask will be filled with water?

Core

1 Which of the organisms in the picture reproduce sexually? Write down the correct letters.

2 Describe briefly how the male and female sex cells are normally brought together in the following organisms:
a plaice **b** cow **c** buttercup.

3 In plants, fertilised ovules develop into seeds. Copy the picture and label:
a the seed
b the swollen ovary.

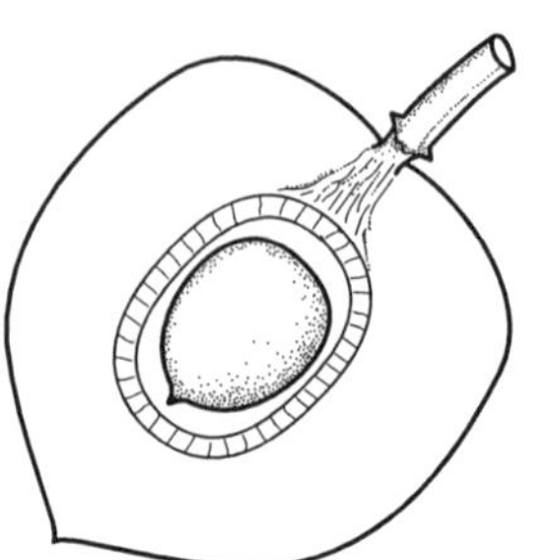

4 Write down two ways in which seeds may be dispersed.

5 During sexual intercourse, the man ejaculates.
a Where in the man do the sperm start their journey?
b Along which tube do they pass out of the man?

6 **a** The diagram shows the human female reproductive organs. Write down the correct letter.
i Where is the egg (ovum) fertilised?
ii Where are the sperm placed during intercourse?
b What happens to the egg immediately after it is fertilised?
c What happens to the egg if it is not fertilised?

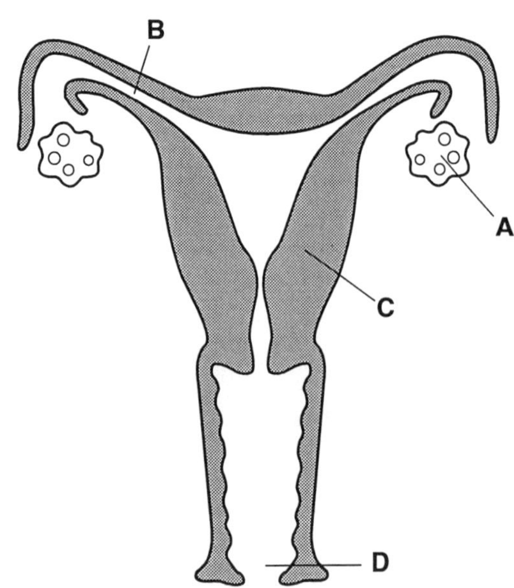

7 This picture shows a fetus growing inside its mother.

a Use the words below to match the labels A, B, C and D.

uterus fetus placenta umbilical cord

b Why is the umbilical cord important? Give two reasons.

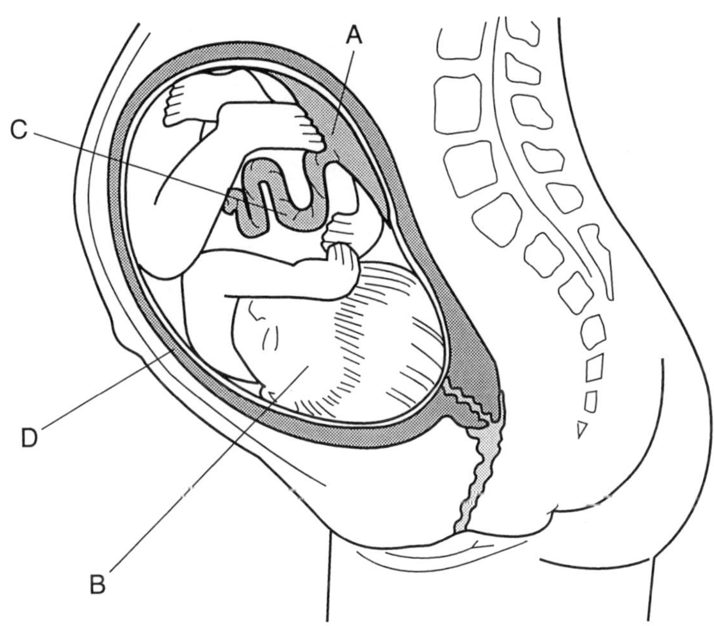

Extension

8 Draw a diagram to show the life cycle of a seagull.

9 **a** How is the fruit in the diagram below dispersed?

b Say how its shape and size help this way of dispersal.

10 Copy and complete the following table. It shows which of the things in the list below are done by the baby inside the uterus, and which are done by a fully developed baby after birth. You can use the same word more than once. An example has been done for you.

drink milk breathe in air digest food hear sounds respire

Inside the uterus	After birth
	drink milk

Core

1 The diagram shows the symbols for some components. Write down the correct letter. Which is the symbol for:

a a switch **b** a battery **c** a motor **d** a bulb?

2 Copy and complete the following table. Write these words in the correct columns.

iron wood glass leather copper aluminium

Electrical conductors	Electrical insulators

3 For each circuit **a** to **d**, say whether **both** bulbs would light when the switch is closed. Write **yes** or **no**.

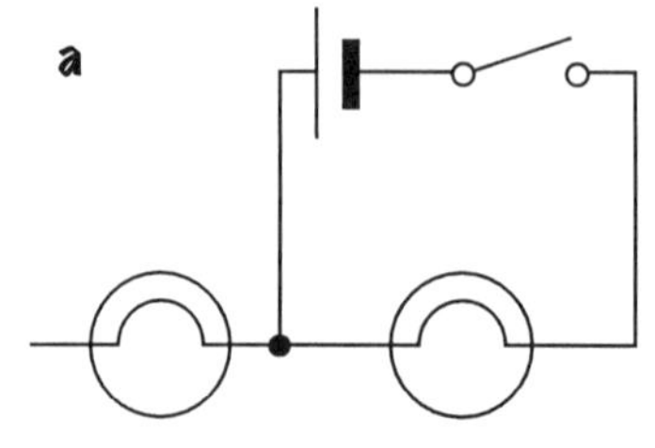

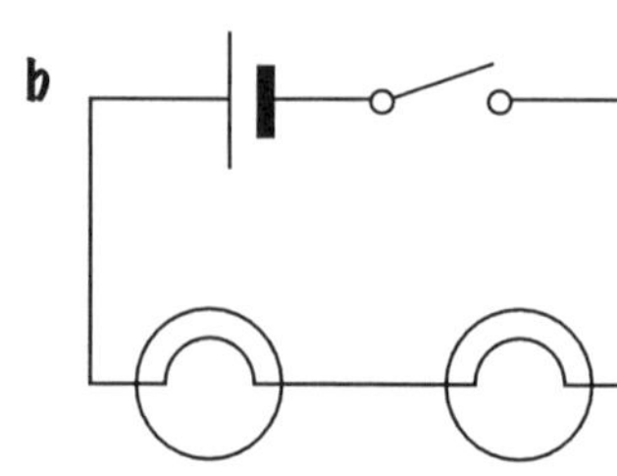

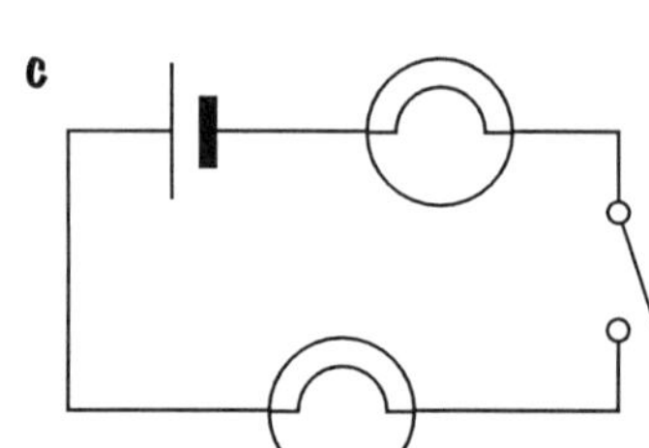

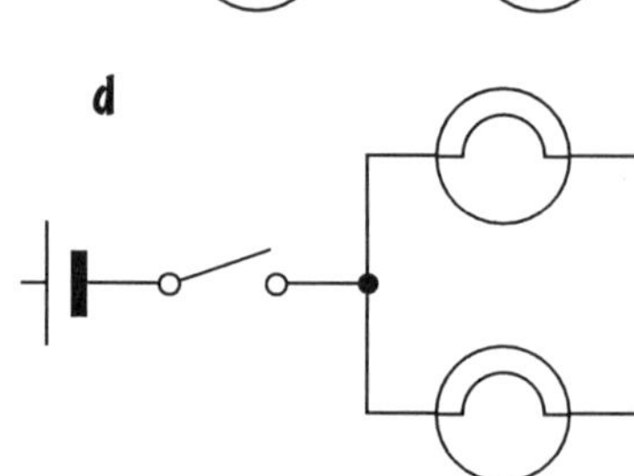

4 The diagram shows two circuits. The bulbs in the circuits are all the same.

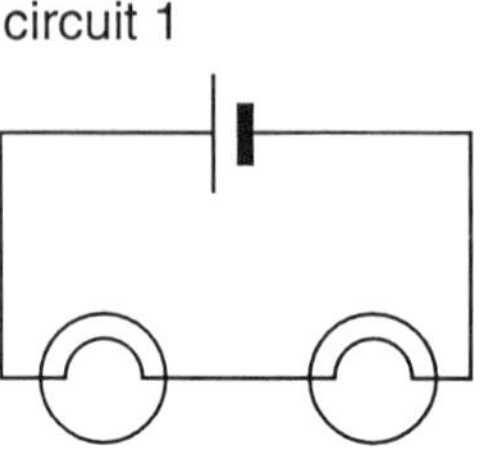

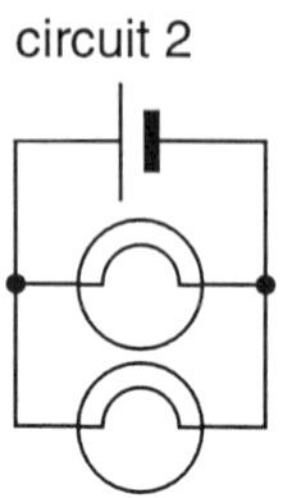

a What is the name given to the way the bulbs are connected together in circuit 1?

b What is the name given to the way the bulbs are connected together in circuit 2?

c In which circuit will the bulbs be brighter?

5 **a** The diagram shows a circuit with two ammeters. If the reading on ammeter 1 is 2 amperes, what will be the reading on ammeter 2? Write down the correct letter.

A more than 2 amperes

B 2 amperes

C less than 2 amperes

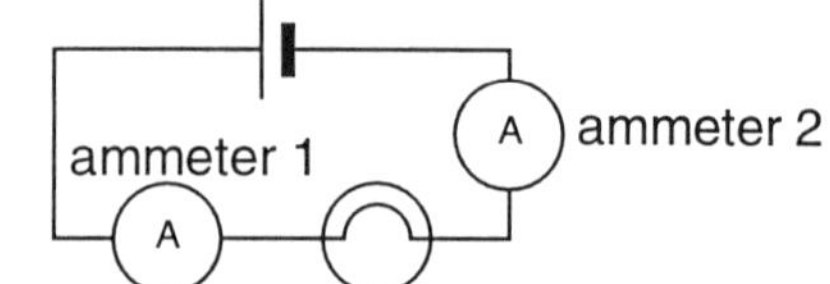

b A new circuit was made. Redraw the circuit to include a switch that will turn both bulbs on and off at the same time.

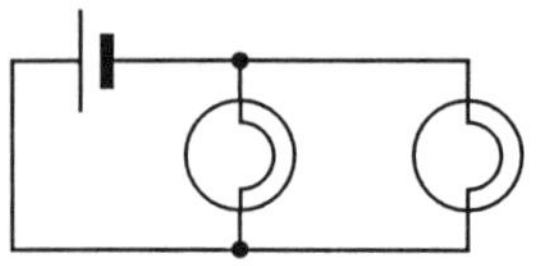

c A more complex lighting circuit was constructed. Copy and complete the following table. Write **on** or **off** in each box.

Switches	Bulb 1	Bulb 2	Bulb 3	Bulb 4
all switches open				
switch A closed, B and C open				
switches A and B closed, C open				

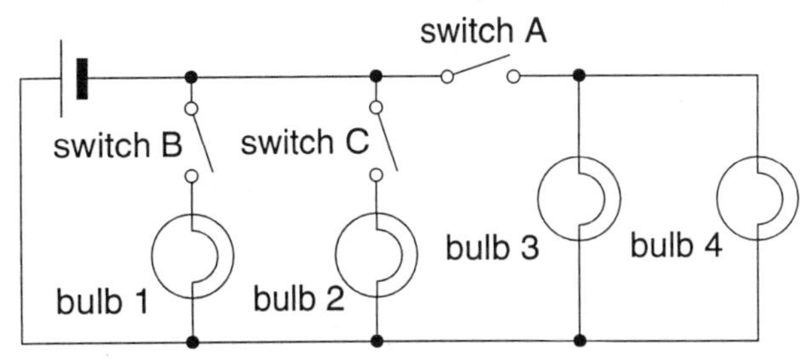

6 **a** Look at the circuit in the diagram. What could you do to make the bulb brighter? Write down the correct letter(s).

A connect an extra battery in series
B increase the resistance of the circuit
C decrease the resistance of the circuit

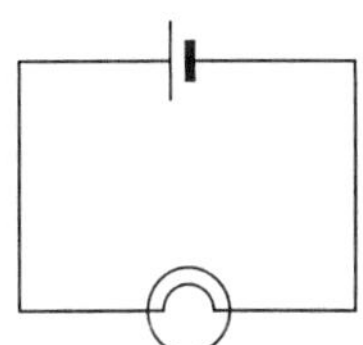

b When the circuit above was left on, the bulb began to go dim. Explain why this happens in terms of energy changes.

Extension

7 This diagram shows two circuits. The same type of bulb has been used in both circuits.

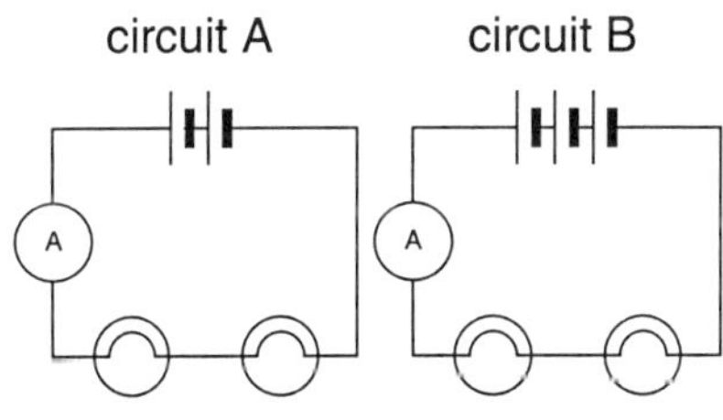

a What difference in brightness would you notice between the two circuits?
b Why is there this difference?
c If you compared the two ammeter readings, what difference, if any, would there be?

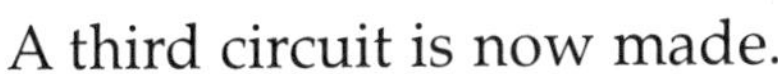
A third circuit is now made.

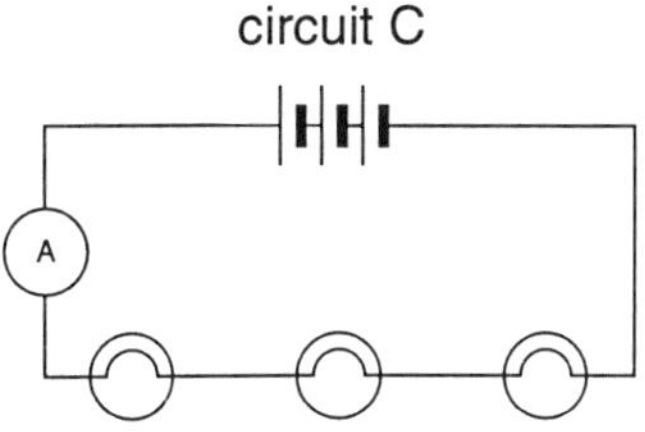

d Of the three circuits A, B and C, which would have the greatest resistance?
e Explain why.
f Draw a circuit diagram including a variable resistor in circuit A to alter the brightness of the bulbs.
g How does adjusting the variable resistor change the brightness of the bulbs?

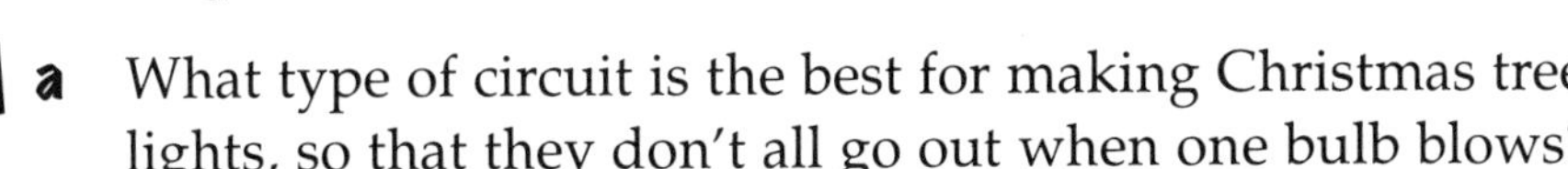
8 **a** What type of circuit is the best for making Christmas tree lights, so that they don't all go out when one bulb blows?
b Give reasons for your answers.
c How could you change the connecting wires to make the bulbs brighter?

Core

1 The diagram shows the layers of the Earth. Choose from the words below to match the labels **a**, **b** and **c**.

core crust mantle solid liquid

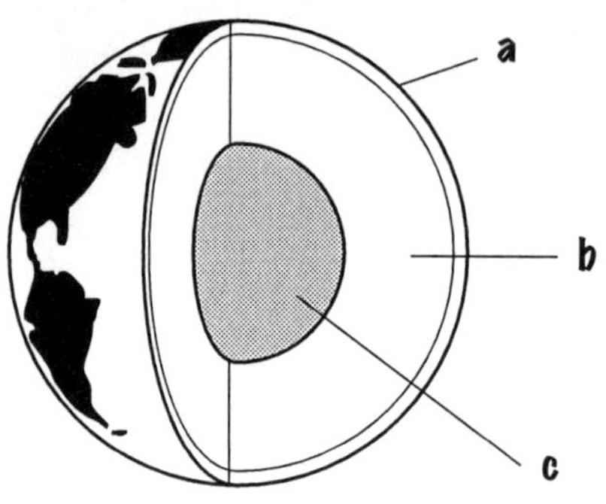

2 **a** The diagram shows a close-up view of two igneous rocks, showing the crystals. Both rocks formed when molten rock cooled. Which of these rocks was formed by slower cooling? Write down the correct letter.
b Give the reason for your answer to **a**.
c Name an igneous rock with large crystals.

A

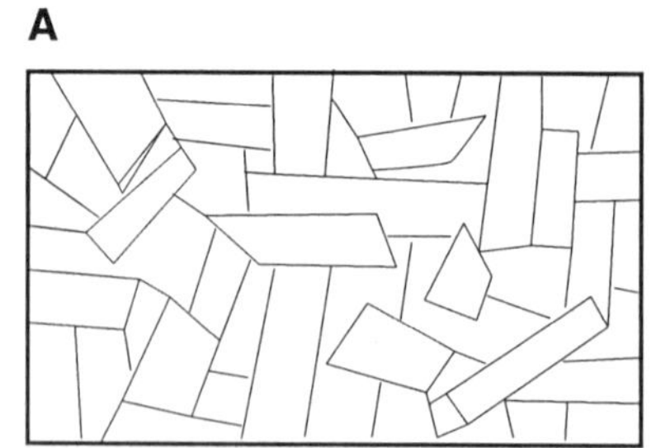

B

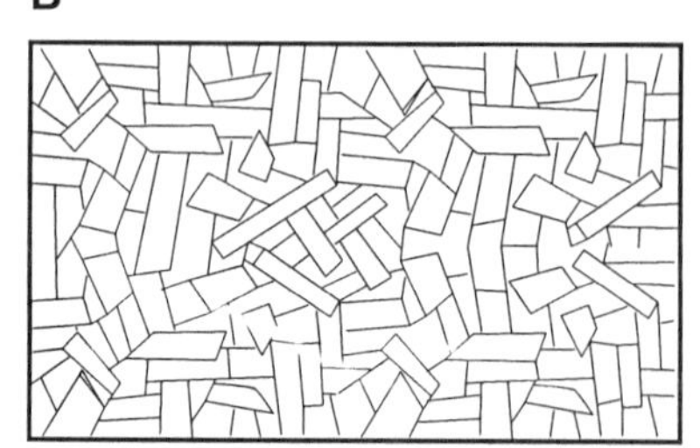

3 Why do rocks in the desert gradually break into small pieces?

4 **a** In mountain streams, rocks at the bottom are often sharp. In rivers near the sea, there are usually rounded pebbles. Why is this?
b Mountain streams have stony beds, whereas rivers near the sea often have soft muddy or sandy beds. Why is this?

5 Igneous rocks can be turned into sand and clay if they are attacked by air and water. What is the name given to this type of weathering?

6 **a** How are sedimentary rocks formed?
b Give an example of a sedimentary rock.

7 The diagram shows a process that happens to rock. Write labels for **a** and **b**.

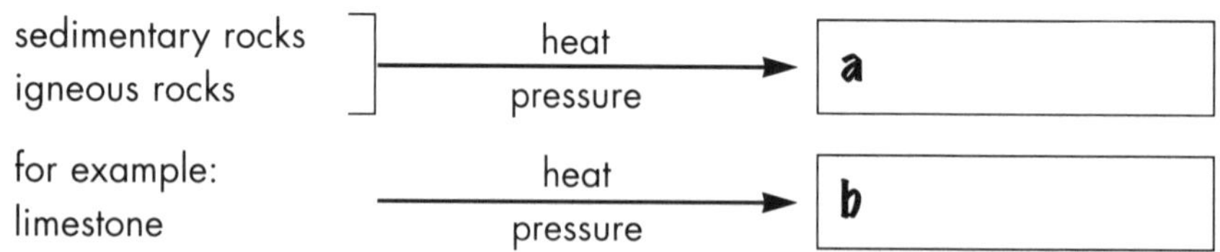

8 Look at this diagram of the rock cycle. Write labels showing the type of rock for **a**, **b** and **c**.

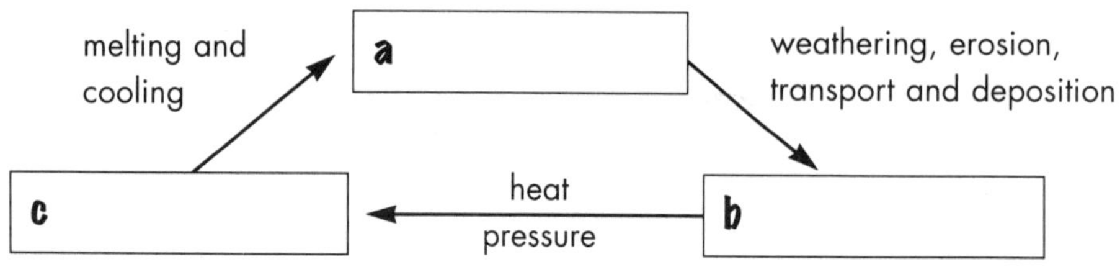

Extension

9 Look at this diagram of the Grand Canyon.

a Which rocks are likely to be the oldest, those at **A**, **B** or **C**? Write down the correct letter.

b Give a reason for your answer to **a**.

c In which layer would you find the hardest rocks? Why?

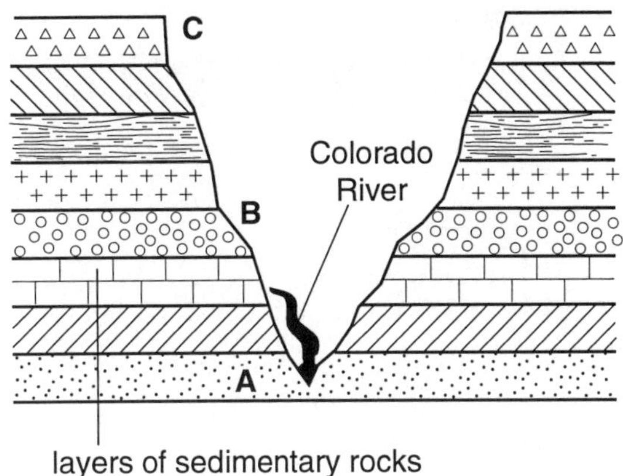

10 On Ingleborough, a hill in Yorkshire, rocks containing fossilised sea shells can be found.

a What does this tell you about where the rocks were formed?

b How did they come to be on a hill in the middle of Britain?

11 Pumice is a volcanic rock formed from lava which has cooled. It has lots of little holes in it, like a sponge. Suggest what might have caused these holes.

12 Identify rocks **A** to **F**, using the key provided. Use the information in the boxes to help you.

Rock A
- does not react with acid
- individual round grains which may rub off
- hard
- may contain fossils

Rock B
- does not react with acid
- hard
- small interlocking crystals visible under a microscope

Rock C
- does not react with acid
- crumbly layers
- very soft
- may contain fossils
- grains too small to see

Rock D
- reacts with acid
- white/grey
- contains fossils or shell fragments
- has fine grains

Rock E
- hard
- large interlocking crystals

Rock F
- reacts with acid
- very tough
- no fossils

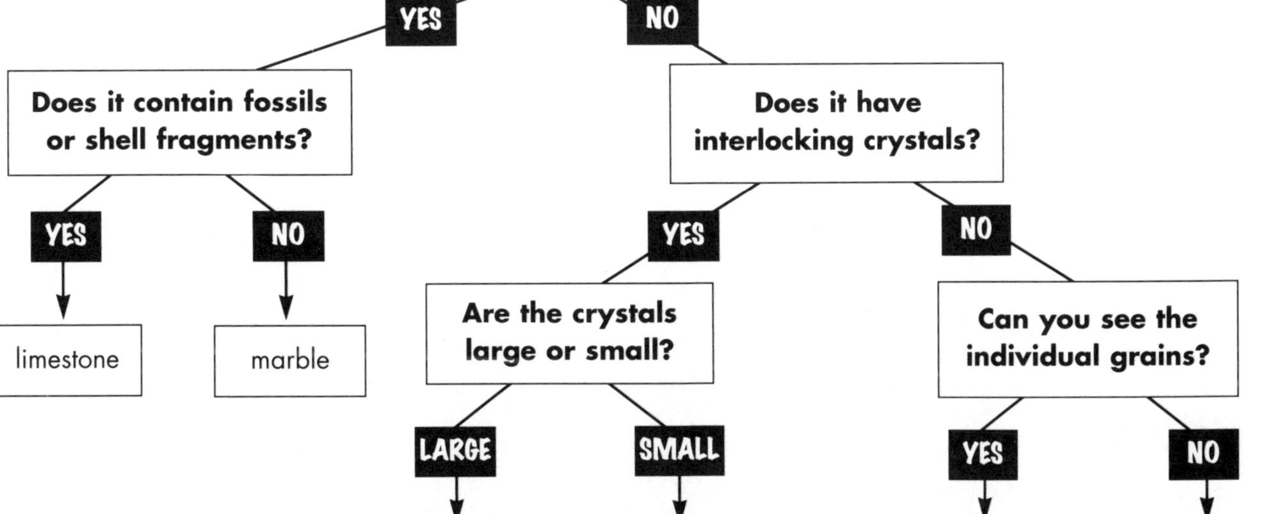

Mark scheme

Answers	Marks
Core	
1 Forces, newtons, balanced.	1 / 1 / 1
2 C.	1
3 B.	1
4 Seal, streamlined *or* smoother *or* aerodynamic *or* sleek.	1 / 1
5 Weight: Earth 750 N Moon less than 750 N Outer space zero Mass: 75 kg for Moon and outer space.	1 1 1 2
6 **a**  gravity ↓ ↑ air resistance *Relative lengths of arrows are not important.* **b** 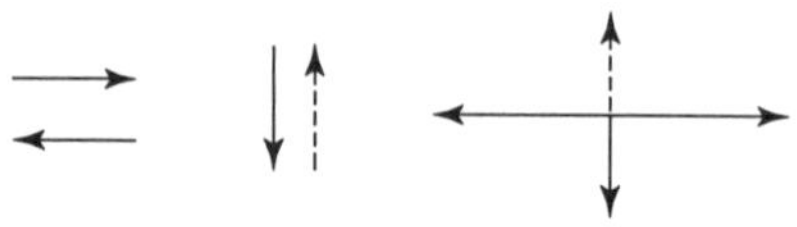 *Reaction force is not expected but accept it if given.*	2 / 3
7 Lever.	1
8 500 N *or* the same as Sameena *or* John's weight is equal to Sameena's.	1
9 **B** *or* nearer the middle *or* between **A** and **C**.	1
Core total marks	20
Extension	
10 Streamlined *or* reduces air resistance/friction.	1
11 **a** A: pushing force *or* engine driving forward. B: friction *or* air resistance *or* opposing force *or* drag. **b** → ← *or* → air resistance, ←, → friction *Arrows to be equal lengths* *or any equivalent answer.* **c** Parachute fills with air and slows car down *or* greater air resistance because of larger surface of parachute. Forward force less than force going backwards *or* driver puts brakes on, therefore increasing force of friction.	1 1 2 1 1
12 **a** Laura's weight = Emma's weight = 350 N. Mass = 350/10 = 35 kg. **b** Anywhere closer to Laura's mum.	1 1 1
Extension total marks	10

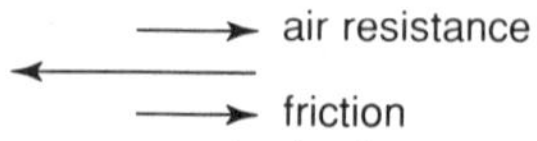

Outcome of test by levels

	Core (out of 20)	Total (out of 30)
Level 3	3–6	4–8
Level 4	7–10	9–14
Level 5	11–14	15–20
Level 6	15 and over	21 and over

Mark scheme

Answers	Marks
Core	
1 D.	1
2 A.	1
3 C.	1
4 *Must use words provided to answer questions.* Dogs can: feed, grow, move, reproduce. Motor bikes can: move. *Accept feed (take in fuel).*	 1 1
5 Amphibian, toad. Reptile, crocodile. Mammal, dog. Bird, penguin. *If pupils substitute an alternative example which is a correct example, the mark should be credited (correct knowledge).*	1 1 1 1
6 **a** Nucleus. **b** Cell wall *or* thick cell wall *or* thick wall. Chloroplasts.* Large vacuoles.* **Not all plant cells have chloroplasts. Also, vacuoles are not universal. These generalisations are common at this level.*	1 *any two* { 1 1 1
7 **a** B. **b** C. **c** A. **d** D.	1 1 1 1
8 **a** W: Wem poplar. **b** X: December fir. **c** Y: lap fir. **d** Z: trench poplar.	1 1 1 1
Core total marks	20
Extension	
9 **c** R: leaf A. **d** S: leaf D. **e** T: leaf C. **f** U: leaf F.	1 1 1 1
10 **a** Conifers. **b** Mosses. **c** Flowering plants or conifers. **d** Ferns.	1 1 1 1
11 **a** Fish, reptiles or birds. **b** If fish: gills/fins. If birds: feathers. If reptiles: dry skin.	1 1
Extension total marks	10

Outcome of test by levels

	Core (out of 20)	Total (out of 30)
Level 3	3–7	4–9
Level 4	8–10	10–14
Level 5	11–14	15–20
Level 6	15 and over	21 and over

Mark scheme

Answers	Marks
Core	
1 Solids: fixed shape and fixed volume. Liquids: no fixed shape, fixed volume. Gases: no fixed shape, no fixed volume.	1 1 1
2 **a** Gas. **b** Solid. **c** Liquid.	1
3 Liquids and gases have no fixed shape *or* can flow.	1
4 **a** Hard. **b** Light. **c** Strong *or* hard.	1 1 1
5 Any figure between 3.5 and 7.	1
6 **a** Shiny. **b** *Any two from*: Conducts heat. Conducts electricity. Bends and stretches. Can be shaped. Make a ringing sound. **c** Any metal example apart from iron (steel), cobalt and nickel.	1 2 1
7 **a** Copper. **b** All have the same volume *or* other factors kept the same.	1 1
8 Carbon dioxide.	1
9 **a** Near each other, but move more easily. **b** Further apart and moving freely. } *Or any equivalent answer.*	1 1
10 **a** Particles in a gas are further apart and they can be squashed together. **b** Particles in a liquid move easily relative to each other *or* particles in a liquid are not in a fixed arrangement.	1 1
Core total marks	**20**
Extension	
11 **a** C and D. **b** Formula (density = mass ÷ volume). Answer = 2.7 grams per cubic centimetre.	1 1 1
12 **a** *One of:* the particles spread out, the particles move more quickly, the particles move from the liquid to the gas. **b** It takes up less space **c** **i** Y. **ii** Strong. Stiff *or* keeps its shape.	1 1 1 1 1
13 Particles become detached from solids (where they are not free to move). Particles move through liquid (where they are free to move relative to each other).	1 1
Extension total marks	**10**

Outcome of test by levels

	Core (out of 20)	Total (out of 30)
Level 3	3–7	3–8
Level 4	8–10	9–13
Level 5	11–14	14–20
Level 6	15 and over	21 and over

Mark scheme

Answers	Marks
Core	
1 Moving bicycle: kinetic energy. Stretched spring: elastic energy. Ringing bell: sound energy. Slice of bread: chemical energy.	1 1 1 1
2 a Elastic energy → kinetic energy. **b** Chemical energy → heat energy + light energy. **c** Electrical energy → light energy + sound energy *or* heat energy.	1 1 1
3 a Coal *or* wood *or* paper *or* peat *or* smokeless fuel *or* charcoal. **b** Oil *or* gas *or* propane *or* LPG *or* kerosene. **c** Electricity.	1 1 1
4 a Kilojoules *or* kJ. *Accept joules or J.* **b i** B. **ii** It has the most energy *or* it contains more joules *or any equivalent answer.*	1 1 1
5 a Electrical energy → light energy. **b** Kinetic (movement) energy → electrical energy **c** Chemical energy → kinetic (movement) energy *or* heat energy. **d i** Dimmer *or* less bright *or* may not be lit. **ii** Less kinetic energy is being changed into electrical energy.	1 1 1 1 1
6 a Heat. **b** Kinetic (movement).	1 1
Core total marks	20
Extension	
7 a i Chemical energy. **ii** Heat energy. **iii** Kinetic (movement) energy. **iv** Electrical energy. **v** Light energy. **b i** Heat. **ii** Friction (surfaces rubbing together) *or* from the warm bulb *or* from the steam engine *or* from the escaping steam. **iii** Spread out into the air *or* to the air.	1 1 1 1 1 1 1 1
8 a 90 kJ. **b** Some energy lost as heat *or* not all of the energy is converted.	1 1
Extension total marks	10

Outcome of test by levels

	Core (out of 20)	Total (out of 30)
Level 3	3–7	4–9
Level 4	8–11	10–15
Level 5	12–14	16–20
Level 6	15 and over	21 and over

Mark scheme

Answers	Marks
Core	
1 a Carbohydrate. Fibre. Minerals *or* vitamins. **b** Growth *or* repair *or* new cells *or* building muscles. Store of energy *or* energy store. Prevents scurvy *or* keeps skin healthy.	1 1 1 1 1 1
2 a i Carbohydrates, oxygen. **ii** Carbon dioxide. **iii** Sunlight. **b** To trap sunlight *or* catch energy for this process.	2 1 1 1
3 a Grass *or* nettles. **b** Rabbit *or* field vole *or* caterpillar. **c** Any of the following: Grass → rabbit → fox. Grass → field vole → fox. Grass → field vole → hawk. Nettle → field vole → fox. Nettle → caterpillar → field vole → fox. Nettle → field vole → hawk. Nettle → caterpillar → field vole → hawk. Nettle → caterpillar → blackbird → hawk. *One mark for each stage, but no marks if the chain doesn't start with a producer, up to a maximum of 3.* **d** The feeding relationships between the plants and animals living in one place.	1 1 3 1
4 Anchorage *or* holding the plant in place. Taking up water. Taking up minerals.	1 1 1
Core total marks	**20**
Extension	
5 a Nitrogen *or* nitrate. **b** To make crops grow better. **c** Root growth. **d** Manure *or* dung *or* sewage sludge *or* slurry *or* leaf mould *or* compost.	1 1 1 1
6 a i They will decrease. **ii** Hunger *or* starvation *or* not enough food for owls and foxes to eat *or* too few mice to go round all the owls and foxes. **b** More blackberries remained uneaten.	1 1 1
7 carbon dioxide + water + sunlight —chloroplasts→ carbohydrate + oxygen $\frac{1}{2}$ *mark for each word placed correctly in box.*	3
Extension total marks	**10**

Outcome of test by levels

	Core (out of 20)	Total (out of 30)
Level 3	3–7	4–10
Level 4	8–10	11–15
Level 5	11–15	16–22
Level 6	16 and over	23 and over

Mark scheme

Answers	Marks
Core	
1 **a** Soluble. **b** Filtration *or* filtering.	1 1
2 **a** Filtration *or* filtering *or* decanting. **b** Large lumps from small lumps *or* large objects from small objects *or* different sized particles *or any equivalent answer.* **c** Chromatography. **d** Magnetic from non-magnetic materials *or* iron and steel from non-iron materials.	1 1 1 1
3 **a** Solvent. **b** Solute. **c** Solution. **d** Saturated. **e** Stir, heat up *or* warm.	1 1 1 1 1 / 1
4 **a** B, C, D. **b** Water. **c** D.	1 1 1
5 **a** Evaporation. **b** Colder surface *or* condensation. **c** Distillation. **d** Flask *or* liquid in flask. **e** Water.	1 1 1 1 1
Core total marks	20
Extension	
6 A: **energy** from the sun is absorbed *or* taken up by the sea *or* warms the sea. B: water **evaporates** and turns into water vapour. C: **water vapour** rises up as **moist air**. D: the air **cools** and water **condenses** because it is colder *or* the **moist air** forms clouds. E: **water vapour condenses** *or* turns to **rain** which falls over the land *or* mountains.	1 1 1 1 1
7 **a** Filtration *or* decanting. **b** Chromatography. **c** 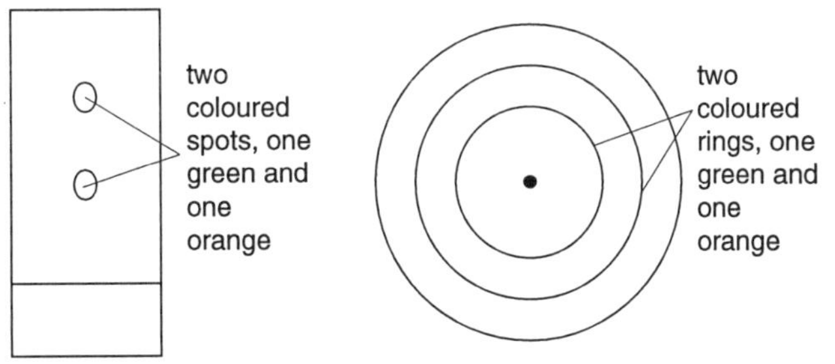 two coloured spots, one green and one orange / two coloured rings, one green and one orange **d** The dyes *or* pigments *or* colours *or* chlorophyll in cabbage are not soluble *or* do not dissolve in water *or* dissolve better in propanone. **e** Distil it *or* evaporate and then condense it.	1 1 1 1 1
Extension total marks	10

Outcome of test by levels

	Core (out of 20)	Total (out of 30)
Level 3	2–5	3–8
Level 4	6–9	9–13
Level 5	10–16	14–22
Level 6	17 and over	23 and over

Mark scheme

Answers	Marks
Core	
1 **a** B. **b** A. **c** D. **d** C.	1 1 1 1
2 **a** The sound will bounce off. **b** Echo. **c** A.	1 1 1
3 A.	1
4 B.	1
5 **a** Her sister. **b** Wear ear plugs *or* ear defenders *or* ear protectors. **c** Sam.	1 1 1
6 Produces light: bonfire, torches, fireworks. Only reflects light: moon, cat, grass, fence.	1 1
7 **a** C. **b** Vibrations. **c** Eardrum *or* small bones *or* fine hairs *(any two).*	1 1 1 / 1
8 **a** Two straight lines drawn from slit, touching sides of ball, to screen. **b** Shadow indicated as area between two lines on screen.	1 1
Core total marks	**20**
Extension	
9 **a** A: eardrum. B: small bones *or* ossicles *or* hammer, anvil, stirrup. C: cochlea *or* fine hairs *or* nerve endings. **b** **i** A. **ii** C. **iii** B.	1 1 1 1 1 1
10 **a** B. **b** D.	1 1
11 **a** Path of ray of light drawn parallel to the sides of the periscope. **b** To see over a crowd *or* to see above the water from a submarine, etc.	1 1
Extension total marks	**10**

Outcome of test by levels

	Core (out of 20)	Total (out of 30)
Level 3	2–5	3–7
Level 4	6–10	8–14
Level 5	11–15	15–21
Level 6	16 and over	22 and over

Mark scheme

Answers	Marks
Core	
1 0 °C, 100 °C.	1
2 a Melts. **b** Condenses. **c** Freezes.	1 1 1
3 a Oxygen. **b** Heat and light. **c** Chemical. **d** Heat or light given out, new substance formed, irreversible *(any two).*	1 1 1 1 / 1
4 a Air *or* oxygen, water.	1
5 a Chemical. **b** Physical. **c** Chemical. **d** Chemical.	1 1 1 1
6 Summer: hotter, expands, gap smaller. Winter: cooler, contracts, gap bigger.	1 1
7 The bridge would buckle *or* bend *or any equivalent answer.*	1
8 a Gas. **b** Solid. **c** Liquid.	1 1 1
Core total marks	**20**
Extension	
9 a The air circulates and carries the water droplets. The water droplets are cooled to –5 °C and freeze. **b** Four *(allow three if explanation correct).* **c** Because there are four sections, and a new coating forms each time.	1 1 1 1
10 a steel; bimetallic strip; brass; central heating system; electrical circuit **b** The strip bends up as the temperature rises. This breaks the circuit and switches off the heating. As the temperature falls, the strip straightens and switches on the heating again.	1 1 1
11 a A new substance (rust) forms *or* they change colour. **b** The oxygen is removed and the water moves in to replace it. **c** One-fifth.	1 1 1
Extension total marks	**10**

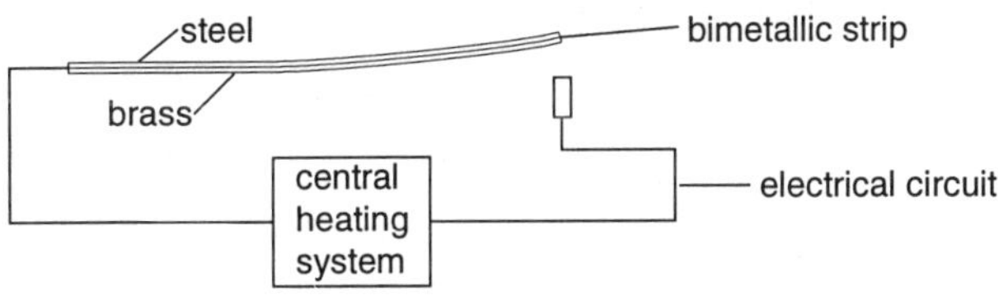

Outcome of test by levels

	Core (out of 20)	Total (out of 30)
Level 3	3–7	4–9
Level 4	8–10	10–13
Level 5	11–16	14–23
Level 6	17 and over	24 and over

Mark scheme

Answers	Marks
Core	
1 A, B, C, D, E *Note that some roses are sterile, so also give the mark if* **D** *omitted.*	1
2 **a** Ova (eggs) and sperm released into (sea) water. **b** (Fertilisation) in the body of the female. **c** Pollen carried by insects.	1 1 1
3 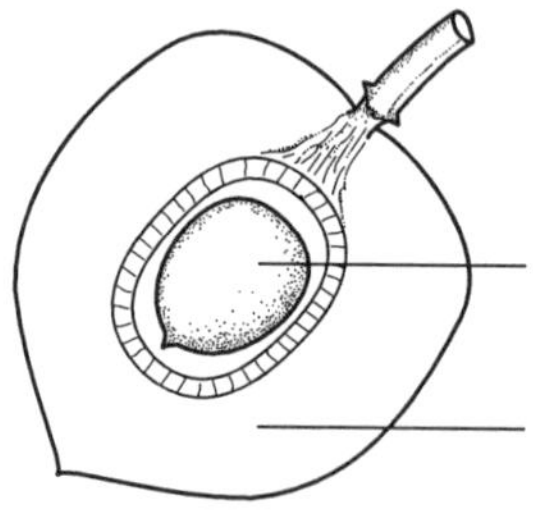 [diagram: seed; swollen ovary] *1 mark for each correct label.*	2
4 *Any two out of:* wind, water, animals' droppings, 'explosive fruits', attached to animals, hidden by animals.	2
5 **a** Testes. **b** Urethra *or* penis.	1 1
6 **a** **i** B. **ii** D. **b** Divides *or* splits *or* implants in the uterus. **c** *Any one of:* menstruation *or* period *or* passes out of body.	2 1 1
7 **a** A: placenta, B: fetus, C: umbilical cord, D: uterus. **b** Carries food and/or oxygen to fetus. Carries waste products and/or carbon dioxide away from fetus.	1/1/1/1 1 1
Core total marks	**20**
Extension	
8 *Indication of:* egg, chick/immature bird, adult, adult producing egg.	1/1/1/1
9 **a** Wind. **b** Large surface area *or* parachute to catch wind *or any equivalent answer.*	1 1
10 In uterus: hear, respire. After birth: (hear, respire) breathe digest.	1/1 1 1
Extension total marks	**10**

Outcome of test by levels

	Core (out of 20)	Total (out of 30)
Level 3	3–6	5–10
Level 4	7–11	11–17
Level 5	12–15	18–23
Level 6	16 and over	24 and over

Mark scheme

Answers	Marks
Core	
1 **a** B. **b** D. **c** A. **d** C.	1/1/1/1
2 Conductors: iron, copper, aluminium. Insulators: wood, glass, leather.	1 1
3 **a** No. **b** Yes. **c** Yes. **d** No.	1 1 1 1
4 **a** Series. **b** Parallel. **c** Circuit 2.	1 1 1
5 **a** B. **b** Switch drawn at any point in series with the cell. **c** (see table below)	1 1
6 **a** A, C. **b** Chemical energy is transferred to electrical energy and then to light and heat energy. The chemical energy is not replaced.	1 1
Core total marks	20
Extension	
7 **a** Circuit B would have brighter bulbs. **b** More cells *or* batteries. **c** Ammeter in circuit B would show a higher current *or* reading. **d** Circuit C. **e** Circuit C has most bulbs in series. **f** Variable resistor drawn in series anywhere in circuit. **g** When the variable resistor is adjusted, the current changes, affecting the brightness of the bulb.	1 1 1 1 1 1 1
8 **a** Parallel. **b** In a parallel circuit, current still flows to the other bulbs. **c** Make them short and thick.	1 1 1
Extension total marks	10

Question 5c:

Switches	Bulb 1	Bulb 2	Bulb 3	Bulb 4	Marks
all switches open	off	off	off	off	1
switch A closed, B and C open	off	off	on	on	1
switches A and B closed, C open	on	off	on	on	1

Outcome of test by levels

	Extension (out of 10)	Total (out of 30)
Level 3	2–5	4–8
Level 4	6–10	9–15
Level 5	11–16	16–23
Level 6	17 and over	24 and over

Mark scheme

Answers	Marks
Core	
1 **a** Crust. **b** Mantle. **c** Core.	1 1 1
2 **a** A. **b** Large crystals result from slow cooling. **c** Granite *or* gabbro.	1 1 1
3 Heating and cooling, expansion and contraction breaks them up.	1 / 1
4 **a** In a river near the sea, the rocks have been rubbing against each other for a long time. **b** Mountain rivers move fast, carrying finer sediment away. Rivers near the sea move slowly, so the finer sediment settles at the bottom.	1 1 1
5 Chemical.	1
6 **a** Sediment settles out in layers. Grains stick together as pressure builds up. **b** Sandstone *or* limestone *or* shale.	1 1 1
7 **a** Metamorphic rocks. **b** Marble.	1 1
8 **a** Igneous rocks. **b** Sedimentary rocks. **c** Metamorphic rocks.	1 1 1
Core total marks	20
Extension	
9 **a** A. **b** Rock was laid down first and others laid on top. **c** A, oldest so under pressure for longest time.	1 1 1
10 **a** They must have formed in the sea. **b** Earth movements.	1 1
11 Gas bubbling out of molten rock as it set.	1
12 **A**: sandstone, **B**: basalt, **C**: shale, **D**: limestone, **E**: granite, **F**: marble. *1 mark if 3 correct, 2 marks if 4 correct, 3 marks if 5 correct, 4 marks if 6 correct.*	4
Extension total marks	10

Outcome of test by levels

	Core (out of 20)	Total (out of 30)
Level 3	2–5	3–8
Level 4	6–9	9–13
Level 5	10–13	14–19
Level 6	14 and over	20 and over